THE POLAR IONOSPHERE AND MAGNETOSPHERIC PROCESSES

THE POLAR IONOSPHERE AND MAGNETOSPHERIC PROCESSES

GUNNAR SKOVLI
Editor

NATO Advanced Study Institute
Kjeller, Norway

GORDON AND BREACH, SCIENCE PUBLISHERS, INC.

New York London Paris

Library of Congress Catalog Card Number: 79-107154

Editorial office for Great Britain:
Gordon and Breach, Science Publishers Ltd.
12 Bloomsbury Way
London W.C. 1

Editorial office for France:
Gordon & Breach
7-9 rue Emile Dubois
Paris 14e

Printed in the United States of America

PREFACE

During the period 9 - 18 April 1969 a NATO Advanced Study Institute was arranged in Norway by the Norwegian Defence Research Establishment. The Institute was devoted to the subject of "Production and Maintenance of the Polar Ionosphere" and sponsored by the Science Committee of NATO.

The meeting was divided into fourteen sections, related to various aspects of the problem, and each section was concluded with discussions.

The present book contains all review papers presented at the Study Institute.

Dr. B.N. Mæhlum, NDRE, was Scientific Director of the Study Institute with Dr. S.J. Bauer, NASA/GSFC, USA, Dr. J.O. Thomas, Imperial College, UK, and Dr. B. Landmark, NDRE, Norway, as Advisory Programme Committee.

Mr. G. Skovli, NDRE, has had the thankless task of editing the proceedings.

Finally, the Publisher is gratefully mentioned for prompt publication of the book.

F. Lied
Director, NDRE

CONTENTS

Session V

PARTICLE IMPACT ON THE POLAR IONOSPHERE

Session VI

THE UPPER POLAR ATMOSPHERE

Session VII

THE POLAR IONOSPHERE THEORETICAL CONSIDERATIONS

Session VIII

THE POLAR F-LAYER 1

Session IX

THE POLAR F-LAYER 2

Session X

IRREGULARITIES IN THE
POLAR F-LAYER ELECTRON DENSITY

Session XI

THE POLAR E- AND D-LAYERS

Session XII

OPTICAL EMISSION OVER THE POLAR CAP

INTRODUCTORY SPEECH

Finn Lied

Director, Norwegian Defence Research Establishment
2007 Kjeller, Norway

1. INTRODUCTION

This meeting is number six in the series of NATO Advanced Study Institute organized by the Norwegian Defence Research Establishment.

One of these, held in 1967, on the topic of radio wave communications in polar regions was organized jointly with the Canadian Defence Research Telecommunication Establishment and one on Plasma Waves held last year jointly with Imperical College of the University of London.

The first 3 meetings, in 1961, 1963 and 1965 were all on the same topic, namely "The electron density distribution in the ionosphere and exosphere".

It was typical of the rapid advances in our field that a meeting on the same topic was needed every second year. For the same reason, this topic would today seem much too wide for a Study Institute. The programme of the present meeting is therefore entirely devoted to the polar, transauroral phenomena in the upper atmosphere.

The first meeting was arranged at this very hotel in 1961 at the dawn of the space age, and very few of the phenomena we are going to discuss during the present conference had been measured quantitatively at that time. In fact, the pure existence of some of the phenomena was not even known.

A variety of irregular, high latitude ionospheric phenomena have been investigated in detail in recent years, among which the "polar cap absorption events" and the "auroral D-layer" are the most prominent members. However, it seems that the regular upper atmosphere within the polar cap has been neglected to a large extent, mainly because of the difficulties involved in making observations in arctic regions.

A number of theories have been forwarded in the literature to account for the maintenance of the regular polar ionosphere. Since only crude estimates of the ion production and loss rates in these layers have been available, most of the theories can only be considered as mere speculations. However, with existing satellite techniques we are now in principle in a position to follow hour-to-hour variations in the sources of ionization for the ionospheric layers. This gives us an opportunity to perform quantitative studies of the upper atmosphere in a way which was impossible only ten years ago.

One of the aims of arranging this meeting is to bring together ionospheric and magnetospheric physicists in order to review our present knowledge on various phenomena in the magnetosphere and their possible impact on the polar ionosphere. The title of the meeting is a more general one, and other important processes in the polar ionosphere will also be covered by various speakers. However, the agenda clearly shows that the programme is biassed towards the problem of interaction between the upper atmosphere and magnetospheric phenomena.

We will start with lectures on the magnetosphere and the earth's magnetic tail, and proceed towards the polar ionosphere over a 9 days period.

2. THE MAGNETOSPHERE

The shape and dynamics of the magnetosphere has been the subject for numerous papers in recent years. During the first half of this century it was common to refer to the earth's magnetic domain as an axial symmetrical donut-shaped region. However, studies of trapped particles in the Van Allen belts as well as more recent studies of particles and fields by deep space probes have revealed an asymmetric shape of the magnetosphere, giving rise to significant local time variations. Furthermore, because of the changes in relative position of the geomagnetic axis to the solar wind, important universal time controlled variations also occur, mainly in the geomagnetic tail region.

At present we do not have a self-consistent analytical description of the magnetosphere, mainly because the complex time variations in the magnetopause location is not fully known. For this reason the true particle trajectories in a realistic geomagnetic field cannot be computed, particularly not for high L-shells. This is very unfortunate when one wants to compare ionospheric and magnetospheric phenomena.

3. LOW ENERGY ELECTRON OBSERVATIONS

One interesting aspect of the physics of the magnetosphere is the low energy plasma region located close to the neutral sheeth in the geomagnetic tail. It has been suggested by various authors that this region in some way is connected to the polar ionosphere, and one should therefore anticipate phenomena to be correlated in these two regions.

Low altitude, polar orbiting satellites have been observing precipitated electrons of energy less than some 10 keV over the polar caps in recent years. Some authors have concluded that the energy spectrum of these electrons display certain similarities with the electron energy spectrum in the neutral sheeth, but it is probably fair to say that we still lack a convincing proof that the two regions have the same origin.

Most recently, low altitude satellite observations of low energy electrons have clearly demonstrated that the auroral zone is not the only region where electron precipitation occurs at high latitudes. In fact, the narrow region of electron precipitation defined as the auroral zone where mostly electrons of energy greater than say 1 keV are important, is embedded in a broad region of very soft electron fluxes, which extends almost all the way to the pole. Investigation of these soft regions has only just been started, and there seems to be some discrepancies as to how the electron fluxes vary in time and space within the polar cap. This problem will be discussed by a series of papers at this meeting.

4. THE POLAR IONOSPHERE

It has been known for the last decennium that significant time variations occur in the F-region exitation and ionization over the winter pole. Most of these studies are based on observations obtained in the Antarctic, but the Arctic region has recently become accessible through the top-side sounder programme.

Speculations have been presented in the literature as to whether the time variations in the winter F-layer can be ascribed to a time varying source of ionization or whether they are caused by redistribution within the layer. During this meeting we shall hear a number of papers concerned with various aspects of this problem, and we will be in a position to discuss quantitatively at least some of the sources for the time variations in the light of recent observations.

5. CONCLUDING REMARKS

In conclusion, I would like to express the hope that through: (i) a number of review lectures presented by experts in various fields, and (ii) the subsequent discussions, we shall together be able to obtain a more coherent picture of the interaction between the magnetosphere and polar ionospheric phenomena.

When looking through the programme of this meeting, which is entitled "The production and maintenance of the polar ionosphere" it is tempting to refer to Mr. J. A. Ratcliffe's classical definition of the ionosphere as being "the region of space which is of interest to an ionospheric physicist at a given time". Apparently, during this meeting the ionosphere fills a substantial fraction of the near space.

With this brief introduction to our subject I would like to open this NATO Advanced Study Institute.

THE MAGNETIC FIELD OF THE MAGNETOSPHERE AND TAIL

D. H. Fairfield

NASA-Goddard Space Flight Center
Greenbelt, Maryland 20771, USA

Abstract. Earth orbiting spacecraft have made extensive measurements of the solar wind-compressed magnetic field in the sunward magnetosphere and the extended magnetic field of the geomagnetic tail. Analysis of the measurements indicate that in an average magnetosphere, field lines crossing the earth above $78^o \pm 3^o$ and $69^o \pm 2^o$ geomagnetic latitude in the noon and midnight meridians, respectively, are extended into the geomagnetic tail. Field lines in the northern and southern hemispheres of the tail are generally oriented parallel or anti-parallel to the earth sun line with the two regions being separated by a neutral sheet or current sheet less than 1 R_E thick. The average field magnitude is 16 γ at 20 R_E and 7 γ at 80 R_E except in a region within about 6 R_E of the neutral sheet where the field is depressed by a factor of 2. The observation of tail-like fields almost 1000 R_E behind the earth by the Pioneer 7 spacecraft sets a lower limit on the radial extent of tail-associated effects. Departures from the average magnetosphere field configuration occur at the times of magnetic storms and bay events. Storms are associated with higher tail fields and the extension of additional field lines to the tail. Magnetospheric substorms are associated with a relaxation toward a more dipole like configuration. The appearance of electrons near the plasma sheet after bays and the increase in the trapping boundary indicate that field lines may be closed up as high as 75° latitude near midnight for brief intervals after substorms.

1. INTRODUCTION

Low-latitude and mid-latitude geomagnetic lines of force are confined by the solar wind to a region within about 15 R_E (earth radii) of the earth known as the magnetosphere. High-latitude lines of force are swept downstream from the earth by the solar wind to form the geomagnetic tail which extends some 100's of earth radii in the anti-solar direction. Numerous spacecraft have made measurements in the magnetosphere and tail and the historical development of the subject has been adequately reviewed (e.g., Ness, 1969).

This paper attempts to consolidate current knowledge of the configuration of the magnetic fields of the magnetosphere and tail. Emphasis is placed on the question "what are the paths traced out by magnetic lines of force from different latitudes and longitudes?" Section 2 reviews work which has integrated large quantities of data and produced an average picture of the magnetosphere and tail. Section 3 surveys developments which show that variations from the average configuration are quite large and significant. Section 4 reviews the high-energy particle measurements which yield information about the paths followed by high-latitude lines of force. Section 5 summarizes the findings and Section 6 outlines outstanding problems.

2. AVERAGE MAGNETIC FIELD CONFIGURATION

2.1 The Inner Magnetosphere

Simple theoretical models which include the compressional effects of the solar wind have had considerable success in predicting the configuration of the magnetosphere in the sunward hemisphere (e.g., Mead, 1964). General hydromagnetic theory has provided some insight into the formation of the geomagnetic tail (review by Dungey, 1968) but the specific processes taking place at the magnetopause are not sufficiently well understood to input into theoretical model calculations. Simple quantitative model calculations have not had

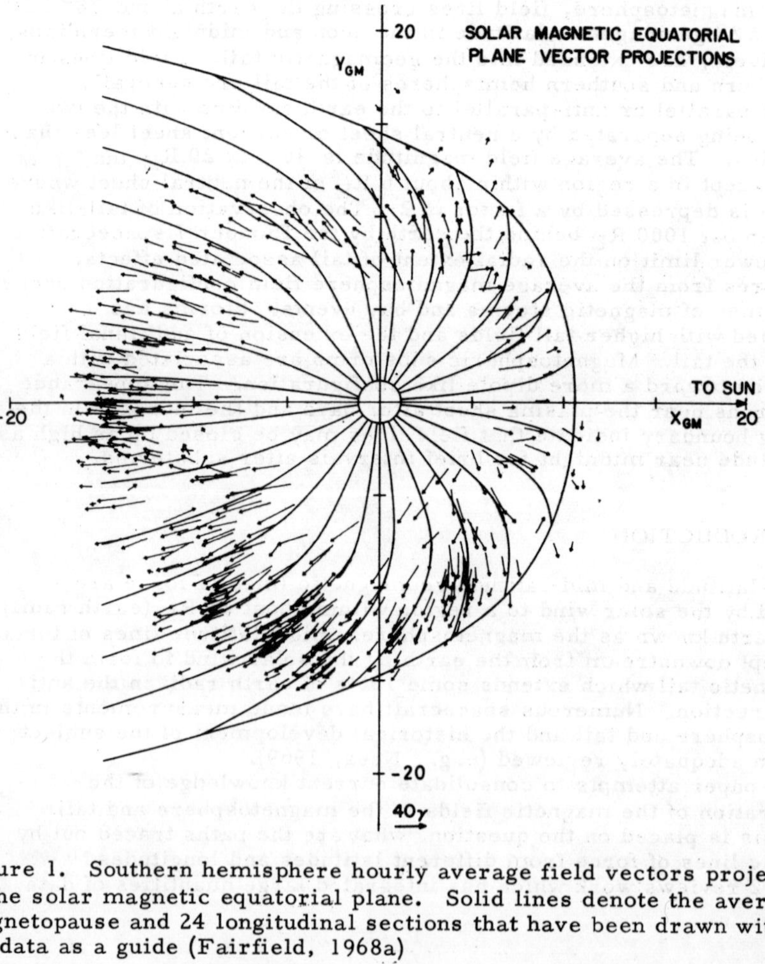

Figure 1. Southern hemisphere hourly average field vectors projected in the solar magnetic equatorial plane. Solid lines denote the average magnetopause and 24 longitudinal sections that have been drawn with the data as a guide (Fairfield, 1968a)

success in predicting the geomagnetic tail. These inadequancies of the theory mean that experimental measurements must take the lead in determining the configuration of the magnetosphere and tail. For this reason the present paper restricts attention to experimental results on the configuration of the magnetosphere and tail. It should be pointed out, however, that quantitative field models have been devised to approximate the experimental results (Williams and Mead, 1965; Taylor and Hones, 1965). Such models are valuable for studies of trapped particles and related effects observed on the ground.

The geomagnetic field inside of 5 earth radii is not greatly distorted by current sources outside the earth's surface except during magnetic storms. Cahill's results (Cahill, 1966) indicate that departures from the dipole field are of the order of 20 γ at 5 R_E on quiet days. The results of Cummings et al. (1968) at 6.6 R_E on quiet days show a $25\gamma \pm 10\gamma$ compression of the field in the sunward hemisphere and a $5\gamma \pm 10\gamma$ weakening in the night hemisphere relative to an ideal dipole.

With this minimal distortion of the inner magnetosphere in mind, Fairfield (1968a) used the IMP 1, 2 and 3 data taken between 5 and 18 R_E to obtain an experimental determination of the outer magnetosphere. Much of these data from the southern hemisphere are shown projected in the geomagnetic equatorial plane in Figure 1. This figure illustrates the manner in which the magnetic field is swept back toward the tail by the solar wind. The vectors in any one region agree among themselves quite well considering that (i) the data were taken by three spacecraft over a three and one-half year period; (ii) data from disturbed intervals were not removed; (iii) the measurements were made over a range of latitudes of about $\pm 70^\circ$. Since lines of force from the same longitude but different latitudes appeared to be distorted in the same manner it was possible to draw the heavy lines in Figure 1 tangent to the vectors and assume that flux passing through the earth at longitudes between a pair of lines also crossed the equatorial plane between the lines. After constructing experimental contours of constant field strength in the equatorial plane, it was possible to equate dipole flux at the earth's surface to the experimentally determined flux through the equatorial plane. Assuming a dipole line of force crossing the earth's surface at $63^\circ 61'$ latitude and intersecting the equatorial plane at 5 R_E, integration northward in latitude at the earth's surface and outward in the equatorial plane yielded the relation between the latitude of a field line and its equatorial crossing point. This procedure was carried out in each of the 15° longitude intervals in the dawn hemisphere and the results are shown in Figure 2. In this view of the magnetosphere equatorial plane the solid lines represent the latitude of the origin of the field lines passing through the equatorial plane at that point. The dashed lines are reproduced from Figure 1 and they represent the longitude of the origin of the field line.

The results of Figure 2 can be used to map phenomena along field lines from the high-latitude ionosphere to the equatorial plane. Fairfield mapped the auroral oval determined from ground observations by Feldstein (1963), the region of low-altitude magnetic fluctuations observed on a low-altitude spacecraft by Zmuda et al. (1967) and a greater than 40 keV electron intensity contour of Frank et al. (1964). These data are shown in Figure 3. It can be seen that the auroral oval maps to the region of the magnetopause in the sunward hemisphere. In the midnight meridian the last field line that closes within 10 R_E runs to approximately 68° latitude. This average magnetosphere suggests that aurora often occur on closed field lines.

Views in meridian planes are shown here as Figure 4. These views incorporate all of the available data from the 15° meridian section of Figure 1 which were taken when the sun was below the geomagnetic

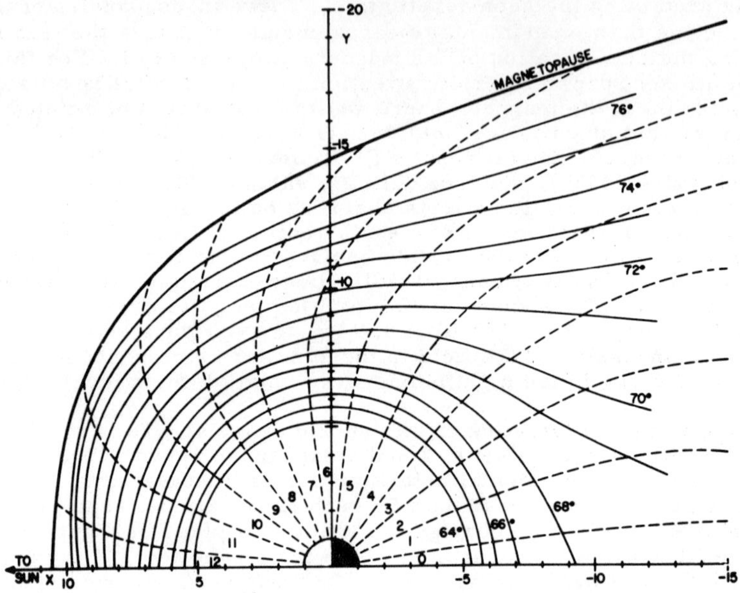

Figure 2. Contours in the equatorial plane designating the latitude and local time of the earth intersection point of the field line (Fairfield, 1968a).

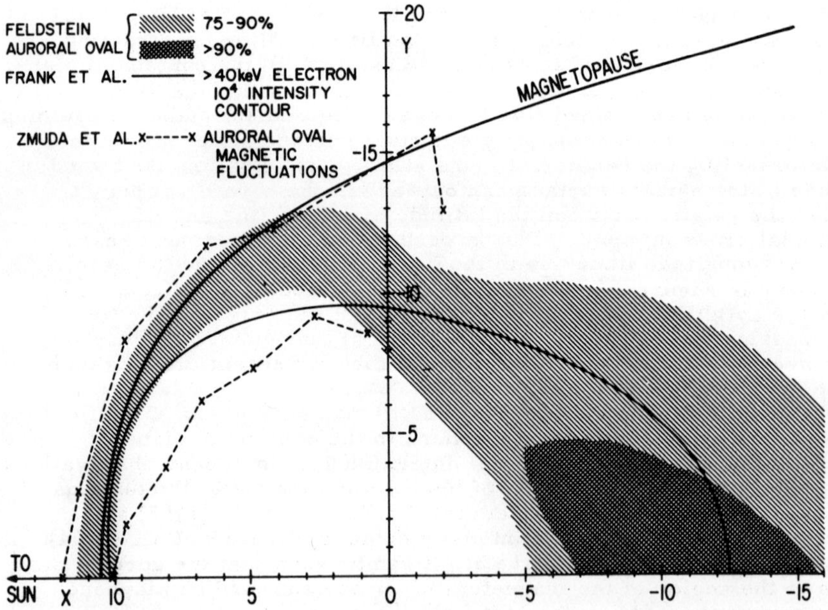

Figure 3. Auroral oval and associated phenomena have been projected along field lines to the equatorial plane using the information of Figure 2 (Fairfield, 1968a).

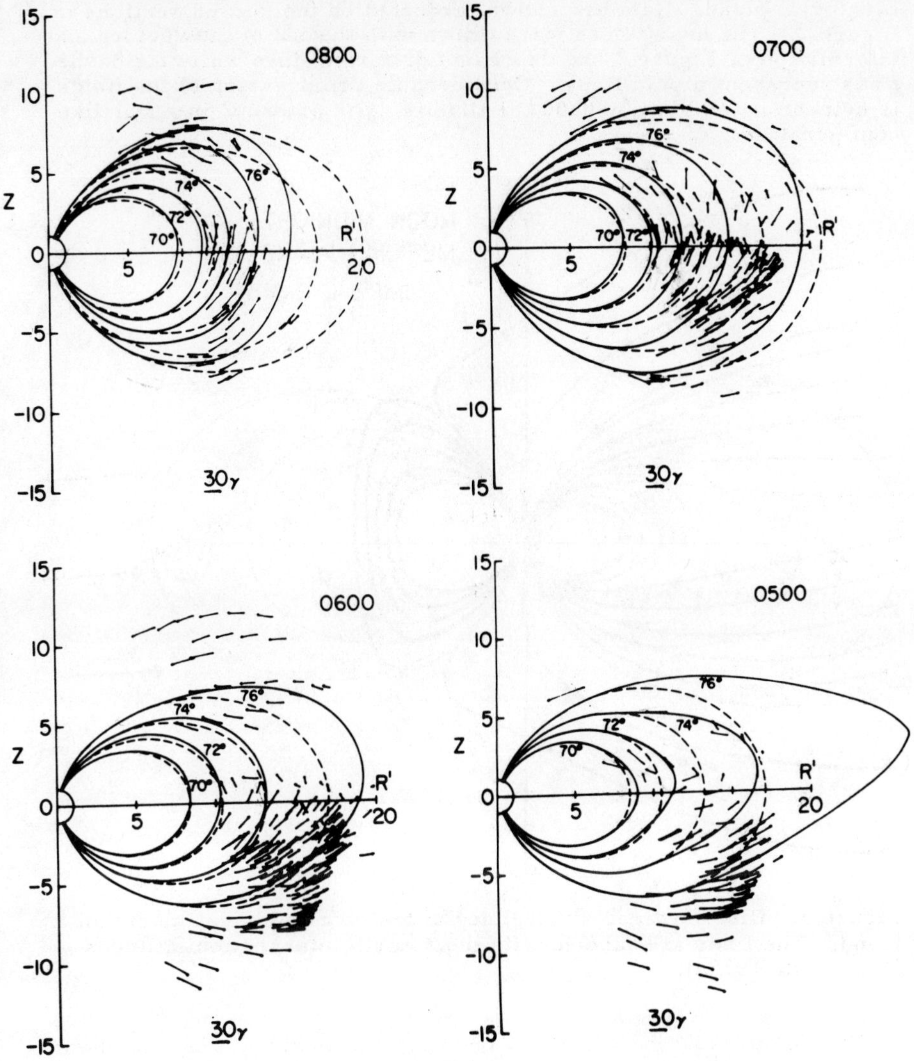

Figure 4. Vectors projected in the curved meridian sections of Figure 1 in the 0800-0500 local time regions for hours when the sun is south of the solar magnetic equatorial plane. Solid lines represent distorted field lines which have been drawn with the data and the information of Figure 2 as a guide. Dashed lines represent undistorted dipole field lines corresponding to each distorted line. The lines are labeled with their earth intersection latitudes (Fairfield, 1968a).

equatorial plane. Data are shown projected on the curved sections of Figure 1. The heavy lines were drawn with the aid of the vectors and information of Figure 2 and represent distorted lines while the dashed lines represent dipole lines. Considerable compression of the fields is evident at 0800 but by 0500 the field is quite extended in a tail-like configuration.

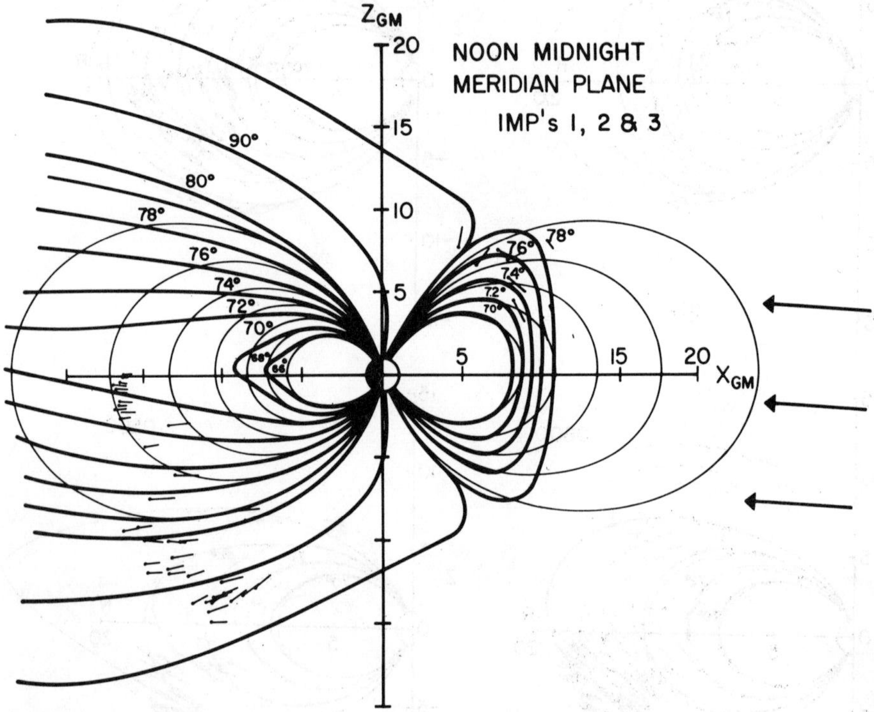

Figure 5. Distorted and dipole lines of the noon-midnight meridian plane. The lines are labeled with their earth intersection latitudes (Fairfield, 1968a).

The comparable view of the noon-midnight meridian plane is shown as Figure 5. The outermost closed field line in the noon meridian is seen to cross the earth at about 78° latitude with higher latitude lines going back into the tail. One hourly average vector near X = 5, Z = 8 illustrates this turning back of the field.

2.2 The Geomagnetic Tail Within 80 R_E

Early observations of the geomagntic tail within 32 R_E of the earth with the IMP 1 spacecraft (Ness, 1965) have now been extended beyond the orbit of the moon at 60 R_E (Ness et al., 1967a; Behannon, 1968; Mihalov et al., 1968) and even to 1000 R_E (Ness et al., 1967b; Wolfe et al., 1967). Figure 6 reproduces figures of Behannon (1968) showing Explorer 33 hourly average vector measurements in the tail region out to 80 R_E. In the top view vectors are projected in the solar magnetospheric XZ plane (X axis along the earth-sun line, Z in the plane formed by the X axis and the dipole axis and Y forming a right hand orthogonal system) with the sun at the left and the northern hemisphere at the top of the page. Field lines originating from the

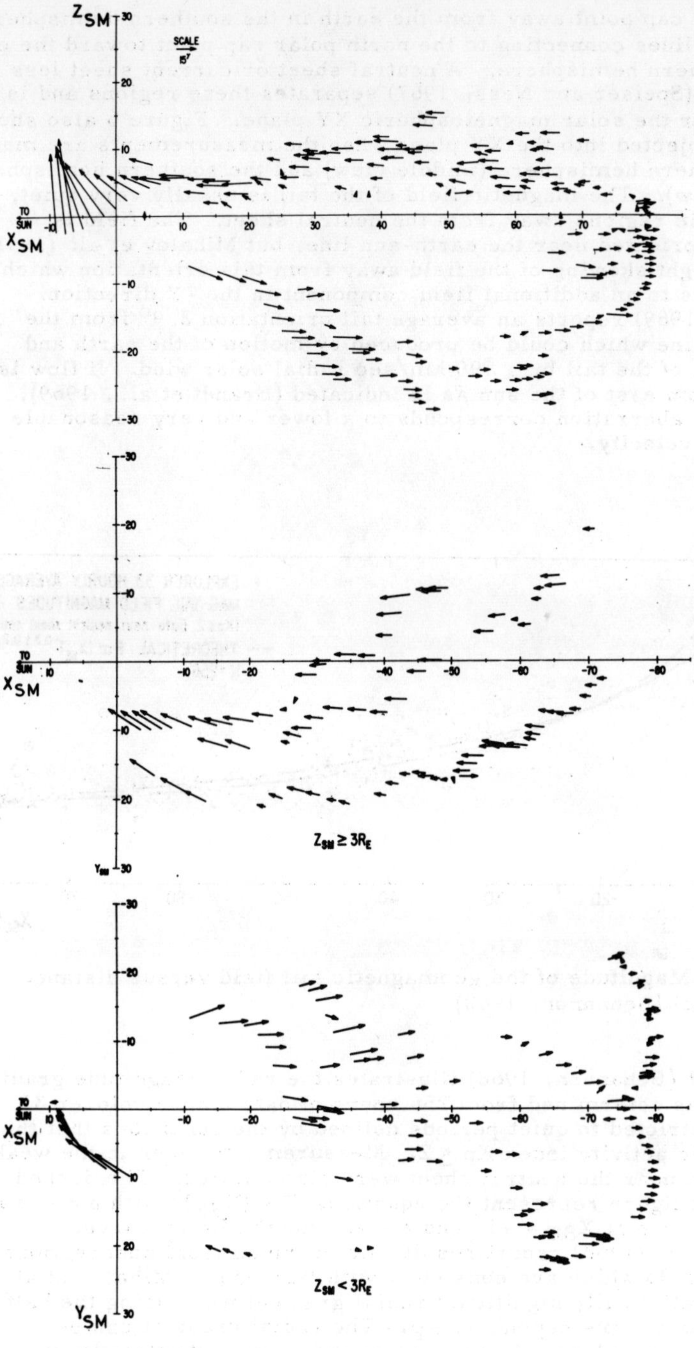

Figure 6. Explorer 33 hourly average vectors projected in solar magnetospheric coordinates. The top view shows vectors projected in the non-midnight meridian plane while the other view shows vectors projected in the equatorial plane when the measurements are taken in the northern hemisphere (middle) and southern hemisphere (bottom). Axes are labeled in earth radii (Behannon, 1968).

south polar cap point away from the earth in the southern hemisphere while field lines connecting to the north polar cap point toward the earth in the northern hemisphere. A neutral sheet or current sheet less than 1 R_E thick (Speiser and Ness, 1967) separates these regions and is located near the solar magnetospheric XY plane. Figure 6 also shows vectors projected into the XY plane when the measurements are made in the northern hemisphere (middle view) and the southern hemisphere (bottom view). The magnetic field of the tail is usually very quiet, especially in regions away from the neutral sheet. The field is invariably oriented near the earth-sun line, but Mihalov et al. (1968) detect a slight skewing of the field away from this orientation which corresponds to an additional field component in the -Y direction. Behannon (1969) reports an average tail orientation 2.9° from the sun-earth line which could be produced by motion of the earth and collimation of the tail by a 590 km/sec radial solar wind. If flow is actually from east of the sun as is indicated (Brandt et al., 1969), Behannon's aberration corresponds to a lower and very reasonable solar wind velocity.

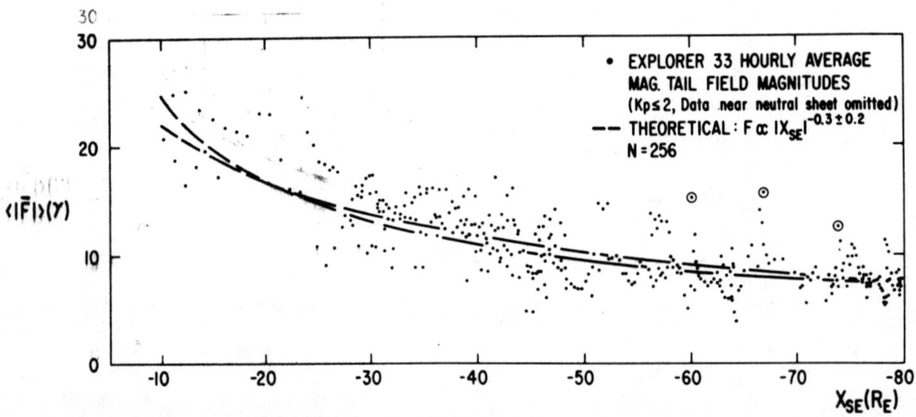

Figure 7. Magnitude of the geomagnetic tail field versus distance from the tail (Behannon, 1968).

Figure 7 (Behannon, 1968) illustrates the radial magnitude gradient of the tail as determined from 256 hours of data from Explorer 33. Data is restricted to quiet periods defined by the conditions that the geomagnetic activity index $Kp \leq 2$. Measurements taken in the weak field region near the neutral sheet were also omitted. The dashed lines in the figure represent the equations $F \sim (X_{SE})^a$ with a = -.5 for the upper curve at $X_{SE} = -10$ and a = -.1 for the lower curve. Mihalov et al. (1968) report results also derived from observations on Explorer 33 which are consistent with Behannon. Mihalov et al. found no statistically significant radial gradient when using the half of their data sample beyond 58 R_E. The radial gradient can be explained by field lines closing across the neutral sheet or by an expansion in the cross-section of the tail. Behannon (1968) suggests that both effects are important. Mihalov et al. investigate the field component across the neutral sheet at the time of field reversals and find cases of both northerly and southerly directed fields. The northerly components are more frequent and the southerly components occur almost exclusively beyond 35 R_E.

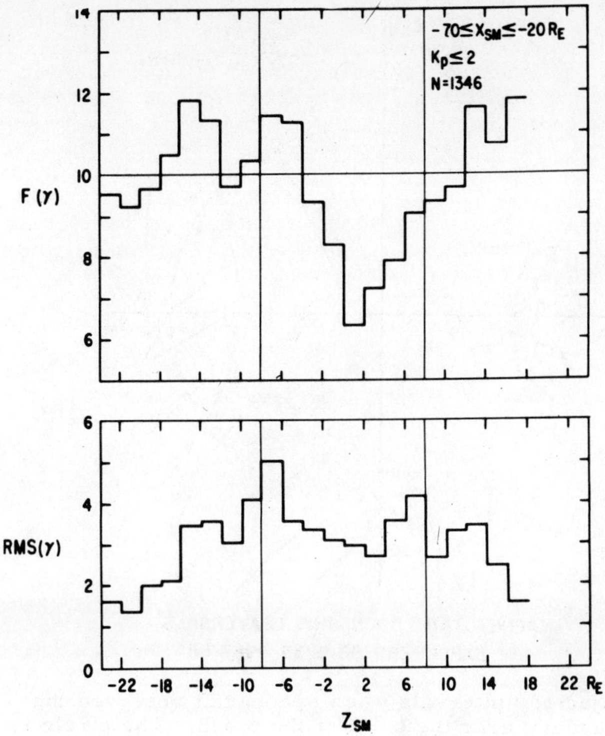

Figure 8. Magnitude of the tail field versus distance from the solar magnetospheric XY plane. Data show a broad region of depressed fields near $Z_{SE} = 0$. The RMS values denote the scatter of the hourly averages which were averaged to give the top curve (Behannon, 1969).

Figure 8 illustrates recent work by Behannon (1969) demonstrating the existence of a region of depressed field magnitude near the solar magnetospheric equatorial plane. The data consist of 1346 hourly average field magnitudes from Explorers 33 and 35 from 1966-1968 observed at positions $-70 \leq X_{SM} \leq -20$ at times with $Kp \leq 2$. Individual hourly averages were computed by averaging instantaneous measurements of the field magnitude. The average of the hourly values are plotted in the top portion of the figure for each 2 R_E interval of Z_{SM}. The RMS deviation computed from the hourly average values is shown in the lower portion of the figure. The fact that the RMS values are not higher in the depressed field region suggests that there exists a broad depressed field region several R_E thick rather than a thinner low field region which moves throughout the interval of low Z_{SM}. This broad depressed field region agrees quite well with the expected plasma sheet which has been measured by Vela spacecraft near 17 R_E (Bame et al., 1967). The existence of a region of depressed field near the equatorial plane and increased field at high-latitudes in the midnight sector between 5 and 11 R_E has been emphasized by Heppner et al. (1967).

Behannon has also obtained the most comprehensive information on the area of the cross-section of the tail in the downstream region (Behannon, 1968, 1969). A figure from his later work is presented as Figure 9. The line segments in this figure represent the trajectory of Explorer 33 and 35 during intervals when they made observations

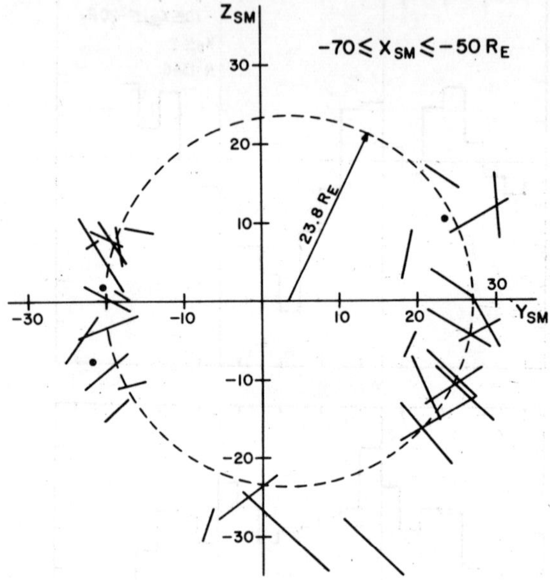

Figure 9. Trajectory intervals when spacecraft observed the magnetotail boundary near the orbit of the moon. The circle is drawn for reference; whereas, the measurements suggest an elliptical cross-section for the tail (Behannon, 1969).

of the tail boundary in the region $-70 \leq X_{SM} \leq -50\ R_E$. A dashed circle of 23.8 R_E radius has been drawn with its center at Y = +3.3 which corresponds to the line through the center of the tail making an angle of 3.1^{o} with the earth-sun line. The observance of boundary positions 30 R_E south of the solar magnetospheric equatorial plane suggests that the cross-section of the tail is elliptical with the semi-major axis oriented along the solar magnetospheric Z axis. This elongation of the tail is not surprising if one considers the tail as formed by two approximately circular bundles of flux from the polar caps aligned adjacent to one another.

2.3 The Extended Geomagnetic Tail

Observations of the geomagnetic tail have been obtained by Pioneer 7 passing behind the earth near 1000 R_E (Ness et al., 1967b; Wolfe et al., 1967; Fairfield, 1968b) and Pioneer 8 near 500 R_E (Mariani and Ness, 1969). The observations at these distances are similar to one another in that tail-like fields are observed for short intervals of a few minutes to a few hours. Pioneer 7 at 1000 R_E observed the tail approximately 7% of the time during an 11 day interval while Pioneer 8 near 500 R_E observed the tail approximately 25% of the time over a 9 day period. Tail observations at these large distances contrast to the measurements nearer the earth primarily in that (i) the intervals of observation are much shorter in the far downstream region; and (ii) the plasma measured in the far downstream tail does not disappear to the extent that it does nearer the earth (Wolfe et al., 1967).

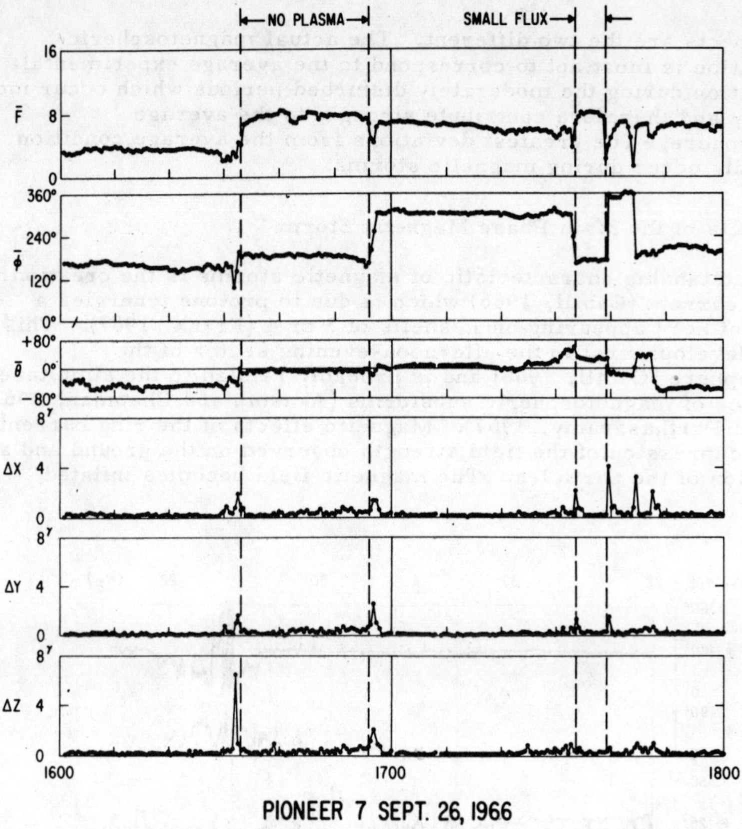

Figure 10. Intervals of tail like magnetic field observed by Pioneer 7 almost 1000 R_E from the earth. A neutral sheet crossing is seen at 1739 (Ness, 1969).

An example of 2 hours of Pioneer 7 data are illustrated in Figure 10 (Ness, 1969). Two intervals 1633-1656 and 1734-1744 are shown where the field aligned itself with the earth-sun line (ϕ = 180 or 360 and θ = 0°). During these two intervals a preliminary comparison with the plasma experiment of the Massachusetts Institute of Technology showed decreased or absent plasma as noted. At 1739 a neutral sheet-like crossing occurred with the field reversing polarity in the manner frequently seen nearer the earth.

The nature of the tail hundreds of R_E behind the earth is not entirely clear from these measurements of the field at isolated points. Mariani and Ness (1969) interpret the short duration of the tail-like intervals as evidence that the tail is broken up into isolated filaments. Fairfield (1968b) suggests that motion of a well collimated coherent tail in response to the varying direction of the solar wind is an equally good explanation of the observations.

3. MAGNETOSPHERE TIME VARIATIONS

Although a rather good description exists for an average magnetosphere and tail, it is important to inquire into how often this average configuration corresponds to an actual configuration, and in what respects are the two different. The actual magnetospheric configuration is most apt to correspond to the average experimental configuration during the moderately disturbed periods which occur most frequently and therefore contribute strongly to the average magnetosphere. The greatest deviations from the average condition undoubtedly occur during magnetic storms.

3.1 Effects of the Main Phase Magnetic Storm

The outstanding characteristic of magnetic storms is the creation of a ring current (Cahill, 1966) which is due to protons (energies a few 10's of keV) appearing on L-shells of 3 or 4 (Frank, 1967). This current develops first in the afternoon-evening sector of the magnetosphere (Cahill, 1966) and is probably related to the simultaneous occurrence of magnetospheric substorms (Akasofu and Chapman, 1963; Davis and Parthasarathy, 1967). Magnetic effects of the ring current create a depression of the field strength observed on the ground and at the location of the particles. The magnetic field becomes inflated

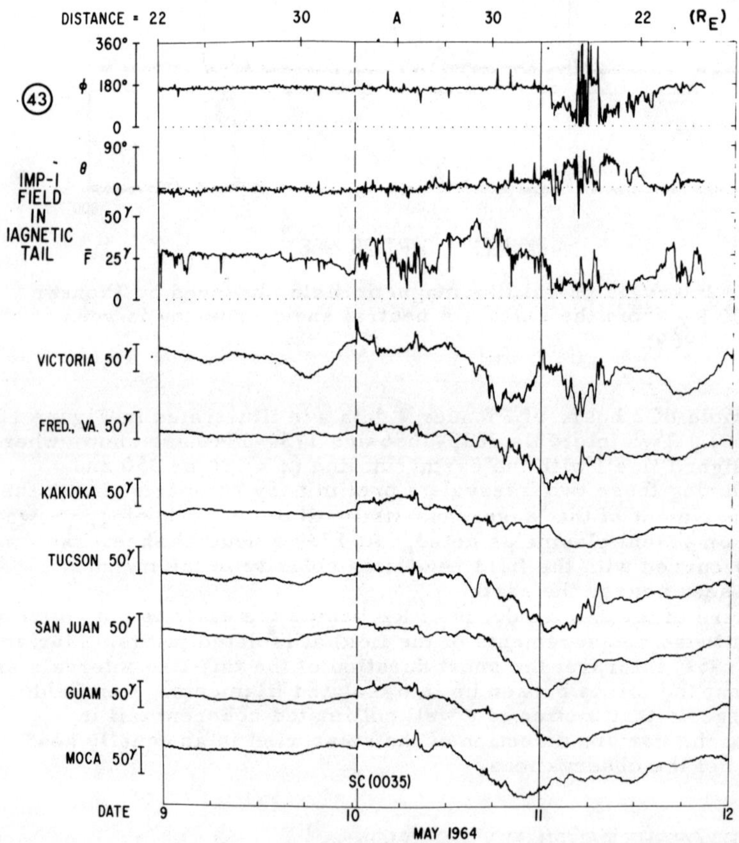

Figure 11. Three days of geomagnetic tail data and mid-latitude magnetograms. Enhanced tail magnitudes are seen until the spacecraft approached the weak field region surrounding the neutral sheet on May 11 (Behannon and Ness, 1966).

since dipole lines which would normally cross the equatorial plane near the earth are forced further out into the magnetosphere.

This addition of field lines to the outer magnetosphere will result in either (i) extending the magnetopause and keeping the field strength constant; (ii) increaseing the field strength and keeping the magnetopause position constant; (iii) causing a readjustment in the magnetosphere configuration so that additional lines of force go into the geomagnetic tail; or (iv) a combination of these 3 alternatives. No clear evidence exists for movement of the magnetopause boundary during periods of high Kp (Patel and Dessler, 1966) but there are observations of unusually distant boundary crossings during the recovery phase of storms (Cahill and Patel, 1967).

Evidence does exist for additional lines of force moving to the geomagnetic tail during storms. Behannon and Ness (1966) demonstrated that the field of the tail tended to increase during periods of high Kp. Figure 11 from Behannon and Ness shows the IMP 1 measurements of the magnitude F, latitude angle θ and longitude angle ϕ in the tail during a three day storm period in May, 1964. The tail field increases abruptly at the time of the storm sudden commencement at 0035 UT on May 10 and tends to remain high until the spacecraft approaches the neutral sheet near the end of May 10. These authors use a tail boundary crossing abnormally close to the X axis and flux conservation arguments to show that only part of the enhanced tail magnitude is due to compression of the tail with the remainder being explained by the addition of lines of force to the tail. Sugiura et al. (1968) have used precise knowledge of the arrival time of the interplanetary shock associated with the sudden commencement of July 8, 1966, to argue that an observed increase in the tail field at OGO 1 was due to additional field lines moving into the tail.

Williams and Ness (1966), Ness and Williams (1966) have used simultaneous measurements of the trapping boundary for electrons > 280 keV measured on the low-altitude polar orbiting satellite 1963 38C and the IMP 1 magnetic field measurements in the tail to gain further evidence that field lines are added to the tail during magnetic storms. Figure 12 from Williams and Ness shows three-hour averages of the tail magnetic field along with the high-latitude cutoff, Λ_c, and the geomagnetic activity index Kp for two days in April. The circles represent the observed low-altitude cutoff for passes of 1963 38C over the polar cap. Open and solid circles designate alternative definitions of the cutoff point. The solid line for Λ_c is obtained by inserting the observed tail field magnitude into the Williams-Mead model of the magnetic field (Williams and Mead, 1965) and noting the latitude of the outermost field lines which could support trapped particles. A sudden commencement storm at 1525 hours on April 1 is reflected in increased values of F and Kp. The observed particle cutoff exhibits a characteristic decrease from pre-storm values of 66° to post SC values near 60°. The value of Λ_c predicted from the field observations undergoes a similar change in latitude even though the absolute values are higher. The difference in absolute value is attributed primarily to inadequacies of the model and the observed changes are attributed to the addition of field lines to the tail. The lower fields and larger differences between the two methods on April 2 are due to IMP 1 measuring the weak fields associated with the equatorial region.

3.2 Effects of Substorms

Information on the behavior of the magnetospheric magnetic fields during substorms has been obtained from the OGO 1 and ATS spacecraft. Figure 13 (Heppner et al., 1967) shows 33 hours of data

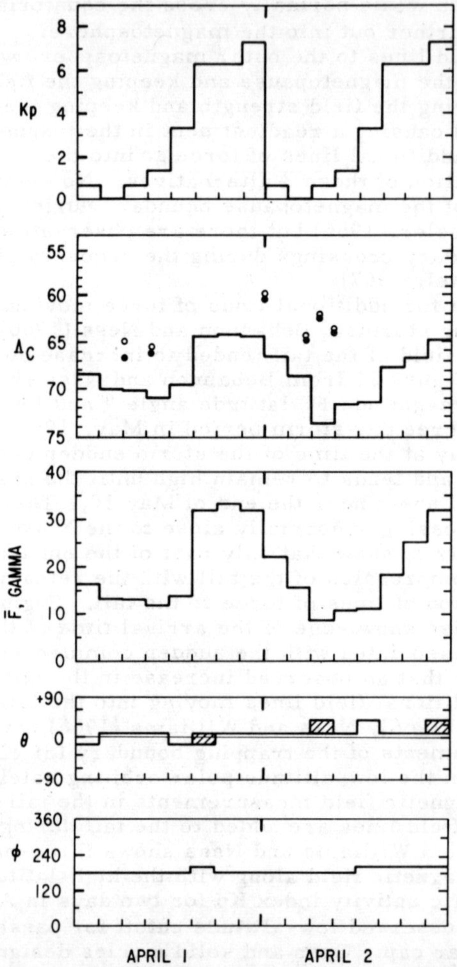

Figure 12. Magnetic field parameters, geomagnetic index Kp and trapped particle cutoff latitude Λ_c for a two day storm period. A decrease in Λ_c after the rise in Kp is seen in both the direct observations (circles) and the values calculated by putting the observed tail magnitudes in a field model (Williams and Ness, 1966).

from an inbound pass of OGO 1 through the tail region. The top portion shows the XY component of the tail field and the Z component while the lower portion of the figure shows two auroral zone magnetograms along with Kp. Changes in field magnitude at 1937 and 2139 in the tail are seen to be associated with magnetic bays observed on the ground near the longitude of the spacecraft. The ground perturbations have earlier onset times of 1922 and 2125. Decreases in the field magnitude at the time of a bay occur when the spacecraft is in a region where the field is greater than that predicted by spherical harmonic reference fields. If the spacecraft is in the region near the equatorial plane between 6 and 11 R_E where the field is normally below the reference field, the spacecraft sees an increase at the time of a bay. In either case the tendency is for the field to evolve toward a more dipole-like configuration following the bay (e.g., Sugiura et al., 1968).

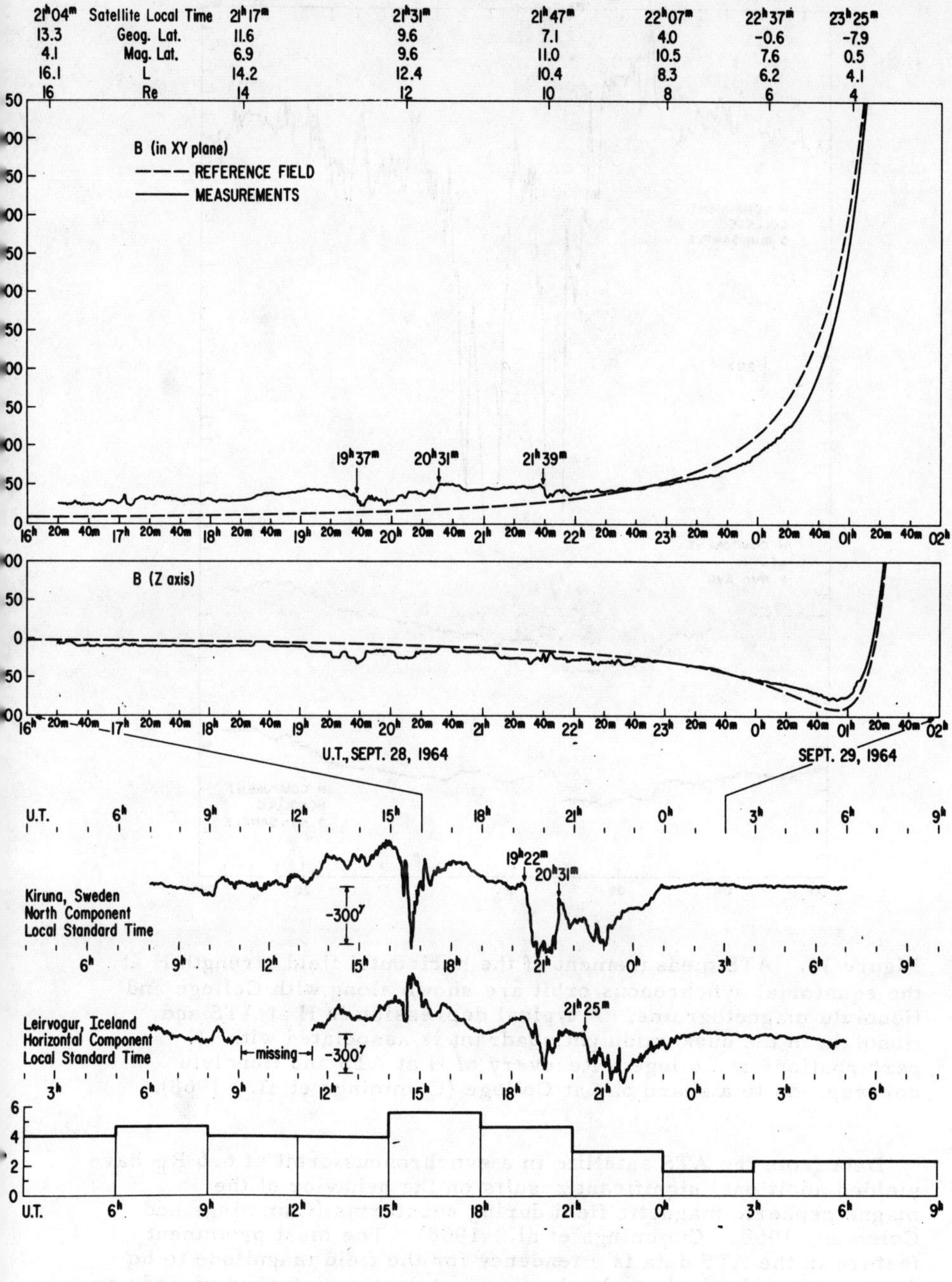

Figure 13. OGO 1 measurements in the geomagnetic tail illustrating decreases in the tail magnitude which are associated with auroral zone bays observed at earlier times (Heppner et al., 1967).

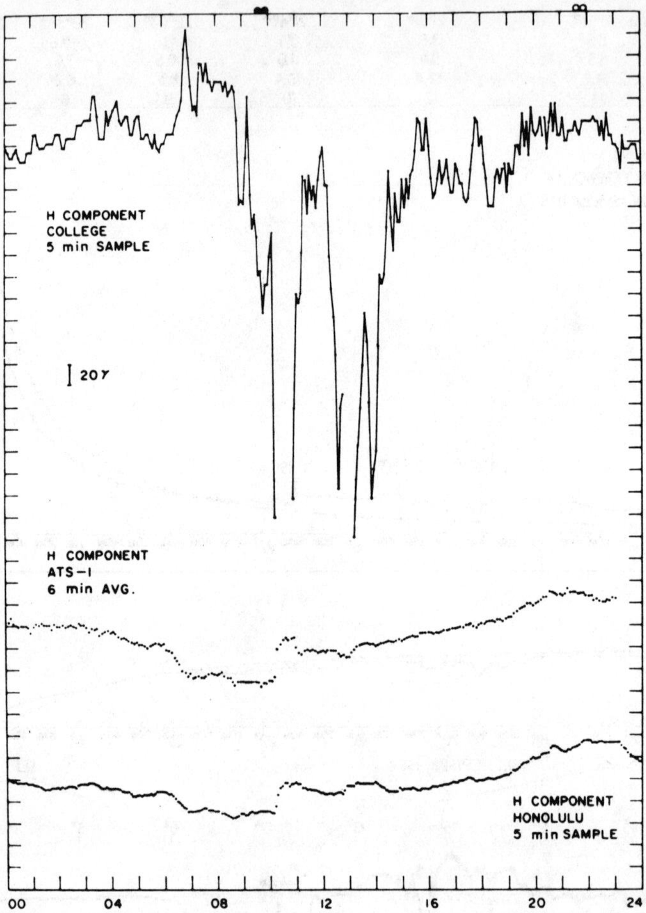

Figure 14. ATS measurement of the horizontal field strength H at the equatorial synchronous orbit are shown along with College and Honolulu magnetograms. A typical depression in H at ATS and Honolulu in the dusk-midnight quadrant is associated with +H perturbations at College. Recovery of H at ATS and Honolulu corresponds to a sharp bay at College (Cummings et al., 1968).

Data from the ATS satellite in a synchronous orbit at 6.6 R_E have yielded additional significant results on the behavior of the magnetospheric magnetic field during substorms (Cummings and Coleman, 1968a; Cummings et al., 1968). The most prominent feature in the ATS data is a tendency for the field magnitude to be depressed in the dusk-midnight quadrant during disturbed periods as is illustrated in Figure 14 (Cummings et al., 1968). This depression in H at ATS (0400-1000 UT = 1800-2400 LT) is typically accompanied by a very similar depression on the Honolulu H magnetogram (bottom trace Figure 14) near the sub satellite point and a +H perturbation at College which is located near the foot of the ATS field line. The expansive phase of a substorm then corresponds to a recovery of the field at ATS in a manner similar to that described by Heppner et al., 1967. This recovery can be seen near 1000 UT (local midnight) in

Figure 14 and is accompanied by a very similar change at Honolulu. The striking correspondence between the observations at ATS and Honolulu refutes the traditional interpretation that bay effects observed on mid-latitude surface magnetograms are due to return currents from the auroral electrojet which flow only in the ionosphere between the ground and the spacecraft. An explanation now seems to require partical ring currents flowing down field lines and closing through the ionosphere (Cummings et al., 1968; Meng and Akasofu, 1968; Schield et al., 1969).

The plasma and energetic particle detectors on the Vela spacecraft have yielded data which is relevant to the question of the magnetic field configuration (Hones et al., 1967; Hones and Singer, 1968). These workers find that if a Vela spacecraft is in the plasma sheet at the time a bay develops, the electrons are decreased in energy or tend to disappear. At the peak of the bay or during the recovery of the bay, the plasma and electrons reappear. Even when a spacecraft is several earth radii from the position of the plasma sheet the electrons and plasma tend to be seen during the recovery of the bay. The authors interpret their results as indicating that substorms originate within the orbit of Vela and only during the recovery do plasma and particles reach the tail.

Rothwell and Wallington (1968) use measurements from the IMP 2 spacecraft with a more eccentric orbit to verify and extend the results of Hones et al. Rothwell and Wallington find that electron intensities in the outer magnetosphere between 5 and 9 R_E increase simultaneously with auroral zone bays while the electrons appearing beyond these distances are seen after the bays and with greater time delays at greater distances. The authors suggest the Kennel-Petchek (1966) mechanism is instrumental in the deflation of tubes of force. This region would propagate outward with the magnetoacoustic velocity which could explain the observed delay times.

Rao (1969) has investigated Injun 4 data to determine the effects of substorms on the low-altitude outer trapping boundary of > 40 keV electrons, which usually is located between 64° and 69° invariant latitude. He finds that during the expansive phase of a substorm the intensity of the electrons increases but the cutoff latitude tends to remain constant. During the recovery phase of the substorm, however, the cutoff latitude typically moves up to 75°. Lin et al., (1968) observe similar high-latitude boundaries during substorms on Alouette 1 and find a corresponding northward movement in the riometer absorption. Although Injun 4 particles are not all trapped, it is probably necessary to have closed field lines up near 75° at these times. This situation clearly represents a magnetosphere configuration grossly different from the average picture represented in Figures 2 and 5.

The common feature of all these measurements during substorms is that they point to the existence of a more dipolar-like magnetospheric configuration immediately after a substorm with significantly more flux crossing the equatorial plane near the earth. This relaxed configuration of the magnetosphere after substorms appears to favor theories which suggest storage of the energy for a substorm in the tail (Axford, 1965; Siscoe and Cummings, 1969) but the work of Hones et al. (1967; 1968) and Heppner et al. (1967) seems to indicate that the initial effects of the storm take place nearer the earth and not in the tail.

4. INDIRECT CONFIGURATION MEASUREMENTS

The question of whether the magnetic tail lines of force eventually connect across the neutral sheet or connect with the interplanetary field is of considerable interest. Energetic particles from the sun may provide the best answers to this question. The suggestion is that energetic solar particles will either (i) tend to follow field lines and gain access to the magnetic tail and polar cap almost instanteneously if tail field lines connect to the interplanetary field (the open model); or (ii) take many hours to diffuse slowly across the field lines if they are not allowed immediate access. To distinguish between these hypotheses simultaneous particle measurements are required from one spacecraft in the tail or over the polar cap and a second spacecraft measuring ambient flux in the interplanetary medium. The time delay between a change in the ambient flux and the corresponding change observed in the tail or over the polar cap is the relevant quantity to be measured.

Williams and Bostrom (1967) found differences in the interplanetary proton flux at Mariner 4 and the flux averages for passes over the polar cap by Injun 4 and 1963 38C. They attributed these to a changing magnetosphere configuration which alternately improved or detracted from access to the polar cap. Krimigis et al. (1967) found good agreement between the fluxes observed at Explorer 33 and Injun 4 over the polar cap with a delay less than $\frac{1}{2}$ hour. These authors interpreted their results as supporting the open model unless diffusion were much more rapid than that predicted by simple particle diffusion theory.

Blake et al. (1968) studied protons over the polar cap and found higher fluxes at lower latitudes nearer the cutoff point. They warn against using polar cap averages and feel their results argue against interplanetary connection. Lanzerotti (1968) compares Explorer 34 interplanetary data with ATS data and finds that solar protons have quick access to 6.6 R_E. He suggests that if particles can diffuse to 6.6 R_E so readily, little can be learned about magnetosphere topology from low-energy particles. Williams and Bostrom (1969) suggest that the particles have easy access to the neutral sheet and therefore, it is not surprising that they can diffuse rapidly to 6.6 R_E.

The best data pertinent to this question of access to the tail are the recent data of Montgomery et al. (1969). These authors consider simultaneous Vela measurements which clearly show that flux changes in the tail are delayed by times ranging from 15 minutes to 2 hours. Often sharp flux increases outside the tail were broadened in the tail but the delays were not energy dependent. The fluxes were often anisotropic outside and away from the tail and isotropic near and within the tail. These measurements fail to support the open magnetosphere model in that a time delay is observed and the particles are not anisotropic. The measurements also fail to support diffusion theory in that the delay is not as long as predicted and no energy dependence is found. After considering these results which do not fully support either theory, the authors suggest an intermediate interpretation of multiple scattering in and near the tail with merging of tail and interplanetary fields downstream.

Williams and Bostrom (1969) have compared proton intensities from individual passes of 1963 38C over the polar cap with IMP-4 measurements in the interplanetary medium. They often find enhanced fluxes of 1.2 - 8.5 MeV protons near the auroral regions following increases in the interplanetary proton fluxes. They interpret these observations as indicating diffusion into each bundle of polar flux. At other times they find polar cap profiles which may be consistent with direct access to the polar cap (Williams, personal communication). Their suggestion is that both means of entry are possible at different times with the exact method depending on the boundary conditions imposed on the magnetosphere by the interplanetary medium.

Less data is available on the entry of electrons to the tail and polar cap. Early measurement by Lin and Anderson suggested that the delay time between a solar flare and the observations of the associated electrons was not strongly dependent on whether or not the spacecraft was in the tail. This observation was cited as favoring the open magnetosphere model, but if diffusion is at all important this would not be so. Van Allen and Ness (1968) use particle shadowing by the moon to demonstrate that the diffusion velocity of solar electrons across the tail is less than 100 km sec^{-1}. Lin (1968) uses similar particle measurements near the moon and concludes that the electrons have direct access to the tail beyond the orbit of the moon.

The net result of all these particle measurement is that they have been unable to supply definitive information on whether field lines in the tail connect to those of interplanetary space. Diffusion appears to play a role in the entry of particles to the tail but it appears to be much more rapid than previously predicted (Michel and Dessler, 1965). Direct access along field lines connecting directly to the interplanetary medium may also occur at certain times.

Connection of magnetosphere field lines to those of the interplanetary medium may be tested directly by studying magnetopause crossings and searching for a field component normal to the boundary. Sonnerup and Cahill (1967) studied Explorer 12 data and concluded that sometimes there was evidence for such connection. Cummings and Coleman (1968b) studied rare boundary crossings at the 6.6 R_E synchronous distance and concluded there was no connection.

Fairfield and Cahill (1966) have shown that geomagnetic activity tends to be higher when the magnetosheath field has a southward component. Since a southward interplanetary field should correspond to the greatest connection, the most rapid magnetosphere convection, and the highest geomagnetic activity, this result has been interpreted as indirect evidence for the connection of the interplanetary field to that of the magnetosphere. This experimental evidence associating a southward field with geomagnetic activity has been well verified (reviewed by Hirshberg and Colburn, 1968) but an alternate interpretation of the results is still a possibility.

5. SUMMARY AND CONCLUSIONS

Mid-latitude dipole field lines are simultaneously compressed by the solar wind in the sunlit hemisphere, extended in the night hemisphere and distorted from the meridian planes. High-latitude field lines are swept back to form the geomagnetic tail. The outermost closed field line orginates near 78° on the day side of the earth and near 68° on the night side. Lines from the south polar cap directed away from the earth form the lower half of the geomagnetic tail while lines from the north polar cap directed toward the earth form the upper half of the tail. These two halves are separated by a neutral sheet less than 1 R_E thick which is itself contained within a broad region of depressed fields approximately 12 R_E thick. The average magnitude of the tail field decreases from a value of 20 γ at 15 R_E to a value of 8 γ at 60 R_E. Any magnitude gradient beyond 60 R_E is very small. The cross-section of the tail is elliptical with typical transverse dimensions of 45 R_E east-west and 60 R_E north-south at the orbit of the moon. The average tail-field direction makes a small angle with the earth-sun line which is consistent with the aberration associated with the earth's motion about the sun. The magnetic tail has been observed for brief intervals (minutes to hours) at 500 R_E and 1000 R_E. These observations are interpreted as being caused by either filamentary structure of the tail or by rapid movement of the tail past the spacecraft in response to changing solar wind directions.

Important changes in the magnetospheric magnetic field configuration occur at the time of magnetic storms and substorms. The ring current associated with main phase magnetic storms inflates the inner magnetosphere forcing the equatorial crossing point of field lines out further into the magnetosphere. At the same time the outer boundary of trapped particles collapses and the field strength in the tail increases. These results are interpreted as an increase in the number of field lines forming the tail.

Magnetospheric substorms correspond to a relaxation of the magnetosphere towards a more dipolar-like state with more lines of force closing near the earth. This relaxation is suggested by direct measurements of the magnetic field and supported by the observed increase in the trapped particle boundary to 75° after substorms. The appearance of plasma and particles at 17 R_E after substorms is further evidence supporting this relaxation. The fact that magnetospheric substorms cause gross readjustments in the magnetosphere configuration should be taken into consideration when using an average magnetosphere configuration. Roederer et al. (1968) have partially circumvented the problem of time variations by using the measurements of a synchronous satellite as input to a field model which defines the field at all points.

The question of connection of the magnetosphere lines of force to the interplanetary lines has not been fully resolved. Access of energetic particles to the tail and polar cap appears to be easier than that expected for diffusion but lower than the expected for direct access. The experimental evidence relating the interplanetary field to geomagnetic activity has been confirmed. This result has been interpreted as evidence for connection; however, another interpretation is possible.

6. FUTURE WORK

The general properties of a time average magnetosphere are rather well known. Numerous spacecraft have probed the outer magnetosphere in latitudes below 60° but the higher latitude regions are still unexplored. Although Figure 1 does not reveal any obvious dawn-dusk asymmetry, the ATS measurements suggest that such effects do exist. This question can be studied in the future with additional equatorial measurements in the dusk hemisphere.

The most important magnetosphere problems lie in the area of time variations. Adjustments in the field configuration at the time of magnetic storms and substorms have recently been investigated but additional work is called for. Field changes must be studied as a function of local time and related to changes in the magnetospheric plasma and particles. Further studies of trapped particles are needed to give information on the latitude of the outermost closed field lines near midnight and how they vary with time.

Detailed studies are needed to determine whether the magnetopause is (i) a tangential discontinuity indicating complete separation of geomagnetic and interplanetary field lines, (ii) a rotational discontinuity with geomagnetic field lines connecting to the interplanetary field, or (iii) tangential and rotational discontinuities at different times. The above investigation is relevant to the very important questions of how field lines are dragged back to form the tail and how energy is transferred from the solar wind to the magnetosphere.

The exact nature of the tail at 500 - 1000 R_E and its extent beyond 1000 R_E remain unanswered questions. Finally the development of improved quantitative models of the magnetosphere and tail which include time variations should be a future goal.

Acknowledgements. The author would like to thank Mr. K. W. Behannon and Drs. N. F. Ness, D. J. Williams and F. Mariani for helpful discussions and access to their data prior to publication.

REFERENCES

Akasofu, S.-I. and S. Chapman, Magnetic storms: the simultaneous development of the main phase (DR) and of polar magnetic substorms (DP), J. Geophys. Res., 68, 3155-3158, 1963.

Axford, W. I., Magnetic storm effects associated with the tail of the magnetosphere, paper presented at ESRO Conference on the Magnetosphere, November 1965.

Bame, S. J., J. R. Asbridge, H. E. Felthauser, E. W. Hones, and I. B. Strong, Characteristics of the plasma sheet in the earth's magnetotail, J. Geophys. Res., 72, 113-129, 1967.

Behannon, Kenneth W., Mapping of the earth's bow shock and magnetic tail by Explorer 33, J. Geophys. Res., 73, 907-930, 1968.

Behannon, Kenneth W., Geometry of the geomagnetic tail, NASA-GSFC preprint, April 1969.

Behannon, K. W. and N. F. Ness, Magnetic storms in the earth's magnetic tail, J. Geophys. Res., 71, 2327-2351, 1966.

Blake, J. B., G. A. Paulikas, and S. C. Freden, Latitude-intensity structure and pitch-angle distributions of low-energy solar cosmic rays at low altitude, J. Geophys. Res., 73, 4927-4934, 1968.

Brandt, John C., Charles Wolff, and Joseph P. Cassinelli, A calculation of the angular momentum of the solar wind, Interplanetary Gas XVI, preprint, 1969.

Cahill, Laurence J., Jr., Inflation of the inner magnetosphere during a magnetic storm, J. Geophys. Res., 71, 4505-4519, 1966.

Cahill, L. J., Jr. and V. L. Patel, The boundary of the geomagnetic field, August to November 1961, Planetary Space Sci., 15, 997-1033, 1967.

Cummings, W. D. and P. J. Coleman, Jr., Simultaneous magnetic field variations at the earth's surface and at synchronous, equatorial distance. Part 1, bay associated events, Radio Sci., 3, 758-761, 1968a.

Cummings, W. D. and P. J. Coleman, Jr., Magnetic fields in the magnetopause and vicinity at synchronous altitude, J. Geophys. Res., 73, 5699-5718, 1968b.

Cummings, W. D., J. N. Barfield, and P. J. Coleman, Jr., Magnetosphere substorms observed at the synchronous orbit, J. Geophys. Res., 73, 6687-6698, 1968.

Davis, T. N. and R. Parthasarathy, The relationship between polar magnetic activity DP and growth of the geomagnetic ring current, J. Geophys. Res., 72, 5825-5836, 1967.

Dungey, J. W., Theory of the quiet magnetosphere, in Solar-Terrestrial Physics, edited by J. W. King and W. S. Newman, Academic Press, New York, pp. 91-106, 1968.

Fairfield, D. H. and L. J. Cahill, Jr., Transition region magnetic field and polar magnetic disturbances, J. Geophys. Res., 71, 155-169, 1966.

Fairfield, D. H., Average magnetic field configuration of the outer magnetosphere, J. Geophys. Res., 73, 7329-7338, 1968a.

Fairfield, D. H., Simultaneous measurements on three satellites and the observation of the geomagnetic tail at 1000 R_E, J. Geophys. Res., 73, 6179-6187, 1968b.

Feldstein, Y. I., Some problems concerning the morphology of auroras and magnetic disturbances in high latitudes, Geomagnetism and Aeronomy, 3, 183-192, 1963.

Frank, L. A., J. A. Van Allen and J. D. Craven, Large diurnal variations of geomagnetically trapped and of precipitated electrons observed at low altitudes, J. Geophys. Res., 69, 3155-3167, 1964.

Frank, L. A., On the extraterrestrial ring current during geomagnetic storms, J. Geophys. Res., 72, 3753-3767, 1967.

Heppner, J. P., M. Sugiura, T. L. Skillman, B. G. Ledley, and M. Campbell, OGO A magnetic field observations, J. Geophys. Res., 72, 5417-5471, 1967.

Hirshberg, J. and D. S. Colburn, Interplanetary field and geomagnetic variations - a unified view, to be published Planetary Space Sci., 1969.

Hones, E. W., Jr., J. R. Ashbridge, S. J. Bame, and I. B. Strong, Outward flow of plasma in the magnetotail following geomagnetic bays, J. Geophys. Res., 72, 5879-5892, 1967.

Hones, E. W., Jr. and S. Singer, Simultaneous observations of electrons ($E > 45$ keV) at 2000 kilometer altitude and at 100,000 kilometers in the magnetotail, J. Geophys. Res., 73, 7339-7359, 1968.

Kennel, C. F. and H. E. Petschek, Limit on stably trapped particle fluxes, J. Geophys. Res., 71, 1-28, 1966.

Krimigis, S. M., J. A. Van Allen and T. P. Armstrong, Simultaneous observations of solar protons inside and outside the magnetosphere, Phys. Rev. Letters, 18, 1204-1206, 1967.

Lanzerotti, L. J., Penetration of solar protons and alphas to the geomagnetic equator, Phys. Rev. Letters, 21, 929-933, 1968.

Lin, R. P., Observations of lunar shadowing of energetic particles, J. Geophys. Res., 73, 3066-3071, 1968.

Lin, R. P. and K. A. Anderson, Evidence for connection of geomagnetic tail lines to the interplanetary field, J. Geophys. Res., 71, 4213-4217, 1966.

Lin, W. C., I. B. McDiarmid, and J. R. Burrows, Electron fluxes at 1000 km altitude associated with auroral substorms, Can. J. Phys., 46, 80-83, 1968.

Mariani, F. and N. F. Ness, Observation of the geomagnetic tail at 500 earth radii by Pioneer 8, NASA-GSFC preprint X616-69-54, 1969.

Mead, G. D., Deformation of the geomagnetic field by the solar wind, J. Geophys. Res., 69, 1181-1195, 1964.

Meng, C. I. and S.-I. Akasofu, A study of polar magnetic substorms-II three dimensional current system, Preprint, Geophysical Institute, University of Alaska, 1969.

Michel, F. C. and A. J. Dessler, Physical significance of inhomogenities in polar cap absorbtion events, J. Geophys. Res., 70, 4305-4311, 1965.

Mihalov, J. D., D. S. Colburn, R. G. Currie, and C. P. Sonett, Configuration and reconnection of the geomagnetic tail, J. Geophys. Res., 73, 943-959, 1968.

Montgomery, Michael, D. and S. Singer, Penetration of solar energetic protons into the magnetotail, J. Geophys. Res., 74, 2869-2880, 1969.

Ness, N. F., The earth's magnetic tail, J. Geophys. Res., 70, 2989-3005, 1965.

Ness, Norman F., The geomagnetic tail, Rev. of Geophys., 7, 97-127, 1969.

Ness, N. F., K. W. Behannon, S. C. Cantarano, and C. S. Scearce, Observations of the earth's magnetic tail and neutral sheet at 510,000 kilometers by Explorer 33, J. Geophys. Res., 72, 927-933, 1967a.

Ness, Norman F., Clell S. Scearce, and Sergio C. Cantarano, Probable observations of the geomagnetic tail at 10^3 R_E by Pioneer 7, J. Geophys. Res., 72, 3769-3776, 1967b.

Ness, N. F. and D. J. Williams, Correlated magnetic tail and radiation belt observations, J. Geophys. Res., 71, 322-325, 1966.

Patel, V. L. and A. J. Dessler, Geomagnetic activity and size of magnetospheric cavity, J. Geophys. Res., 71, 1940-1942, 1966.

Rao, C. S. R., Some observations of energetic electrons in the outer radiation zone during magnetic rays, J. Geophys. Res., 74, 794-801, 1969.

Roederer, Juan G., Warren D. Cummings, Paul J. Coleman, and Marilyn F. Robbins, Determination of magnetospheric parameters from magnetic field measurements at synchronous altitudes, University of Denver, Dept. of Physics, Denver Research Institute Report, 1968.

Rothwell, P. and V. Wallington, The polar substorm and electron "Islands" in the earth's magnetic tail, Planetary Space Sci., 16, 1441-1451, 1968.

Schield, M. A., J. W. Freeman, and A. J. Dessler, A source for field-aligned currents at auroral latitudes, J. Geophys. Res., 74, 247-256, 1969.

Siscoe, G. L. and W. D. Cummings, On the cause of geomagnetic rays, UCLA preprint, Februar 1969.

Sonnerup, B. U. Ö. and L. J. Cahill, Jr., magnetopause structure and attitude from Explorer 12 observations, J. Geophys. Res., 72, 171-183, 1967.

Speiser, T. W. and N. F. Ness, The neutral sheet in the geomagnetic tail: its motion, equivalent currents, and field line reconnection through it, J. Geophys. Res., 72, 131-141, 1967.

Sugiura, Masahisa, Results of magnetic surveys of the magnetosphere and adjacent regions, NASA-GSFC, preprint X612-69-12, January 1969. Presented at the IAGA Symposium on Description of the Earth's Magnetic Field, Washington, D. C., October 1968, to be published in the Report of the World Magnetic Survey.

Sugiura, M., T. L. Skillman, B. G. Ledley, and J. P. Heppner, Propagation of the sudden commencement of July 8, 1966, to the magnetotail, J. Geophys. Res., 73, 6699-6709, 1968.

Taylor, Harold E. and Edward W. Hones, Jr., Adiabatic motion of auroral particles in a model of the electric and magnetic fields surrounding the earth, J. Geophys. Res., 70, 3605-3628, 1965.

Van Allen, J. A. and N. F. Ness, Particle shadowing by the moon, J. Geophys. Res., 74, 71-93, 1969.

Williams, D. J. and C. O. Bostrom, The February 5, 1965 solar proton event 2. Low energy proton observations and their relation to the magnetosphere, J. Geophys. Res., 72, 4490-4506, 1967.

Williams, D. J. and C. O. Bostrom, Proton entry in the magnetosphere on May 26, 1967, J. Geophys. Res., 74, 3019-3026, 1969.

Williams, Donald J. and Gilbert D. Mead, Nightside magnetosphere configuration as obtained from trapped electrons at 1100 kilometers, J. Geophys. Res., 70, 3017-3029, 1965.

Williams, Donald J. and Norman F. Ness, Simultaneous trapped electron and magnetic tail field observations, J. Geophys. Res., 71, 5117-5128, 1966.

Wolfe, J. H., R. W. Silva, D. D. McKibbin, and R. H. Mason, Preliminary observations of a geomagnetic wake at 1000 earth radii, J. Geophys. Res., 72, 4577-4581, 1967.

Zmuda, A. J., F. T. Heuring, and J. H. Martin, Dayside magnetic distrubances at 1100 kilometers in the auroral oval, J. Geophys. Res., 72, 1115-1117, 1967.

LOW ENERGY PARTICLE FLUXES IN THE GEOMAGNETIC TAIL

Vytenis M. Vasyliunas

Department of Physics and Center for Space Research
Massachusetts Institute of Technology
Cambridge, 02139 Massachusetts, USA

1. INTRODUCTION

Much observational work in space research within the past few years has been concerned with what are commonly called "low energy particles". That term is somewhat ambiguous: 100 MeV may be low energy to a cosmic-ray physicist, while 10 eV is probably a rather high energy to many ionospheric physicists. Hence we should begin by clarifying the meaning of "low energy" in the present context. Most of the early particle observations in the magnetosphere were obtained with instruments (such as Geiger tubes and scintillation counters) capable of observing electrons only above an energy of ~40 keV and protons above ~100 keV. Extensive data on particles above these energy limits had been accumulated by 1964 and the gross features, at least, of their distribution could be mapped out, while knowledge of lower energy particles within the magnetosphere was still rather meager by comparison. Since then, however, numerous observations with Faraday cups and curved-plate analyzers have extended our detailed knowledge of magnetospheric particles down to energies of some tens of eV; these are the so-called "low energy" observations - the energy is low compared to what was being measured some years ago.

However, these low energy observations represent more than a mere improvement in experimental technique. They are qualitatively different from the observations hitherto possible and provide a new way of looking at magnetospheric plasma. This can be illustrated with Figure 1, which shows schematically a typical electron energy spectrum in the magnetosphere. The older "high energy" observations were limited to the small fraction of particles whose energies greatly exceeded the average energy of the entire plasma, i. e., to the "high energy tail" of the particle distribution. The measured quantity usually was the total flux of particles with energies above some threshold set by the instrument and otherwise having no particular physical significance, plus a power-law exponent or some similar measure of the steepness of the spectrum; no information on the collective properties of the entire plasma could be obtained. By contrast, once the measurements were extended to energies of some hundred eV and below, a well-defined peak in the energy spectrum was found; below the peak the particle flux decreases rapidly with decreasing energy. Thus the

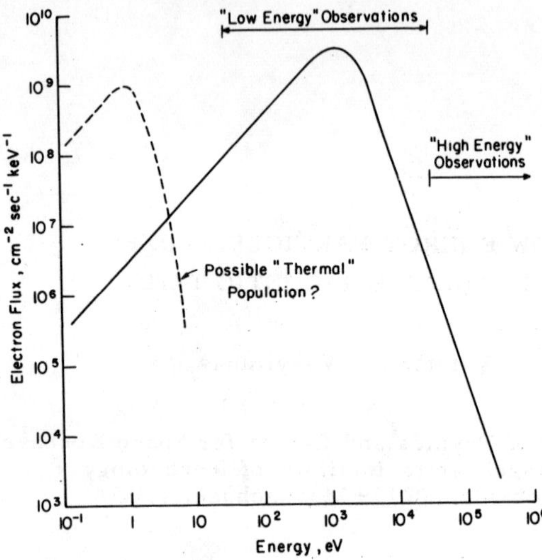

Vasyliunas

Figure 1. Solid line: Sketch of a typical electron energy spectrum (differential flux per unit energy) observed in the outer magnetosphere and the magnetotail. Dashed line: Spectrum of a hypothetical unobserved "thermal" electron population (here drawn with a total density 10 times that of the observed low energy electron population).

magnetosphere contains a population of particles with an average energy in the ~ 100 eV to ~ 10 keV range, and the present "low energy" observations encompass a major fraction of this population. It is now possible to integrate the flux vs. energy curve (the rapid decrease of flux at energies below the peak allows one to extrapolate to zero energy, as sketched in Figure 1, with little effect on the value of the integral) and to compute quantities referring to the entire population, such as total density, total energy density, or total energy flux. The observational results can now be presented using terms of plasma physics, rather than terms borrowed from cosmic ray studies.

As indicated in Figure 1, there may also exist another population of (so-called "thermal") particles at still lower energies, below the range of present "low energy" observations. No reliable direct observations of such low energy particles in the outer magnetosphere and the magnetotail are presently available, to my knowledge; some indirect evidence will be discussed later.

This paper reviews certain aspects of low energy particle observations in the magnetosphere. First we will describe the over-all spatial distribution of the low energy particles. Then we will concentrate on the geomagnetic tail and discuss properties of the particle population within the plasma sheet, evidence bearing on the question of a "thermal" population, and present knowledge of particles outside the plasma sheet. Finally we will present some recent results on a topic of special interest to this meeting, namely, the connection between particle structures in the geomagnetic tail and auroral phenomena in the ionosphere. A major topic that will not be covered is time variations and their relation to geomagnetic substorm activity; the reader is referred to the work of Hones (1968a; 1968b; 1969). The entire field of low energy observations has recently been reviewed by Gringauz (1969). Techniques of measurement and analysis will not be described; they are discussed in many of the original papers, and a recent review has been given by Vasyliunas (1969a).

2. SPATIAL DISTRIBUTION OF LOW ENERGY PARTICLES IN THE MAGNETOSPHERE

Our present conception of the average distribution of low energy electrons in the equatorial region of the magnetosphere is sketched in Figure 2. Intense fluxes of electrons extend across the entire

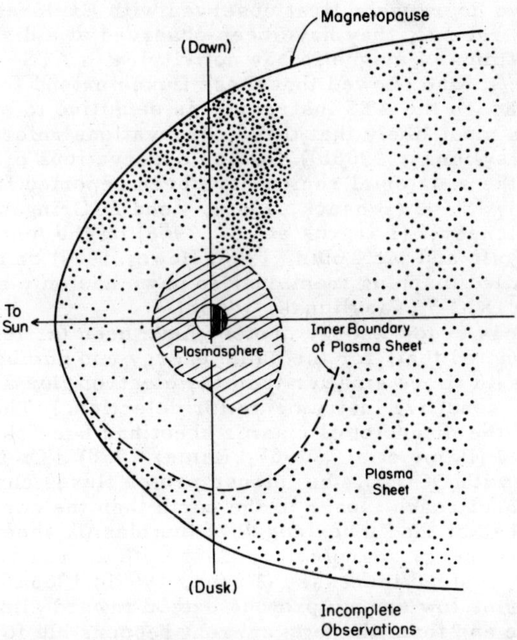

Figure 2. The average low energy electron distribution in the equatorial region of the magnetosphere, viewed from the north pole. The density of dots indicates qualitatively the flux of electrons with energies from 100 eV to several keV. The shaded region in the predawn sector has not yet been studied in detail; it is, however, known that strong electron fluxes are present within it.

magnetotail, forming the plasma sheet, first detected by the Luna 2 space probe in 1959 (Gringauz et al., 1960a; 1960b; Gringauz, 1961) and since 1964 extensively studied with instruments aboard the Vela and OGO satellites (Bame et al., 1966; 1967; Bame, 1968; Hones, 1968a; 1968b; 1969; Vasyliunas, 1968a; Frank, 1967a; 1967b; 1968a; 1968b). On the evening side of the magnetosphere these electron fluxes terminate at a well-defined boundary, called the inner boundary (or inner edge) of the plasma sheet. This boundary lies at an equatorial radial distance of ~ 11 R_E during magnetically quiet periods and 6 - 8 R_E during times of magnetic bay activity, and has been traced from near the midnight meridian through dusk to a local time near noon, where it meets the magnetopause (Vasyliunas, 1968a; 1968b). Earthward of the inner boundary, at local times between dusk and midnight the electron flux in the 100 eV to a few keV energy range is very weak in comparison with that within the plasma sheet (as will be discussed later, this is not a decrease in electron density but a decrease in their mean energy to values well below 100 eV). As one proceeds from dusk toward earlier local times, appreciable electron fluxes appear between the plasmapause and the inner edge of the plasma sheet, grow in intensity and acquire a marked outward radial gradient in the morning side, and become very intense (exceeding typical plasma sheet fluxes by an order of magnitude) at local times near and before dawn (Vasyliunas, 1968b). Electrons in the morning side of the magnetosphere were the first low energy particles to be detected in outer space, with an instrument aboard Luna 1 (February 1959) (Gringauz et al., 1960b; Gringauz, 1961; 1964). The very intense electron fluxes characteristic of the predawn hours were first observed with Explorer 12 by Freeman (1964); more recently they have been observed at a distance of 6.6 R_E (only during times of magnetic bay activity) with ATS-1 by Freeman and Maguire (1967), who showed that these fluxes extend from dawn to midnight (although the ATS instrument is sensitive to both electrons and protons, it is most likely that these observations refer primarily to electrons (Vasyliunas, 1968b)). Other observations of low energy electrons in the equatorial region have been reported from IMP 1 (Serbu, 1964), IMP 2 (Binsack, 1966), Zond 2 (Gringauz and Khokhlov, 1965), and Pioneer 7 (Lazarus et al., 1968) on the morning side, and Pioneer 6 (Wolfe and McKibbin, 1968; Howe, 1969) on the evening side; the first detailed mapping from dusk to dawn has only recently been obtained with OGO 3 (Vasyliunas, 1968b).

Observations of low energy protons have been far less numerous. (It should be noted that, for the same energy and number density, the proton flux is 43 times smaller than the electron flux and thus may require more sensitive instruments for detection.) The proton component of the magnetotail plasma sheet has been observed with the Vela satellites (Bame et al., 1967; Bame, 1968). On the evening side (as observed with OGO 3), the intense proton fluxes characteristic of the plasma sheet reach closer to the earth than the corresponding electron fluxes; at the "inner edge" of the plasma sheet, where the electron mean energy sharply decreases with decreasing distance, the proton energy instead increases (Frank, 1967b; 1968a; 1968b). During magnetic storms low energy protons extend inward almost to the plasmasphere and form the ring current responsible for the storm main phase (Frank, 1967c; 1968a). Very recently Frank (1968c) has reported that the largest proton fluxes occur in the noon-to-dusk quadrant.

So far we have discussed particles in the equatorial region. Figure 3 illustrates schematically the distribution of low energy electrons in a meridian plane on the night side of the earth. (Also shown are typical orbits of some of the principal satellites that have provided data on low energy electrons.) In the geomagnetic tail, plasma sheet particles are confined in the north-south direction to within a few R_E of the magnetic neutral sheet, as first established by Vela measurements (Bame et al., 1966; 1967; Bame, 1968; Hones, 1968a; 1968b; 1969) and confirmed by OGO 1 and 3 (Vasyliunas, 1968a). (This planar geometry was the reason for the term "plasma sheet", introduced by Bame et al., 1966.) A seasonal variation in the north-south position of the plasma sheet has been observed (Bame et al., 1967; Vasyliunas, 1968a) that reflects the north-south displacement of the neutral sheet as the earth's dipole - solar wind angle varies (Speiser and Ness, 1967; Russell and Brody, 1967). Thus the plasma sheet appears to be always at least roughly centered about the magnetic neutral sheet; however, its width of several earth radii greatly exceeds the ~ 10^3 km width of the actual magnetic field reversal region (Ness, 1965; Speiser and Ness, 1967).

Even before the geometry of the tail plasma sheet had been observationally established, intense low energy electron fluxes were

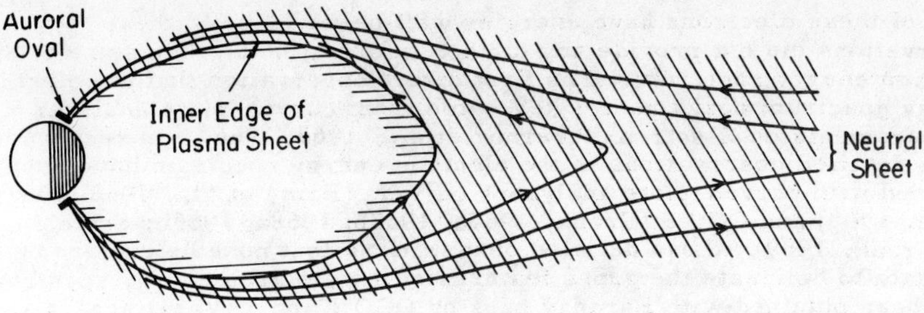

Figure 3. The average low energy electron distribution in a meridian plane on the night side of the magnetosphere. The sun is to the left, the magnetotail to the right. The satellite orbits are shown rotated into a common meridian plane, without regard to local time (the Vela observations cover all local times, the OGO 1, OGO 3, and Mars 1 observations are mostly between dusk and midnight, and the Electron 2 observations are near midnight and near dawn).

observed close to the earth at high latitudes by Mars 1 in 1962 (Gringauz et al., 1964) and Electron 2 in 1964 (Vernov et al., 1965; 1966). These fluxes occurred in a narrow band of latitudes, roughly at invariant latitudes 65 - 70°, i.e., in the neighborhood of the auroral zone. Gringauz et al. (1964) at once suggested that these electron fluxes were part of the same low energy particle region that had earlier been observed by Luna 2 near the equatorial plane and which we now know to be the plasma sheet. Vasyliunas (1968a) traced the inner boundary of the plasma sheet from the equator up to latitudes of ~ 40° and concluded that the inner boundary coincides with a shell of magnetic field lines. All these observations suggest the picture sketched in Figure 3: plasma sheet electron fluxes extend from the magnetotail toward the earth in a pair of narrow "horns", reaching the ionosphere at or near the auroral zones. (Pictures of this kind were earlier presented and the connection with the aurora suggested by Dessler and Juday (1965), O'Brien (1967), Piddington (1967a), and possibly others.) Recently the secular change of the OGO 1 orbit has made it possible to investigate the precise connection between the plasma sheet and the auroral oval; we shall discuss this topic in Section 4.

Outside the plasma sheet, particle fluxes in the magnetotail are very weak and have only recently been detected by Vela 4 (Bame, 1968); they were generally below the sensitivity and/or energy limits of earlier instruments (Bame et al., 1967; Vasyliunas, 1968a). This absence of strong fluxes at great heights above the neutral sheet persists right up to the magnetopause (Vasyliunas, 1968a; Bame, 1968); the early suggestion by Dessler and Juday (1965) of a θ-shaped plasma sheet has not been borne out by the data.

3. PROPERTIES OF LOW ENERGY PARTICLE POPULATION IN THE MAGNETOTAIL

The early observations by Gringauz and co-workers indicated that within what we now call the plasma sheet the total flux of electrons with energies above ~ 200 eV is of the order of $2 \cdot 10^{8}$ cm^{-2} sec^{-1} and that most of these electrons have energies well below 50 keV; these observations did not provide any further information on the nature of the electron energy spectrum. The first direct observation that the electron energy spectrum peaks near 1 keV came from curved-plate analyzer measurements on Electron 2 (Vernov et al., 1965; 1966). In recent years detailed observations of the electron energy spectrum have been obtained with curved-plate analyzers on Vela (Bame et al., 1966; 1967; Bame, 1968) and OGO 3 (Frank, 1967a; 1967b; 1968a; 1968b; Schield and Frank, 1969). Less detailed observations that nonetheless are adequate to delineate the gross features of the electron energy spectrum have been obtained with Faraday cups on OGO 1 and 3 (Vasyliunas, 1968a), Pioneer 7 (Lazarus et al., 1968), and Pioneer 6 (Howe, 1969). The typical energy spectrum of electrons within the plasma sheet is found to be rather similar to the idealized sketch of Figure 1: there is a broad maximum at an energy that can range from ~100 eV to several keV, and a flux decrease at both low and high energies. Well above the peak the differential flux decreases roughly as an inverse power of the energy, with exponents typically in the range 3 - 5, and extrapolates more or less smoothly into the power law spectra observed at high (50 - 100 keV) energies (Montgomery et al., 1965; Montgomery, 1968). On the other hand, there is some evidence from occasional observation of a double-peaked energy spectrum (Frank, 1967a) and from detailed study of time variations of electron fluxes at different energies (Montgomery, 1968) that electrons above some 10 keV energy may not be merely the high energy tail of the main plasma sheet electron population but possibly a separate population. The angular distribution of electrons within the plasma sheet appears to be approximately isotropic at all energies up to hundreds of keV (Vernov et al., 1966; Bame et al., 1967; Hones et al., 1968; Retzler and Simpson, 1969).

A smooth, broad, single-peaked, isotropic particle distribution can be quantitatively characterized by two parameters, which may be chosen as the number density n and the energy density U; in terms of the particle velocity distribution function $f(\underline{v})$ these are

$$n = \int d\underline{v}\, f(\underline{v})$$

$$U = \int d\underline{v}\, \tfrac{1}{2} m v^2 f(\underline{v})$$

One may then define the mean particle energy $E \equiv U/n$ (if the distribution function is Maxwellian with a temperature T, then $E = 3/2\, kT$). (Other sometimes useful quantities such as total omnidirectional particle or energy flux may be defined in a similar manner.) These quantities may be deduced from the measurements either by approximating the integrals by sums and estimating the required values of $f(\underline{v})$ from the measured fluxes, or by assuming a specific functional form for $f(\underline{v})$ and varying parameters to obtain a best fit to the measurements; these methods have been reviewed by Vasyliunas (1969a).

Two major surveys of the quantitative properties of the plasma sheet electron population have been published, as of the time of this writing. Vasyliunas (1968a) used OGO 1 Faraday cup measurements at local times from ~17 to ~22 hours and distances up to 24 R_E. Figure 4 shows the distribution of electron parameters obtained from a sample of these measurements; the electron properties were found to be more or less uniform over the entire surveyed region of the plasma sheet. Hones (1968a; 1969) used curved-plate analyzer measurements from the Vela satellites to construct maps showing the average values of electron density, energy density, and mean energy as function of latitude and local time at a constant distance of 17 R_E. The maps indicate that the electron mean energy is ~1 keV near the center of the plasma sheet,

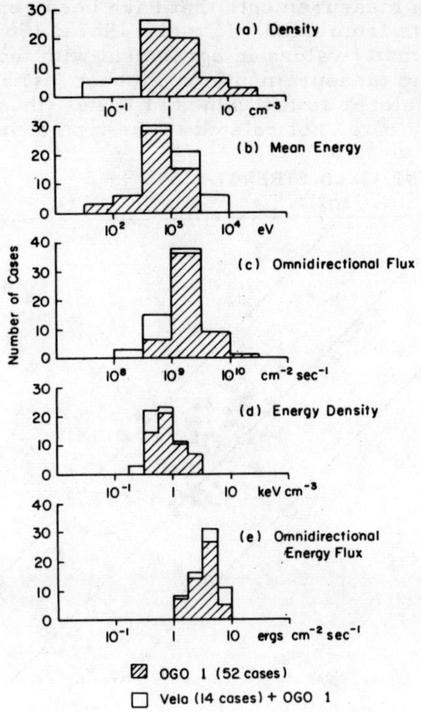

Figure 4. Frequency of occurrence of various electron parameters within the plasma sheet at local times from dusk to midnight from observations with OGO 1 (plus a few observations with Vela). The histograms are constructed at intervals of 1/2 decade in (a), (b), and (c), at 1/4 decade in (d) and (e). The Vela observations were taken from sample values published by Bame et al. (1967) and may not be representative of the entire set of Vela observations.

decreases with increasing latitude to ~ 0.3 - 0.5 keV near the boundaries of the plasma sheet, and shows no significant dependence on local time. The density and energy density show a minimum near midnight; in the center of the plasma sheet the density is ~ 0.1 - 0.3 cm^{-3} near midnight and ~ 0.3 - 0.4 cm^{-3} near the flanks of the magnetotail, while the energy density in the same regions is ~ 0.1 - 0.2 and ~ 0.3 - 0.4 keV cm^{-3}, respectively. The local time dependence of all quantities is roughly symmetric about midnight; the striking dawn-dusk asymmetry observed in high energy (E > 45 keV) electrons (Montgomery et al., 1965; Bame et al., 1967; Hones, 1968b; Montgomery, 1968) apparently does not persist at lower energies.

Comparing the numbers given by Hones with those in Figure 4, it is apparent that both sets of measurements agree on the electron mean energy but there may be some discrepancy on density and energy density: the "most probable" values given by Vasyliunas are a factor of ~ 3 higher than the "average" values of Hones for the same local time range. Whether the discrepancy is significant is difficult to say at present; the difference is no larger than the spread of individual values in Figure 4, and may perhaps be due in part to sampling statistics.

Of the other electron measurements that have been reported, the curved-plate analyzer results from OGO 3 (Frank, 1967a; 1967b; Schield and Frank, 1969) give density values in agreement with those of Vasyliunas, while the Faraday cup measurements on Pioneer 7 (Lazarus et al., 1968) are somewhat closer to the values of Hones (thus the discrepancy appears to be, at any rate, not related to the type of instrument used).

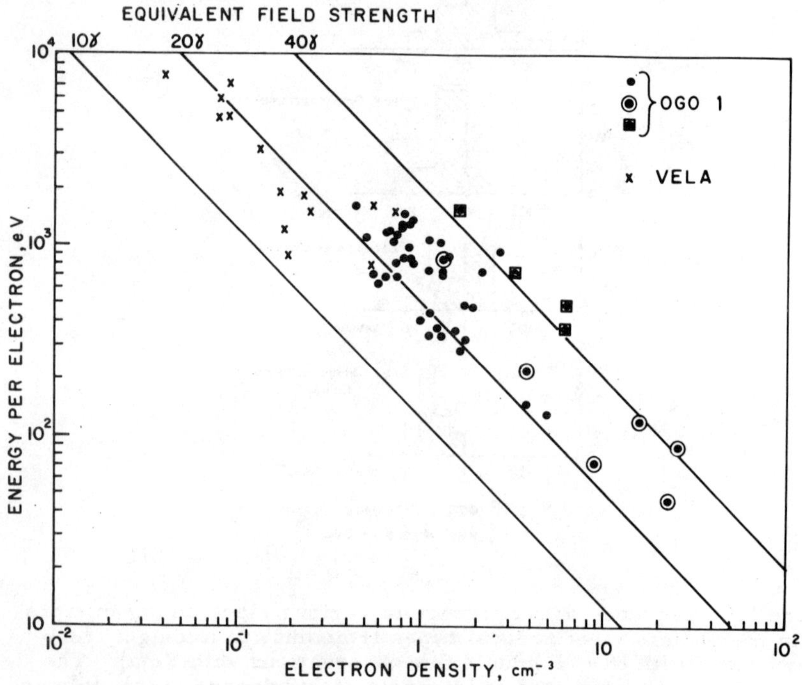

Figure 5. Plot of electron mean energy vs. density, for the same set of data as Figure 4. Points enclosed in circles refer to measurements during the October 3, 1964, magnetic storm; those enclosed in squares refer to the November 15, 1964, storm.

Although one may discuss "typical" or "average" values, it is also clear from Figure 4 that observed individual values of electron density and mean energy vary over a range of several decades. High densities, however, tend to be accompanied by low mean energies and vice versa, so that the electron energy density shows relatively little variation (Vasyliunas, 1968a). This is illustrated in Figure 5, in which mean energy is plotted against density (from the same data as Figure 4); lines of constant energy density have been drawn and labeled by the magnetic field strength corresponding to twice the energy density. The individual points in Figure 5 come from a wide range of spatial locations and conditions of geomagnetic activity (note that very high densities and low energies usually occur during magnetic storms). Figure 5 thus primarily represents gross, large-scale spatial and temporal variations in plasma sheet properties; it shows that the electron energy density is little affected by such variations and on the average is of the same order of magnitude as the magnetic field energy density.

The relation between the instantaneous values of magnetic and plasma energy density (as opposed to the over-all average behavior just discussed) was investigated by Lazarus et al. (1968), using magnetic

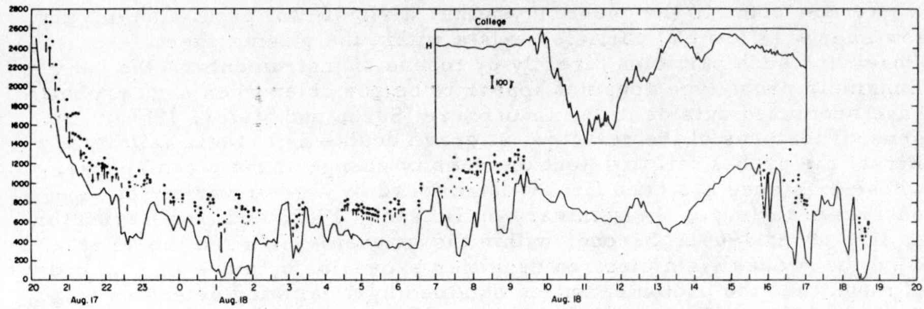

Figure 6. Magnetic field pressure (solid line) and total pressure (data points), assuming equal proton and electron pressures, measured during the magnetosphere passage of Pioneer 7. The spacecraft crossed the magnetotail at a nearly constant local time of ~2 hours; its distance was 11 R_E at the beginning of the figure (the high field values reflect the proximity to the earth) and 32 R_E at 10 hours on August 18 (After Lazarus et al., 1968).

field and electron measurements during the single magnetotail traversal of Pioneer 7. Figure 6, taken from their paper, shows the magnetic field pressure and the sum of it and twice the electron pressure. (Magnetic field pressure is equal to the energy density; particle pressure is 2/3 of the energy density.) As can be seen, major decreases in magnetic pressure are generally compensated by increases in electron pressure, so that the total pressure remains approximately constant; this is especially apparent in Figure 6 during the intervals 22 - 23 h, 1 - 2 h (which contains a magnetic neutral sheet crossing), and 5 - 9 h.

It appears, then, that plasma sheet electrons undergo at least two types of variations. First, their energy density may change, but at the same time the magnetic field also changes so that the total (particle plus field) pressure remains constant; thus these variations preserve, at least approximately, a condition of hydrostatic balance. Second, the electron density and mean energy may both change by as much as an order of magnitude, but the energy density remains approximately constant; these variations preserve what may loosely be called equipartition of energy between electrons and magnetic field.

The proton component of the plasma sheet has been detected with curved-plate analyzers (Bame et al., 1967; Frank, 1967b; 1968a; 1968b; Bame, 1968; Hones, 1968a), but our knowledge of it is as yet far less detailed than that of electrons. The proton energy spectrum has a smooth, broad, single-peaked shape, qualitatively similar to the electron spectrum. The proton angular distribution is approximately isotropic, but slight (~10%) anisotropies possibly indicative of a bulk flow (with speeds $\lesssim$ 50 km sec^{-1}) have been observed (Hones, 1968a). The number density of protons equals that of the electrons, to within the observational uncertainties. The proton mean energy appears to exceed the electron mean energy, but by how much is not yet entirely clear: the ratio of proton to electron mean energy is ~5 according to Hones (1969) and $\gtrsim$ 10 according to Bame (1968), while Frank (1967b) reports examples of proton and electron spectra in which the <u>mean</u> energies of the two are approximately the same (although the proton spectrum has a much broader shape), and Lazarus et al. (1968) infer by an indirect argument based on pressure balance that the ratio is ~1 to 3.

We now come to the question whether a significant population of very low energy ($\lesssim$ few eV) particles exists within the plasma sheet. Observing such particles directly by means of instruments of the Langmuir probe type does not appear to be possible; when such probes have been used outside of the ionosphere (Serbu and Maier, 1966), several features of the results cast grave doubts as to their validity. First, the probes fail to detect the density change at the plasmapause, whose existence has been firmly established by a great variety of observations (e.g., the summary by Binsack, 1967, and also Carpenter et al., 1968; 1969). Second, within the magnetosheath and the solar wind the probes yield electron densities exceeding by more than an order of magnitude the proton densities obtained by solar wind detectors (e.g., Axford, 1968). (It should be kept in mind that even if a so-called "thermal" component existed in the solar wind, it would have to move with the solar wind velocity, at least as far as motion across magnetic field lines is concerned; thus the "thermal" protons would have energies of hundreds of eV and would be readily observable with the usual solar wind detectors.) Since Langmuir probes apparently do not work properly either near the plasmapause or beyond the magnetopause, there is little reason to suppose that they work in the outer magnetosphere and the magnetotail. The difficulty appears to lie in the large fluxes of ~ 1 eV electrons that can be produced at the surface of the satellite by photoemission and secondary emission; Whipple and Parker (1969a; 1969b) have recently demonstrated that these secondary fluxes can account for the anomalous probe results in the solar wind.

Very low energy particles thus must be studied by indirect techniques, the most important of which is observing the effect of the particles on electromagnetic wave propagation. Dispersive effects in radar signals reflected from the moon were used by Howard et al. (1965), Yoh et al. (1966), and Howard (1967) to infer the total electron content per unit area in a column between the earth and the moon. After removing the contribution of the ionosphere, values of the order of 10^{17} electrons m^{-2} were obtained in the direction of the magnetotail. If spread uniformly between the earth and the moon, as was done by Howard et al. and Yoh et al., this columnar content would correspond to a density of ~ 200 - 300 cm^{-3}. However, the electron distribution along the column is certainly not uniform, and it was shown by Howard (1967) that the entire observed content can within experimental uncertainties be accounted for by electrons within the plasmasphere; for the average electron density between the plasmapause and the moon, these observations yield only an upper limit of approximately 30 cm^{-3}.

Electron densities within the plasmasphere and out to several earth radii beyond it have been extensively studied by the well-known whistler technique (e.g., Carpenter and Smith, 1964; Helliwell, 1965; Carpenter, 1966; Angerami and Carpenter, 1966; and many other papers). These measurements do not extend far enough out to provide data on densities within the plasma sheet. Nevertheless, a comparison between the electron density values as measured beyond the plasmapause by the whistler technique and within the plasma sheet by low energy particle detectors may have some significance, since the low energy electron density appears not to change appreciably across the inner boundary of the plasma sheet (Vasyliunas, 1968a). Figure 7, for example, shows an inner boundary crossing observed by OGO 1; the electron energy is observed to drop precipituously as the satellite moves earthward, but the density shows little variation. This result has recently been confirmed by Schield and Frank (1969), using curved-plate analyzer measurements with high spectral resolution from OGO 3. The inner boundary of the plasma sheet thus appears to be a discontinuity in electron energy but not density.

With this in mind, Figure 8 shows, as a function of radial distance, the electron density in the equatorial region, measured by the two

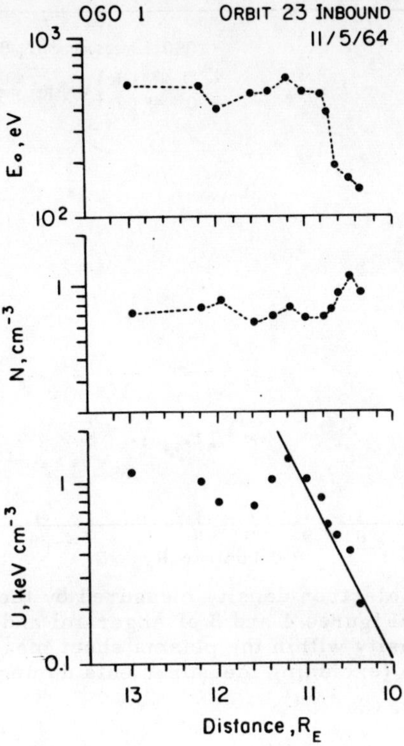

Figure 7. Typical variation of electron properties across the inner boundary of the plasma sheet. E_o is the energy corresponding to the most probable speed and in the present case is approximately 1/2 of the electron mean energy; N is the electron number density and U the energy density. The local time is approximately 19 hours.

techniques. The whistler measurements, taken from Angerami and Carpenter (1966), represent the <u>total</u> plasma density, including electrons of all energies. The OGO 1 measurements represent the density of the observed plasma sheet electron population (the points at distances less than 10 R_E come from periods of bay activity, when the inner boundary of the plasma sheet is displaced earthward). Near 4 R_E the densities obtained from whistler measurements show the familiar drop, from several hundred to ~ 10 cm^{-3}, that constitutes the plasmapause; then they decrease gradually and near 6 - 7 R_E join more or less smoothly onto the $\lesssim 1\ cm^{-3}$ densities measured in the plasma sheet. This is consistent with the supposion that a very low energy ("thermal") electron population either does not exist within the plasma sheet or, if it does exist, its total density is at most comparable to the density of the electron population that has already been observed. (The argument, however, is not conclusive since it is based on comparison of two sets of data taken in different years, at different local times, and under a variety of geomagnetic activity conditions.) The observation mentioned earlier that within the plasma sheet the measured electron and proton number densities are equal is also consistent with this supposition.

In summary, (i) lunar radar observations impose an upper limit of some tens cm^{-3} on the total density of very low energy particles, below

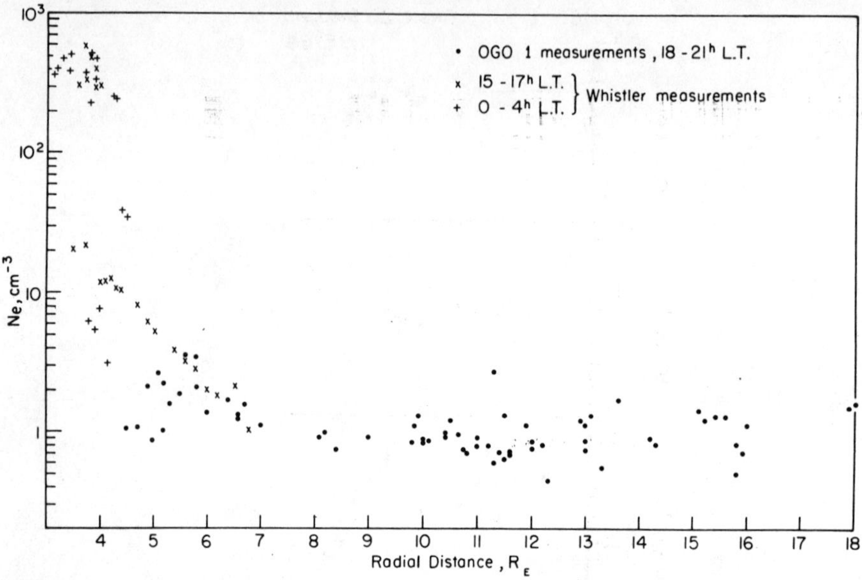

Figure 8. Equatorial electron density measured by the whistler technique (taken from Figures 4 and 8 of Angerami and Carpenter, 1966) and electron density within the plasma sheet measured with a Faraday cup on OGO 1 (excluding measurements during magnetic storms).

the energy range of present detectors, in the magnetotail; (ii) comparison of whistler and plasma sheet observations, as well as the observed equality of proton and electron densities suggest (although do not necessarily prove) that within the plasma sheet the density of such particles does not significantly exceed the density already measured; (iii) there is at present, to my knowledge, no observational evidence that a distinct population of such very low energy particles actually exists within the plasma sheet.

In regions of the magnetotail outside of the plasma sheet, i.e., more than a few earth radii above or below the magnetic neutral sheet, particle fluxes were until recently undetectable with available instruments, as already mentioned in Section 2. Vasyliunas (1968a) considered the upper limits on the electron population that could be deduced from the absence of a measurable flux in the OGO 3 Faraday cup. He concluded that (i) if the electron mean energy was $E \sim 1$ keV, their density was $n \lesssim 10^{-2}$ cm^{-3}; (ii) if $E \sim 50$ eV, then $n \lesssim 5 \cdot 10^{-2}$ cm^{-3}; (iii) if one insisted that the density be high, $n \sim 1$ cm^{-3}, the mean energy had to be very low (to place most of the electron distribution at energies below the range of the detector), $E < 10$ eV. With curved-plate analyzers on Vela 4, Bame (1968) has actually detected electrons outside of the plasma sheet and found that they have an average energy near or under 100 eV and a density that is rather close to the upper limit of Vasyliunas; an example published by Bame shows an electron density $\lesssim 4 \cdot 10^{-2}$ cm^{-3} and an energy density of $\lesssim 2$ eV cm^{-3}, hence a mean energy of ~ 50 eV.

4. RELATION OF THE PLASMA SHEET TO THE AURORAL OVAL

That the auroral oval and the plasma sheet are in some way connected is strongly suggested by the meridian plane configuration of the plasma sheet sketched in Figure 3, as already discussed in Section 2. Hitherto available observations, however, have not been adequate to establish the nature of the connection (due largely to uncertainties in attempting to trace field lines from the position of the observations to the ionosphere) and a number of theoretical suggestions have been made that can be classified into three major groups:

(i) Aurora occurs on all magnetic field lines threading the plasma sheet. This is essentially the concept argued by O'Brien (1967). The equatorward edge of the auroral oval corresponds, in this interpretation, to the inner boundary of the plasma sheet which, as discussed by Vasyliunas (1968a), lies on field lines that are definitely closed and in fact still of a roughly dipolar (although distorted) shape; the poleward edge of the oval corresponds to the outer boundary of the plasma sheet and thus lies on (so-called "open") field lines extending far into the magnetotail.

(ii) Aurora occurs on field lines that come from the immediate vicinity of the magnetic neutral sheet (and thus are "open"). This suggestion has been made by a number of authors, among them Dessler and Juday (1965), Coppi et al. (1966), Speiser (1967a; 1967b), Piddington (1967a; 1967b), and has often been coupled with a theory of particle acceleration at the neutral sheet. Since the inner edge of the plasma sheet, as just discussed, coincides with closed field lines that must lie well earthward of the magnetic neutral sheet, this suggestion implies that the equatorward edge of the auroral oval occurs poleward of the field lines passing through the inner edge of the plasma sheet.

(iii) It has recently been proposed by Schield (1968; also Schield et al., 1969) that aurora occurs at the inner boundary of the plasma sheet and that the auroral oval corresponds to the layer ($\sim 1\ R_E$ thick at the equator) within which the change in the electron energy spectrum associated with the inner boundary occurs (Vasyliunas, 1968a). If the precise location of the inner boundary is taken to be the point at which the electron spectrum just begins to change from its plasma sheet character, as done consistently by Vasyliunas, then Schield's model implies that the poleward edge of the auroral oval coincides with field lines through the inner boundary of the plasma sheet.

Recently Vasyliunas has obtained some observational evidence directly bearing on this problem from electron measurements on the OGO 1 satellite. What follows is a brief description of the results; a more detailed account has been given by Vasyliunas (1969b).

The change of the OGO 1 orbital plane inclination from 31° at launch in September 1964 to 45° a year later allowed the satellite, during certain orbits in 1965 which were favorably phased relative to the daily wobble of the earth's dipole, to approach the earth at relatively high geomagnetic latitudes. A typical nearly equatorial orbit of 1964 and a "high latitude" orbit of 1965 have already been sketched in Figure 3. Data from six such "high latitude" orbits are available, and in each of them the narrow "horns" of the plasma sheet as indicated in Figure 3 were clearly observed. In particular, at these high latitudes the plasma sheet still has a clearly discernible inner boundary, which was crossed by the satellite at geocentric distances ranging between 2.5 and 3 R_E. From such close distances, magnetic field lines can be traced to the earth's surface fairly accurately using the earth's main field alone (inclusion of fields due to magnetopause and tail currents changes the latitude and longitude of the field line intersection with the earth by $\sim 1^o$). Thus the observed position of the inner edge of the plasma sheet can be mapped to ionospheric heights and compared with data on the auroral oval. Simultaneous ground observations of the aurora are not

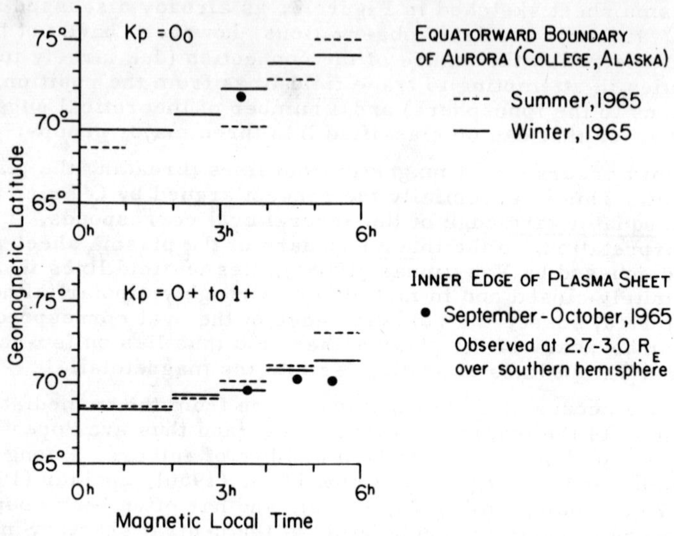

Figure 9. Observed positions of the inner edge of the plasma sheet, mapped to the ionosphere, and the average equatorward boundary of radar aurora as given by Bates (1966).

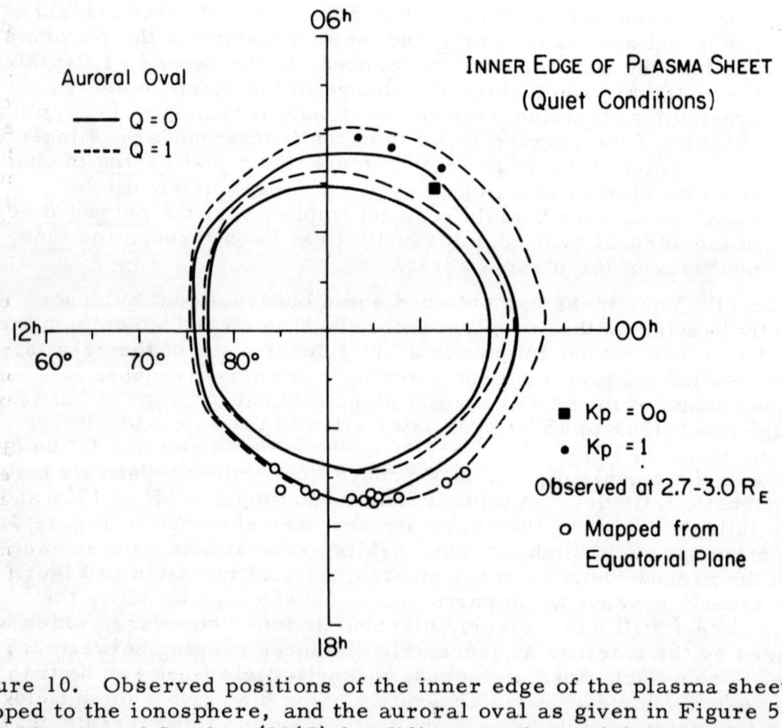

Figure 10. Observed positions of the inner edge of the plasma sheet, mapped to the ionosphere, and the auroral oval as given in Figure 5 of Feldstein and Starkov (1967) for different values of the geomagnetic activity index Q. The observation at Kp = 0o was preceded by an interval of 12 hours with Kp not exceeding 0+, and 42 hours with Kp not exceeding 1o.

available, but according to Feldstein and Starkov (1968) the actual location of the aurora at a particular time is closely approximated by the statistical auroral belt for the corresponding degree of geomagnetic activity (Feldstein and Starkov, 1967).

In Figure 9 the observed positions of the inner edge of the plasma sheet, mapped to the ionosphere as described above, are shown together with the equatorward boundary of the auroral oval as determined from radar backscatter observations (Bates, 1966), for quiet geomagnetic conditions. It has been shown by Bates et al. (1969) that radar and visual oval aurora are closely associated; these particular radar observations are shown here because they were obtained in the same year as the OGO 1 data. In Figure 10 the same observations of the inner edge of the plasma sheet are compared with the auroral oval for quiet geomagnetic conditions as given by Feldstein and Starkov (1967); also included in Figure 10 are the positions of the inner edge observed by OGO 1 in the equatorial plane and mapped to the ionosphere using the field line mapping determined by Fairfield (1968). As can be seen in both figures, the inner edge of the plasma sheet lies on or close to the <u>equatorward</u> boundary of the auroral oval and (in Figure 10) is well removed from its poleward boundary. The same result is even more apparent in Figure 11, which shows observations during periods of magnetic bay activity. The earthward displacement of the inner edge of the plasma sheet during magnetic bays to distances $\leq 6\ R_E$, observed in the equatorial plane (Vasyliunas, 1968a), here is seen to correspond precisely to the well-known equatorward expansion of the auroral oval during magnetically disturbed periods.

The observations thus strongly suggest that the inner edge of the plasma sheet and the equatorward boundary of the auroral oval coincide. This supports suggestion (i) of the three theoretical suggestions described earlier and is in definite disagreement with (ii) and especially with (iii). The position of the poleward boundary of the oval in relation to the plasma sheet is not so clear (the OGO 1 observations of the outer boundary of the plasma sheet occur at larger radial distances where reliable mapping does not appear to be possible). The observation by McDiarmid and Burrows (1968) that the poleward boundary of the auroral

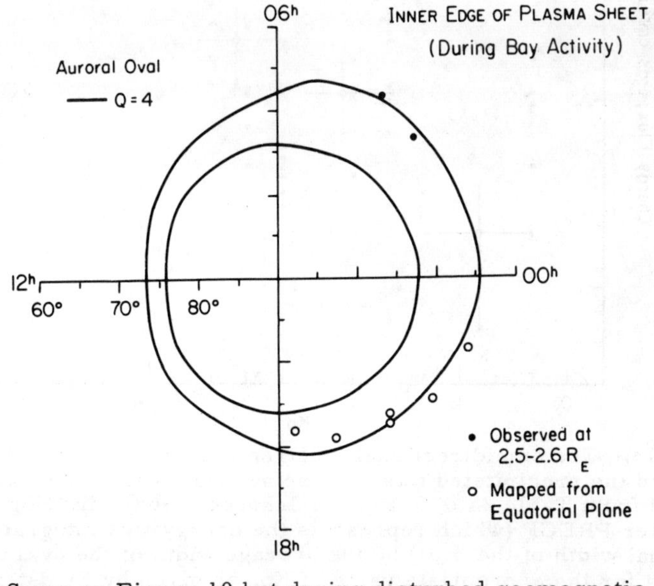

Figure 11. Same as Figure 10 but during disturbed geomagnetic conditions.

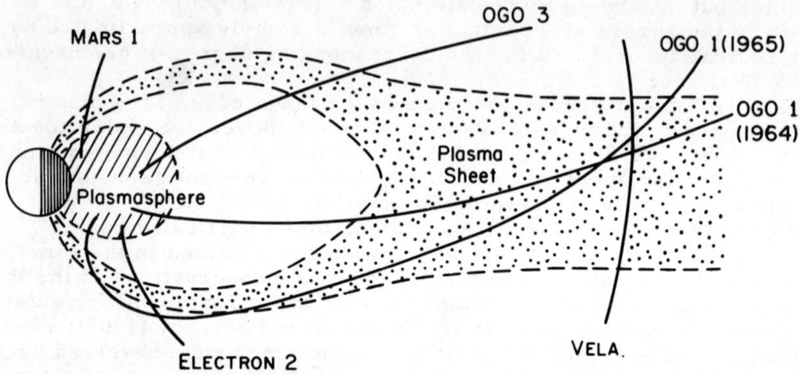

Figure 12. Configuration of the magnetic field lines and the plasma in a meridian plane on the night side of the magnetosphere. The 45° hatching indicates the plasma sheet, which corresponds to the auroral oval according to theoretical suggestion (i) and the present observations; the horizontal hatching indicates the "boundary layer" at the inner edge of the plasma sheet within which the change of electron properties occurs, which would be associated with the auroral oval according to suggestion (iii).

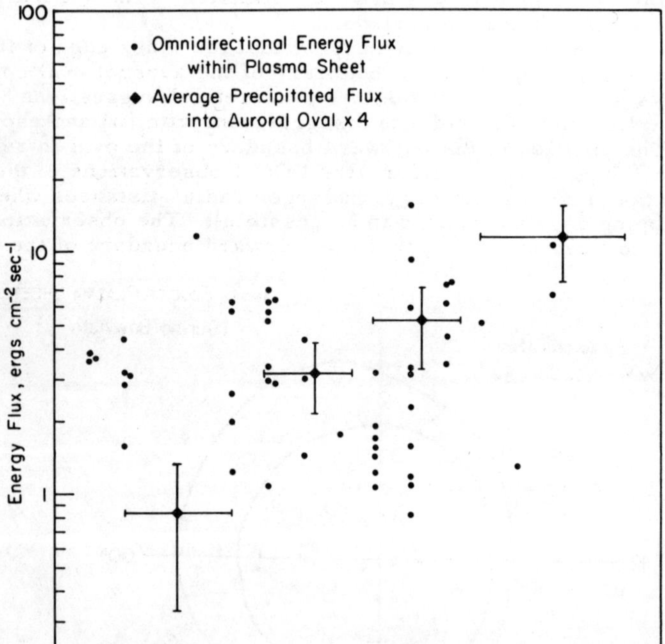

Figure 13. The omnidirectional electron energy flux within the plasma sheet and the precipitated flux into the auroral oval. The latter was obtained from Table 2a of Sharp and Johnson (1968), dividing their parameter PRECP (which represents the energy flux integrated over the latitudinal width of the oval) by the average width of the oval near midnight for the appropriate value of Kp, as given by Feldstein and Starkov (1967).

oval marks the limit of observable high energy ($E > 40$ keV) electron fluxes, together with the fact that high energy electron fluxes observed within the magnetotail are confined to a region (Murayama, 1966; Anderson and Ness, 1966; Montgomery, 1968) that appears to coincide with the plasma sheet, may be taken as (admittedly inconclusive) evidence that the poleward boundary of the auroral oval and the outer boundary of the plasma sheet also coincide. The general configuration of plasma and magnetic fields on the night side suggested by all these observations is sketched in Figure 12: the auroral oval is simply the plasma sheet mapped down to the ionosphere. (The connection between the poleward boundaries of the oval and the plasma sheet, however, is far from being well established, and the relation of either of them to the soft precipitation zone recently observed at low altitudes by Burch (1968) and Hoffman (1969) is not clear.)

It is now of interest to compare the electron energy fluxes observed within the plasma sheet with the energy fluxes precipitated into the auroral oval. Figure 13 shows individual values of the omnidirectional electron energy flux measured by OGO 1 within the plasma sheet (from the same set of data as Figures 4 and 5), as a function of the Kp index. Also shown is the average precipitated electron energy flux into the night side auroral oval, obtained from results published by Sharp and Johnson (1968); this is the total precipitated energy flux averaged over the latitudinal width of the oval and has been multiplied by 4 to make it directly comparable with the equatorial omnidirectional flux (it is readily shown that, if the particle distribution in the equatorial region is isotropic over all directions including the loss cone, then the total precipitated flux observed at the ionosphere is one quarter of the omnidirectional flux observed near the equator). Except possibly at very low Kp values, the equatorial and the precipitated energy fluxes are of the same order of magnitude; averaging all the measurements gives a precipitated flux (times 4) of 5 ± 1 ergs cm^{-2} sec^{-1} and a flux within the plasma sheet of 4 ± 2 ergs cm^{-2} sec^{-1}. This means that if the precipitated particles come from the plasma sheet, the energy flux per steradian within the loss cone is of the same order of magnitude as the energy flux per steradian averaged over all directions. This result is in agreement with the hypothesis of Kennel (1969) and Vasyliunas (1969c), that strong pitch angle diffusion maintains the plasma sheet electron distribution approximately isotropic over the loss cone, and further indicates that the electron precipitation implied by this hypothesis is just adequate to supply the _average_ energy flux observed in the auroral oval. (It should be emphasized that we are here discussing only values averaged over the width of the auroral oval; present results shed no light on the origin of the individual auroral forms.) Independently of theoretical speculation, the observations indicate that the source of auroral particles, whatever it may be, maintains an average energy input to the ionosphere of the same order of magnitude as the maximum energy flux that can be drawn out of the plasma sheet electron population by pitch angle scattering processes.

The present results are consistent with the observation by Chase (1969) that the electron energy spectrum in post-breakup aurora closely resembles, in both shape and intensity, the spectrum within the plasma sheet.

5. SUMMARY

Intense fluxes of low energy particles exist within the geomagnetic tail. They are confined to the equatorial region of the tail and form a _plasma sheet_ several earth radii thick and centered about the magnetic neutral sheet. The mean energy of electrons within the plasma sheet typically ranges from a few hundred eV to a few keV; the mean energy of protons is appreciably higher. The electron and proton number densities are equal. Typical values of the number density ranging from 0.1 to 1 cm^{-3} have been reported; the lowest values occur near midnight. The plasma energy density is comparable to the magnetic field energy density; correlated plasma and magnetic field variations have been observed in which the sum of plasma and magnetic field pressures remains constant. Large correlated changes in electron density and energy, which maintain the energy density constant, have also been observed. There is no direct evidence for a very low energy ("thermal" or "background") particle population within the plasma sheet; if such a population exists at all, its electron density must be less than some tens of particles cm^{-3}, and there are indirect arguments that the density is less than ~1 cm^{-3}.

In the high latitude regions of the magnetotail, above and below the plasma sheet, the observed electron density is $\lesssim 4 \cdot 10^{-2}$ cm^{-3} and the electron mean energy is $\lesssim 100$ eV; thus the energy density is very low in comparison with that within the plasma sheet.

Low energy proton fluxes extend from the magnetotail earthward to well within the outer radiation belt, where during magnetic storms they become very intense and form the ring current responsible for the storm main phase. Low energy electron fluxes extend from the magnetotail along the flanks of the magnetosphere and merge with the complicated (and as yet only partly explored) low energy electron structures on the day side of the magnetosphere. In the evening sector, however, these intense electron fluxes do not extend far into the radiation belts but terminate at the so-called inner boundary of the plasma sheet, which lies at an equatorial distance of 10 R_E during magnetically quiet periods and moves inward to distances ~ 6 R_E during magnetic bays. At this inner boundary the electron density appears not to change but the electron mean energy decreases very rapidly with decreasing distance.

The inner boundary of the plasma sheet has now been traced in latitude to ~ 55° (geocentric distance of ~ 2.5 - 3 R_E) and shown to coincide with magnetic field lines through the equatorward boundary of the auroral oval. It has been suggested that the entire auroral oval is simply the plasma sheet mapped down to the ionosphere, but further study of the poleward boundaries especially is needed. The average precipitated energy flux into the auroral oval appears to be of the same order of magnitude as the electron flux within the plasma sheet.

Acknowledgments. This work was supported by the National Aeronautics and Space Administration under grant NGL 22-009-015.

REFERENCES

Anderson, K. A. and N. F. Ness, Correlation of magnetic fields and energetic electrons on the IMP 1 satellite, J. Geophys. Res., 71, 3705, 1966.

Angerami, J. J. and D. L. Carpenter, Whistler studies of the plasmapause in the magnetosphere, 2, Electron density and total tube electron content near the knee in magnetospheric ionization, J. Geophys. Res., 71, 711, 1966.

Axford, W. I., Observations of the interplanetary plasma, Space Sci. Rev., 8, 331, 1968.

Bame, S. J., Plasma sheet and adjacent regions, in Earth's Particles and Fields, edited by B. M. McCormac, Reinhold, New York, p. 373, 1968.

Bame, S.J., J.R. Asbridge, H.E. Felthauser, R.A. Olson, and I.B. Strong, Electrons in the plasma sheet of the earth's magnetic tail, Phys. Rev. Letters, 16, 138, 1966.

Bame, S.J., J.R. Asbridge, H.E. Felthauser, E.W. Hones, and I.B. Strong, Characteristics of the plasma sheet in the earth's magnetotail, J. Geophys. Res., 72, 113, 1967.

Bates, H.F., The latitude of the dayside aurora, J. Geophys. Res., 71, 3629, 1966.

Bates, H.F., R.D. Sharp, A.E. Belon, and J.S. Boyd, Spatial relationships between HF radar aurora, optical aurora and electron precipitation, Planetary Space Sci., 17, 83, 1969.

Binsack, J.H., Plasma studies with the IMP 2 satellite, Ph.D. thesis, M.I.T., Cambridge, Mass., 1966.

Binsack, J.H., Plasmapause observations with the M.I.T. experiment on IMP 2, J. Geophys. Res., 72, 5231, 1967.

Burch, J.L., Low-energy electron fluxes at latitudes above the auroral zone, J. Geophys. Res., 73, 3585, 1968.

Carpenter, D.L., Whistler studies of the plasmapause in the magnetosphere, 1, Temporal variations in the position of the knee and some evidence on plasma motions near the knee, J. Geophys. Res., 71, 693, 1966.

Carpenter, D.L. and R.L. Smith, Whistler measurements of electron density in the magnetosphere, Rev. Geophys., 2, 415, 1964.

Carpenter, D.L., F. Walter, R.E. Barrington, and D.J. McEwen, Alouette 1 and 2 observations of abrupt changes in whistler rate and of VLF noise variations at the plasmapause - a satellite-ground study, J. Geophys. Res., 73, 2929, 1968.

Carpenter, D.L., C.G. Park, H.A. Taylor, Jr., and H.C. Brinton, Multi-experiment detection of the plasmapause from EOGO satellites and Antarctic ground stations, J. Geophys. Res., 74, 1837, 1969.

Chase, L.M., Evidence that the plasma sheet is the source of auroral electrons, J. Geophys. Res., 74, 348, 1969.

Coppi, B., G. Laval, and R. Pellat, Dynamics of the geomagnetic tail, Phys. Rev. Letters, 16, 1207, 1966.

Dessler, A.J. and R.D. Juday, Configuration of auroral radiation in space, Planetary Space Sci., 13, 63, 1965.

Fairfield, D.H., The average magnetic field configuration of the outer magnetosphere, J. Geophys. Res., 73, 7329, 1968.

Feldstein, Y.I. and G.V. Starkov, Dynamics of auroral belt and polar geomagnetic disturbances, Planetary Space Sci., 15, 209, 1967.

Feldstein, Y.I. and G.V. Starkov, Auroral oval in the IGY and IQSY period and a ring current in the magnetosphere, Planetary Space Sci., 16, 129, 1968.

Frank, L.A., Initial observations of low-energy electrons in the earth's magnetosphere with OGO 3, J. Geophys. Res., 72, 185, 1967a.

Frank, L.A., Several observations of low-energy protons and electrons in the earth's magnetosphere with OGO 3, J. Geophys. Res., 72, 1905, 1967b.

Frank, L.A., On the extraterrestrial ring current during geomagnetic storms, J. Geophys. Res., 72, 3753, 1967c.

Frank, L. A. Recent observations of low-energy charged particles in the earth's magnetosphere, in Physics of the Magnetosphere, edited by R. L. Carovillano, J. F. McClay, and H. R. Radoski, D. Reidel, Dordrecht - Holland, p. 271, 1968a.

Frank, L. A., On the distributions of low-energy protons and electrons in the earth's magnetosphere, in Earth's Particles and Fields, edited by B. M. McCormac, Reinhold, New York, p. 67, 1968b.

Frank, L. A. Observations of low-energy protons and electrons with IMP 4, paper presented at the 7th national fall meeting of the American Geophysical Union, San Francisco, December 1968c.

Freeman, J. W., The morphology of the electron distribution in the outer radiation zone and near the magnetospheric boundary as observed by Explorer 12, J. Geophys. Res., 69, 1891, 1964.

Freeman, J. W., Jr. and J. J. Maguire, Gross local-time particle asymmetries at the synchronous orbit altitude, J. Geophys. Res., 72, 5257, 1967.

Gringauz, K. I., Some results of experiments carried out by means of charged particle collectors carried by Soviet space rockets, Space Res., II, 539, 1961.

Gringauz, K. I., Remarks on papers by J. W. Freeman, J. A. Van Allen, and L. J. Cahill, "Explorer 12 observations of the magnetospheric boundary and the associated solar plasma on September 13, 1961", and by L. A. Frank, J. A. Van Allen, and E. Macagno, "Charged-particle observations in the earth's outer magnetosphere", J. Geophys. Res., 69, 1007, 1964.

Gringauz, K. I., Low-energy plasma in the earth's magnetosphere, Rev. Geophys., 7, 339, 1969.

Gringauz, K. I., V. V. Bezrukikh, V. D. Ozerov, and R. E. Rybchinskii, A study of the interplanetary ionized gas, high energy electrons, and corpuscular radiation from the sun by means of the three-electrode trap for charged particles on the second Soviet cosmic rocket, Dokl. Akad. Nauk SSSR, 131, 1960a. (English translation: Soviet Phys., Dokl., 5, 1961).

Gringauz, K. I., V. G. Kurt, V. I. Moroz, and I. S. Shklovskii, Results of observations of charged particles observed out to R = 1000, 000 km with the aid of charged-particle traps on Soviet space rockets, Astron. Zh., 37, 716, 1960b. (English translation: Soviet Astron. AJ., 4, 680, 1961).

Gringauz, K. I., V. V. Bezrukikh, L. S. Musatov, R. E. Rybchinsky, and S. M. Sheronova, Measurements made in the earth's magnetosphere by means of charged particle traps aboard the Mars 1 probe, Space Res., IV, p. 621, 1964.

Gringauz, K. I. and M. Z. Khokhlov, The outermost zone of charged particles, in Issledovaniya Kosmitcheskogo Prostranstva, edited by G. S. Skuridin et al., Nauka Publishing House, Moscow, p. 467, 1965. (English translation: Space Res.: Transactions of the All-Union Conference on Space Physics, NASA Tech. Transl. TT F-389, p. 633, May 1966).

Helliwell, R. A., Whistlers and Related Ionospheric Phenomena, Stanford University Press, Stanford, 1965.

Hoffman, R. A., Low-energy electron precipitation at high latitudes, J. Geophys. Res., 74, 2425, 1969.

Hones, E.W., Jr., Review and interpretation of particle measurements made by the Vela satellites in the magnetotail, in Physics of the Magnetosphere, edited by R.L. Carovillano, J.F. McClay, and H.R. Radoski, D. Reidel, Dordrecht - Holland, p. 392, 1968a.

Hones, E.W., Jr., Electric fields and plasma convection in the magnetosphere, in Earth's Particles and Fields, edited by B.M. McCormac, Reinhold, New York, p. 403, 1968b.

Hones, E.W., Jr., Observations in the magnetotail and their interpretation, in Aurora and Airglow 1968, edited by B.M. McCormac, Reinhold, New York, 1969.

Hones, E.W., Jr., S. Singer, and C.S.R. Rao, Simultaneous observations of electrons ($E > 45$ keV) at 2000-kilometer altitude and at 100,000 kilometers in the magnetotail, J. Geophys. Res., 73, 7339, 1968.

Howard, H.T., Cislunar electron content as determined by radar group delay measurements, J. Geophys. Res., 72, 2729, 1967.

Howard, H.T., V.R. Eshleman, G.H. Barry, and R.B. Fenwick, Radar measurements of the cislunar electron content, J. Geophys. Res., 70, 4357, 1965.

Howe, H.C., Jr., Pioneer 6 plasma measurements in the magnetosheath, submitted to J. Geophys. Res., 1969.

Kennel, C.F., Consequences of a magnetospheric plasma, Rev. Geophys., 7, 379, 1969.

Lazarus, A.J., G.L. Siscoe, and N.F. Ness, Plasma and magnetic field observations during the magnetosphere passage of Pioneer 7, J. Geophys. Res., 73, 2399, 1968.

McDiarmid, I.B. and J.R. Burrows, Local time asymmetries in the high-latitude boundary of the outer radiation zone for the different electron energies, Can. J. Phys., 46, 49, 1968.

Montgomery, M.D., Observations of electrons in the earth's magnetotail by Vela launch 2 satellites, J. Geophys. Res., 73, 871, 1968.

Montgomery, M.D., S. Singer, J.P. Conner, and E.E. Stogsdill, Spatial distribution, energy spectra and time variations of energetic electrons ($E > 50$ keV) at 17.7 earth radii, Phys. Rev. Letters, 14, 209, 1965.

Murayama, T., Spatial distribution of energetic electrons in the geomagnetic tail, J. Geophys. Res., 71, 5547, 1966.

Ness, N.F., The earth's magnetic tail, J. Geophys. Res., 70, 2989, 1965.

O'Brien, B.J. Interrelations of energetic charged particles in the magnetosphere, in Solar Terrestrial Physics, edited by J.W. King and W.S. Newman, Academic Press, New York and London, p. 169, 1967.

Piddington, J.H., A theory of auroras and the ring current, J. Atmospheric Terrest. Phys., 29, 87, 1967a.

Piddington, J.H., Magnetic field annihilation in current pinches, Planetary Space Sci., 15, 733, 1967b.

Retzler, J. and J.A. Simpson, Relativistic electrons confined within the neutral sheet of the geomagnetic tail, J. Geophys. Res., 74, 2149, 1969.

Russell, C. T. and K. I. Brody, Some remarks on the position and shape of the neutral sheet, J. Geophys. Res., 73, 6104, 1967.

Schield, M. A., The configuration of geomagnetic field lines above the auroral zones, Ph. D. thesis, Rice University, Houston, Texas, 1968.

Schield, M. A., J. W. Freeman, and A. J. Dessler, A source for field-aligned currents at auroral latitudes, J. Geophys. Res., 74, 247, 1969.

Schield, M. A. and L. A. Frank, Inner edge of the plasma sheet, paper presented at the Conference on Electric Fields in the Magnetosphere, Rice University, Houston, Texas, March 1969.

Serbu, G. P., Results from the IMP 1 retarding potential analyzer, in Space Res., V, edited by King-Hele et al., North-Holland Publishing Company, Amsterdam, p. 564, 1965.

Serbu, G. P. and E. J. R. Maier, Low-energy electrons measured on IMP 2, J. Geophys. Res., 71, 3755, 1966.

Sharp, R. D. and R. G. Johnson, Some average properties of auroral electron precipitation as determined from satellite observations, J. Geophys. Res., 73, 969, 1968.

Speiser, T. W., Particle trajectories in model current sheets, 2, Application to auroras using a geomagnetic tail model, J. Geophys. Res., 72, 3919, 1967a.

Speiser, T. W., Auroral particles accelerated in the geomagnetic tail, in Aurora and Airglow, edited by B. M. McCormac, Reinhold, New York, p. 491, 1967b.

Speiser, T. W. and N. F. Ness, The neutral sheet in the geomagnetic tail: its motion, equivalent currents, and field line connection through it, J. Geophys. Res., 72, 131, 1967.

Vasyliunas, V. M., A survey of low-energy electrons in the evening sector of the magnetosphere with OGO 1 and OGO 3, J. Geophys. Res., 73, 2839, 1968a.

Vasyliunas, V. M., Low-energy electrons on the day side of the magnetosphere, J. Geophys. Res., 73, 7519, 1968b.

Vasyliunas, V. M., Deep space plasma measurements, to appear in Methods of Experimental Physics, Vol. IX (Plasma Physics), edited by H. R. Griem and R. H. Lovberg, Academic Press, New York, 1969a.

Vasyliunas, V. M., Relation of the plasma sheet to the auroral oval, to be published, 1969b.

Vasyliunas, V. M., A mathematical description of plasma motions in the magnetosphere, to be published, 1969c.

Vernov, S. N., V. V. Melnikov, I. A. Savenko, B. I. Savin, and T. I. Pervaiya, Recording charged particles with an energy of 0.1 - 10 keV with a spherical electrostatic analyzer, in Issledovaniya Kosmitcheskogo Prostranstva, edited by G. A. Skuridin et al., Nauka Publishing House, Moscow, p. 381, 1965. (English translation: Space Res.: Transactions of the All-Union Conference on Space Physics, NASA Tech. Transl. TT F-389, p. 518, May 1966).

Vernov, S. N., V. V. Melnikov, I. A. Savenko, and B. I. Savin, Measurements of low-energy particle fluxes from the Cosmos and Electron satellites, Space Res., VI, edited by R. L. Smith-Rose, Spartan Books, Washington, D. C., p. 746, 1966.

Whipple, E. C., Jr. and L. W. Parker, Theory of an electron trap on a charged spacecraft, J. Geophys. Res., 74, 2962, 1969a.

Whipple, E. C., Jr. and L. W. Parker, Effects of secondary electron emission on electron trap measurements in the magnetosphere and solar wind, submitted to J. Geophys. Res., 1969b.

Wolfe, J. H. and D. D. McKibbin, Review of Ames Research Center plasma-probe results from Pioneers 6 and 7, in Physics of the Magnetosphere, edited by R. L. Carovillano, J. F. McClay and H. R. Radoski, D. Reidel, Dordrecht-Holland, p. 435, 1968.

Yoh, R., H. T. Howard, B. B. Lusignan, and V. R. Eshleman, Lunar radar measurements of the earth's magnetospheric wake, J. Geophys. Res., 71, 189, 1966.

LOW ENERGY PARTICLE FLUXES OVER THE POLAR CAP

Brian J. O'Brien

School of Physics
University of Sydney, Sydney, Australia

Abstract. The electron and proton fluxes observed at low altitudes ($< 1\ R_E$) and high latitudes ($\Lambda > 60^o$) are reviewed, with particular attention to the polar cap and, indeed, with attention to the very question of whether there is any generically distinct region called the polar cap. It is concluded that because, for example, particle precipitation and auroras occur over the magnetic poles, any distinction of a polar cap is purely for convenience. As such, the cap is then defined at low altitudes as being the region beyond Λ of 74° to 84° during local day, and Λ of 71° to 83° during local night where the flux of precipitated electrons drops abruptly by a factor of ten or so. The wide range of the above latitude edge of the "polar cap" is caused by its very great control by the relative angle between the geomagnetic axis and the ecliptic. At high altitudes, the polar cap field lines are deemed to thread the region between the plasma sheet and the magnetopause. The flux of low-energy (~ 50 eV to a few keV) electrons precipitated over the polar cap is generally about 10^7 eV cm^{-2} sec^{-1} sr^{-1}, with a soft spectrum similar to that at lower latitudes in the auroral regions. The usual flux of precipitated protons with energy of ~ 100 eV to ~ 30 keV, is less than 10^6 eV cm^{-2} sec^{-1} sr^{-1}. Occasionally, very intense "spikes" of electrons are precipitated. A discussion is given of auroral particles only to the extent necessary to delineate dissimilarities and similarities with fluxes in the polar region.

1. INTRODUCTION

In beginning a review of the low-energy particle fluxes over the polar cap, it appears profitable to put this into historical perspective by summarizing the relevant rocket and satellite studies of the past decade. Because, in these proceedings, there will be extensive reviews of other facets of magnetospheric research (e.g., geomagnetic morphology, ionospheric phenomena and the polar wind), we concentrate here on high altitude (> 100 km) direct measurements of non-thermal charged particles.

However, it is generally recognized that too rigid demarcation between such fields of study can lead to very serious errors or omissions. This review then, is intended to be only part of a hopefully comprehensive whole.

It is little more than a decade since two important direct discoveries of energetic-charged particle fluxes were made. With Explorers I and

III, Van Allen et al. (1958) discovered geomagnetically-trapped radiation at high altitudes and low latitudes. About the same time, McIlwain (1960) found that the dominant cause of excitation of bright, visible auroral forms is the bombardment of the atmosphere by electrons with energy of some 1 to 10 keV. McIlwain (1960) also detected smaller fluxes of higher-energy protons. These auroral measurements were made at low altitudes and, of course, at a high latitude.

The glamour of satellite research and other factors led to a relative neglect of high-latitude and auroral rocket research for some years, as satellites and space probes explored the magnetosphere. Extensive reviews of these magnetospheric studies have been given by Hess (1968), O'Brien (1967), and others. From these measurements and from extensive theoretical studies there has come a gradual understanding of many of the high-altitude magnetospheric properties. In the following, we will concentrate on the properties that can be "mapped" down into the earth's atmosphere at high latitudes. Because this mapping is only tentative as yet from high-altitudes (e.g., Taylor and Hones, 1965), we pay particular attention to measurements made with rockets and low-altitude satellites. There is still a very real need for suitably instrumented satellites (such as an IMP) to be put into high-inclination, highly eccentric orbits.

2. DELINEATION OF THE AURORAL ZONE AND THE TRANSAURORAL ZONE

From many recent and not so recent ground-based observations of visible auroral light, the auroral zone or oval has been reasonably well defined (reviewed by Davis, 1965). The region of most frequent occurrence for low values of K_p is at an invariant latitude (O'Brien, 1963) $\Lambda \sim 67^o$ around local midnight and at $\Lambda \sim 75^o$ around local noon. During times of geomagnetic disturbances the auroras may be seen at lower values of Λ, although the actual latitudinal behaviour during auroral substorms is far too complex to summarize here (cf. Akasofu, 1964).

In a series of observations with the satellites Injun 1 and Injun 3, it was shown that:

(i) there exists a high-latitude boundary of "trapping", (O'Brien, 1963),

(ii) statistically, this high-latitude boundary varies with local time and with geomagnetic activity in much the same manner as do visible auroras (O'Brien, 1963; Maehlum and O'Brien, 1963), and finally

(iii) that simultaneous observations from the satellite of auroral light and this boundary show them to be generally in the same location (O'Brien and Taylor, 1964).

These observations therefore suggested that, in some manner, this high-latitude termination of trapping was intimately related to the processes that cause auroras. The above observations suffered from the disadvantage that they were of electrons with energy $E > 40$ keV, whereas it was known that lower-energy $1 < E < 10$ keV excite most of the auroral light. Indeed, observations of such lower-energy electrons on the same satellite showed that they were often found at somewhat higher latitudes than the 40 keV electrons (Fritz and Gurnett, 1965).

Nevertheless, this concept of the auroral zone as a dividing region between the stable magnetosphere with closed field lines and the polar cap threaded by "open" field lines still appears to be a useful one. Although intuition must be used with great care in plasma and magnetospheric research, it does intuitively appear reasonable to

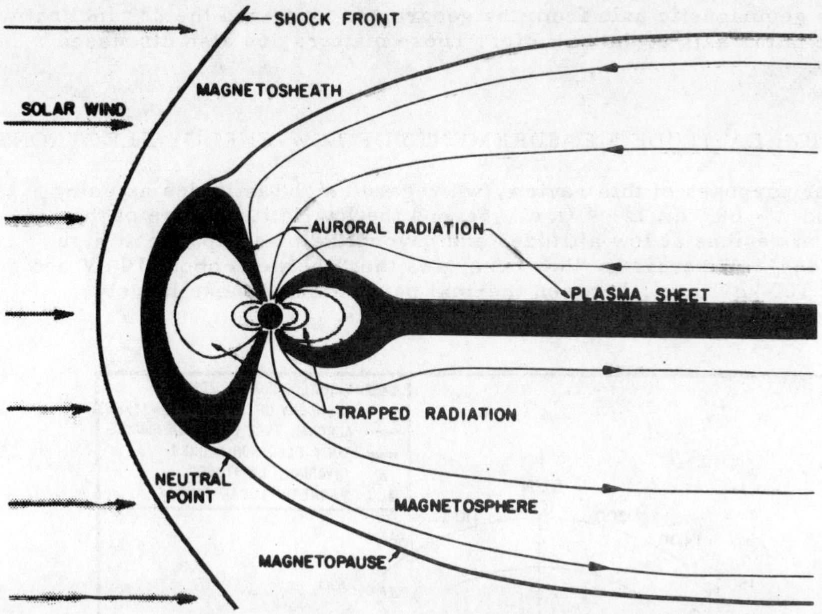

Figure 1. Rough sketch of the energetic particle domains in the magnetosphere (O'Brien, 1967). No attempt is made to show the plasmasphere on the sketch.

continue to regard auroral phenomena as playing a pivotal role in solar-terrestrial relations (O'Brien, 1967). That is the reason for the rather gross extrapolations sketched in Figure 1.

Consider, for example, the ratio (β) of the charged particle kinetic energy density $\frac{1}{2}nmv^2$ (where n is number density of particles of mass m and velocity v) to the magnetic field energy density $B^2/8\pi$. Inside the plasmasphere, $\beta < 1$ and the particles are trapped (Van Allen particles). In the streaming magnetosheath the particles drag the geomagnetic field lines out into a comet-like tail (Ness, 1965) since $\beta > 1$. Now, certainly in auroras at low altitudes and in the plasma sheet at very high altitudes and also just beyond the plasmasphere we can find $\beta \sim 1$, i.e., the intermediate situation. (For details of these several regions, cf. for example, McIlwain, 1960; Ness, 1965; Frank, 1967; Bame et al., 1967; O'Brien, 1967.)

However, it will be noted from the above discussion and also from Figure 1 that we have neglected the vast region of space between the "auroral radiation" zone and the magnetosheath. It is precisely this region, sometimes called the transauroral region or the polar cap, that is the dominant topic of this symposium for the very reason that it is little known and poorly mapped.

It appears therefore, that one of the major aims of this symposium should be to replace Figure 1 and others of its ilk with another sketch that does not so neglect the polar cap. It is the purpose of this review to provide data on low-altitude measurements of charged particles as a necessary part of redrawing such sketches.

We draw attention to a variety of other magnetospheric sketches compiled by O'Brien (1967) and Hess (1968), and also to the fact that, whereas Figure 1 shows the solar wind impinging normally to the geomagnetic axis, very great changes in this sketch are rendered mandatory for the more usual situation, taking account of the 11° offset of the geomagnetic axis from the geographic axis, and the 23° inclination of the latter axis to the exliptic. These matters are also discussed below.

3. HIGH LATITUDE MEASUREMENTS OF LOW-ENERGY ELECTRONS

For purposes of this review, we regard "high" latitudes as being beyond $\Lambda \sim 60°$ or $L \sim 4$ (i. e., beyond the low-latitude edge of the auroral regions at low altitudes and beyond the plasmapause at high altitudes). We treat as "low" energies those between about 10 eV and about 100 keV i. e., between thermal particles and those largely unaffected by magnetospheric electric fields.

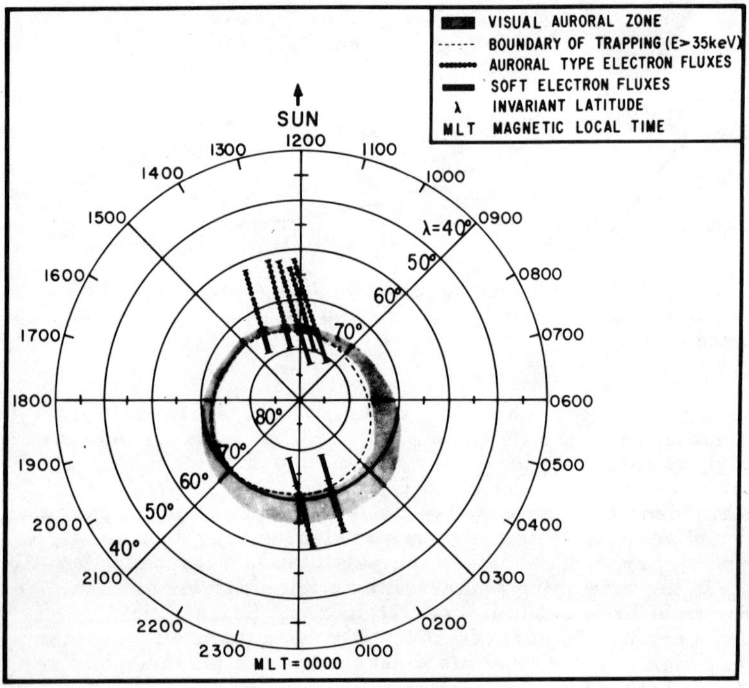

Figure 2. Local-time projection on the polar cap showing the latitude range of electron fluxes, the high-latitude boundary of trapping and the auroral zone (Burch, 1968b).

The exploratory measurements of the latitudinal and diurnal distribution of electrons with E > 40 keV (O'Brien, 1964) were mentioned above. Extensive studies by McDiarmid and Burrows (1968) have mapped out in detail the high-latitude "boundary" of electrons with energies greater than 35 keV. This is shown in Figure 2, together with the location of the visual auroral zone and with some more recent satellite data of Burch (1968a; 1968b) that will be discussed below.

Definitive satellite studies of the lower-energy electrons i. e., those primarily responsible for excitation of auroral light, were begun by the Lockheed group using several low-altitude satellites with lifetime of only a few days.

Johnson et al. (1966) found two apparently distinct zones of electron precipitation on the day-side. They used two integral detectors, one

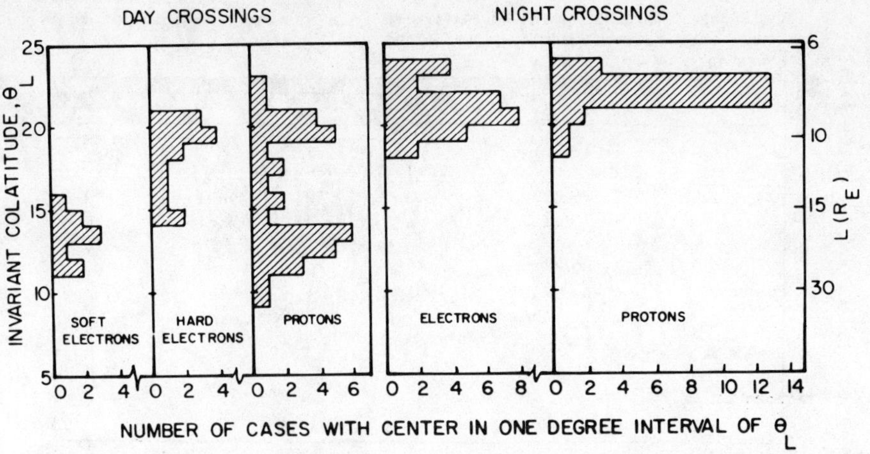

Figure 3. Latitudinal distribution of the "centers" of particle precipitation in the northern and southern hemisphere (Sharp and Johnson, 1968).

sensitive to electrons above 80 eV and the other to electrons with $E > 20$ keV. As shown in Figure 3, the low-energy ($0.08 < E < 20$ keV) electrons seemed grouped around $\Lambda \sim 75^{o}$ while those with $E > 20$ keV were more intense around $\Lambda \sim 70^{o}$. It is clear from Figure 3 that statistics were limited, as was the sensitivity of each detector. Nevertheless, subsequent more exacting measurements e.g., by Burch (1968a) have shown the early Lockheed results to be correct, albeit incomplete.

Johnson et al. (1966) found no such bifurcation in the electron precipitation on the night-side. As discussed by Burch (1968b) this feature (or lack of it) can be explained on the grounds of semantics and experimental deficiencies, and Burch (1968a) finds indeed some such division on the night-side as well.

It is as well to mention a note of caution here. Within the limits of their definition of a "soft" zone i.e., one where the $E > 80$ eV detector counted above background but the $E > 20$ keV detector did not, and their definition of a "hard" zone where both integral fluxes were about the same, the bifurcation on the day-side as found by Johnson et al. (1966) may be taken to be real.

But from an analytical viewpoint the situation is not clearly defined. Such a demarcation of two zones may be largely instrumental in origin. Such, indeed, was once the case with the "inner" and "outer" Van Allen radiation zones. Given a limited number of sensors or points on an energy spectrum which can vary greatly with Λ, one may well obtain clear-cut differences in overall responses without these being due to a clear-cut physical distinction between the regions. For example, a spectrum that became monotonically softer with increasing Λ could very simply produce such instrumental responses. Note that most theories, ranging from classical Størmer studies to Taylor and Hones (1965) study of particle drift in the magnetospheric electric fields, actually do predict some such latitudinal variations of the energy spectra but no bifurcation.

Similar words of caution must be applied to the more detailed work of Burch (1968a; 1968b) or rather to interpretation of its significance. Here we present the results, and discuss its significance in a separate section.

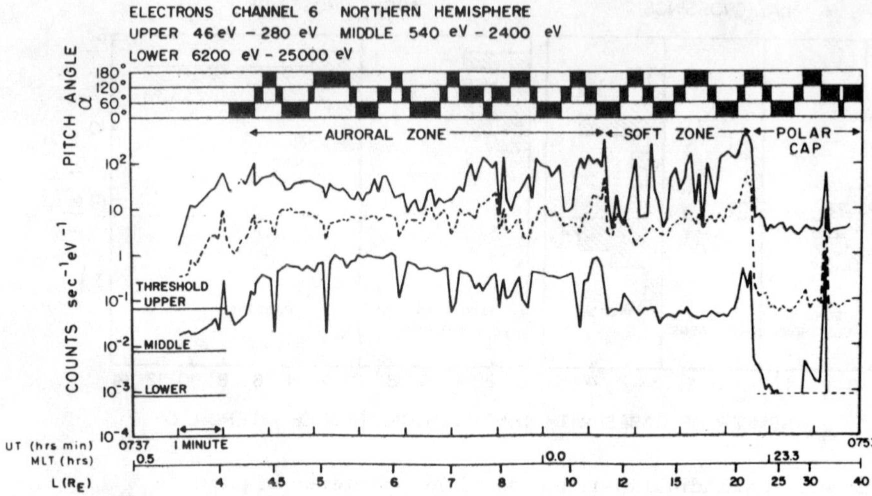

Figure 4. Typical response to electrons of the wide-band channel in the spectrometer on Aurora 1. This pass was at night in the northern hemisphere on June 30, 1967 when $K_p \sim 3$ when the satellite was still tumbling (Burch, 1968b).

With a multichannel electron spectrometer on the Aurora 1 satellite, Burch (1968a) found three high-latitude zones of electron fluxes at $L > 4.5$. These zones were delineated both in daylight (near noon) and in nighttime (near midnight) passes of the near-polar satellite in both hemispheres at an altitude of almost 4000 km. The three zones were defined by Burch (1968a)as being: (i) an "auroral zone", (ii) a "soft zone", and (iii) the "polar cap". The location of these "zones" is shown in Figure 2 for several passes. A typical satellite pass from which such a distinction is made is shown in Figure 4.

According to Burch (1968a) the "auroral zone" is characterized by relatively hard electron spectra and by a tendency to isotropy at least over the upper hemisphere.

The "soft zone" at a higher latitude (Figure 2) is characterized by a softer electron spectrum and by rapid variations in measured fluxes - where the latter effect could be due to any or all of the variations with pitch angle, space or time.

In the "polar cap" the fluxes of electrons in the ranges 46 - 280 eV and 540 eV - 2.4 keV are some 10 to 100 times less than in the "soft zone", while the flux in the 6.2 keV - 25 keV range is below background except for occasional solar UV contamination (Burch 1968b).

The above findings therefore agree with Johnson et al. (1967) and Sharp and Johnson (1968) except that the higher sensitivity of the instrument permits distinction of "hard" (or "auroral") and "soft" zones on both the night and day sides. There may also be an effect due to the higher fluxes in the more magnetically-disturbed conditions of the part of the Aurora 1 flight. It may be noted that Burch (1968b) found that the boundary (Λ_1) between zones 1 and 2 (i.e., the auroral and soft zones) moved to lower Λ during magnetic disturbances, while Maehlum (1968) found that the boundary (Λ_2) between the soft zone and the polar cap varied as a very strong function of UT. Since this topic will be treated by other speakers at this symposium, it will only be

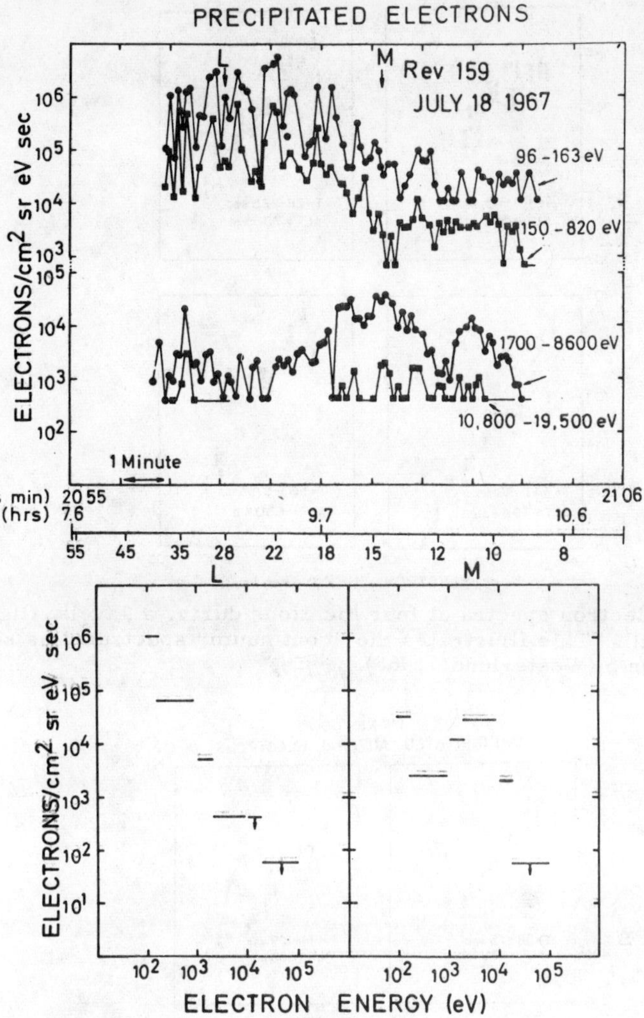

Figure 5. Electron fluxes in selected energy ranges for a day-side high-latitude pass when $K_p \sim 1$. Sample energy spectra are given to compare the "continuum" spectrum (at L) with the continuum plus auroral peak (at M) (Burch, 1968b).

treated briefly in the discussion herein.

When the detailed energy spectra are considered (Figure 5) one may more easily contrast the soft and auroral spectra, as sampled at L and M, respectively.

A very important feature, as shown by the spectrum at M in Figure 5, is that there can occur a second peak in the differential electron spectrum, with the peak lying between 1 and 10 keV. Many such examples are found in the Aurora 1 data, but it is very important to note that occasionally such effects were found sporadically during an auroral bay disturbance even in what was called the "soft zone".

Now, precisely this type of effect has been found in rocket flights through visible auroral displays near Fort Churchill. Figures 6 and 7, from Westerlund (1968) show sample spectra. Other rocket

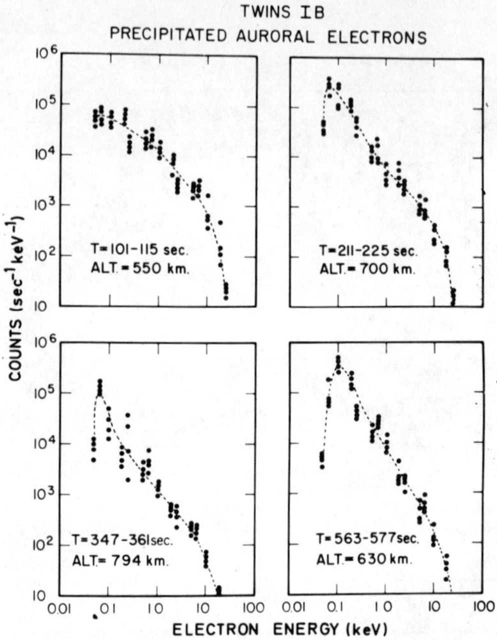

Figure 6. Electron spectra at four locations during a Javelin flight at Fort Chufchill. This illustrates the "continuum" spectrum, as shown at L in Figure 5 (Westerlund, 1968).

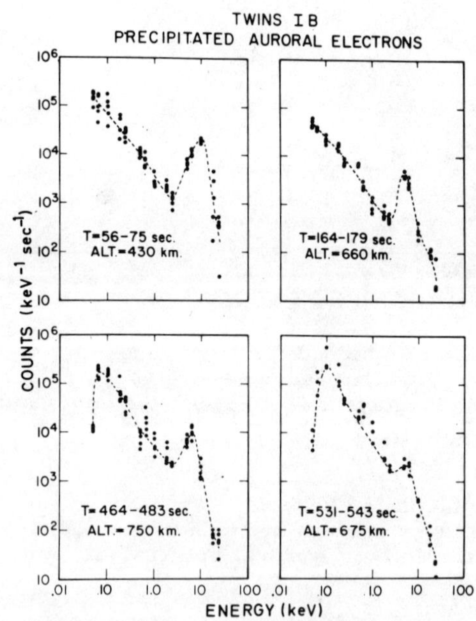

Figure 7. Electron spectra at four other locations during the same rocket flight. The first two measurements were definitely over visible auroral arcs, but seeing conditions made it impossible to make positive statements about the last two spectra. Note the enhanced flux at 1 to 10 keV superimposed on the continuum, and compare this with M in Figure 5. Note also the error in calling the auroral flux "monoenergetic" (Westerlund, 1968).

measurements e.g., those of Evans (1967) and Albert (1967) which measured auroral electrons above some keV, found the peaks between 1 and 10 keV (as shown in Figures 5 and 7) which are erroneously often called "monoenergetic" fluxes.

Analysis of the electron flux under the near-continuum spectra of Figure 6 show that it brings some 6 ergs cm^{-2} sec^{-1} sr^{-1} into the atmosphere while the particles in the "peaks" of Figure 7 bring in an energy flux an order of magnitude greater. These findings correspond very well (Westerlund, 1968) with the instruments passing over bright visible auroral arcs with a diffuse glow between them, and indeed photometric observations of 3914 Å intensity are consistent with the measured particle precipitation (Westerlund, 1968).

Thus, on the basis of both rocket and satellite electron data, we can state that there is a broad zone in which there is precipitation of electrons whose energy spectrum is a continuum which peaks in flux down to an energy of some 100 eV, with superimposed intense fluxes of electrons with energies in the 1 to 10 keV range. These 1 to 10 keV electrons produce bright and discrete auroral forms, while the continuum produces a diffuse glow. The higher probability of discrete auroras in the auroral zone, with occasional excursions to higher latitudes during auroral substorms (Akasofu, 1964), leads us to infer that the distinction between the "auroral" and the "soft" zones (Burch, 1968a) while being useful and informative in many ways, is largely a statistical rather than a generic distinction. Thus both regions are bombarded by the continuum, but this is most often dominated by the 1 to 10 keV auroral electrons at the lower latitudes where auroras are most common.

If the above discussion completed our knowledge of the high-latitude electron fluxes, then we could be tempted to revert to the simple picture of Figure 1 and state that the entire blackened or auroral region is pervaded by the continuum, and that discrete auroras are caused by precipitation of electrons in the 1 to 10 keV range, which are accelerated in this region, most frequently near its innermost edge. Since such precipitation is seen at altitudes of 4000 km (Burch, 1968a) and 7000 km (Bolyunova et al., 1966), the acceleration must occur generally (if not always) at an altitude above about 1 R_E. From simultaneous studies with the high-altitude Vela and the low-altitude Injun 4 satellites, Hones et al. (1968) place the acceleration location between 1 R_E and more than 7 R_E but less than 17 R_E.

However, two important fidings have been omitted from the above discussions of electrons (protons are treated separately below). These facts relate to the pitch-angle distributions and the high-latitude "spikes" of electrons measured by low-altitude satellites.

Pitch-angle distributions have been studied extensively in many rocket and satellite measurements. In general, ignoring the complex (but non-trivial) backscatter effects, the measurements have found the same as O'Brien (1964) i.e., that the flux of electrons is normally peaked at $\alpha \sim 90^o$ (i.e., quasi-trapped) but in intense precipitation events the flux approaches and generally reaches isotropy over the upper hemisphere.

However, this intuitively satisfying picture has been disturbed by recent OGO observations by Hoffman and Evans (1968). They report finding at an altitude of some 600 km and invariant latitudes of 70° to 80°, that the flux of electrons with energies around 2 keV was often found to have a maximum value parallel to the magnetic field. No detailed statistical analysis of the results was presented, but, as the authors point out, "the fact that such events can be found at all is of considerable importance".

No such results have been reported by the Lockheed or Aurora 1 experimenters. However, both had relatively inadequate capability of detecting such events reliably, or at least of verifying the unusual anisotropy, so the "discrepancy" may not be real. However, since only the OGO-4 data were from a dawn-dusk orbit, while the others were from noon-midnight orbits, it is conceivable that the effect is dependant on local time.

Hoffman and Evans (1968) argue from their results that a possible source mechanism for these high-latitude particles is an electric field parallel to the geomagnetic field B at an altitude of several earth radii. They specifically do not apply this concept to be the source of electron bombardment in the auroral zone. However, since such anisotropic bursts were found at $\Lambda \sim 70^{o}$ at dawn, i.e., in the auroral zone, and at dusk "in the region of the auroral zone", but at midnight in the "high-latitude polar region", it would seem as well to await further detailed statistical analysis before attempting a zonal demarcation.

Intense high-latitude "spikes" of precipitating electrons with $E > 35$ keV were singled out as a separate phenomenon by McDiarmid and Burrows (1965). Found primarily in night-time passes, fluxes of up to 10^9 electrons cm^{-2} sec^{-1} sr^{-1} occurred at latitudes generally beyond the usual boundary of trapping, generally with larger fluxes at lower L values, but with intense fluxes up to $\Lambda \sim 83^{o}$. The authors relate the observations to "islands" of electrons found in the magnetospheric tail. However it must be borne in mind that the events are most frequent nearest to the boundary of trapping.

These two discoveries, viz of the "anomalous" pitch-angle distribution and the high-latitude islands, indicate the need for a sketch more complex than Figure 1.

4. HIGH-LATITUDE MEASUREMENTS OF LOW-ENERGY PROTONS

Again, since this is but one of several related reviews, we concentrate on direct rocket and satellite low-altitude measurements of low-energy protons, and deliberately exclude the valuable trans-polar measurements of H_β made in the 1968 Convair 990 expedition and also we exclude discussion of solar cosmic-ray events.

The earliest rocket measurements were by McIlwain (1960) and the first productive satellite measurements were by the Lockheed group (Sharp et al., 1967). McIlwain (1960) found fluxes of protons in the energy range above 50 keV to be more stable in time, more diffuse in space and less intense than the fluxes of electrons with a few keV that caused visible auroral forms.

The satellite data of Sharp et al. (1967) have already been illustrated in Figure 3. The fluxes of protons (many of which had already been neutralized at the low satellite altitude of 200 km) showed a bifurcation into two "zones" on the day-side but not on the night-side. Note in Figure 3 the similar location of the two dayside proton zones and the two electron zones discussed earlier. The measurements of protons are for energies above some 20 keV, and, unlike the two electron zones the two proton zones did not show any marked difference in characteristic energy, with most spectra consistent with an exponential spectrum with an e-folding value with E_o between 10 and 20 keV.

Sharp et al. (1967) did find that the median peak intensity of these precipitated protons was some five times larger at night than during the day just as was the total flux of precipitated electrons (Sharp and Johnson, 1967). Local times of crossing the auroral zones were about 1100 and 2300 hours, and were thus very similar to those of the Aurora 1 satellite. (Studies at 0600 and 1800 hours to study dawn-dusk asymmetries would be extremely valuable.)

However, again as with the electron studies, the higher sensitivity of the detectors on Aurora 1 found the two day-time proton zones discovered with the Lockheed satellite, but also found two night-side zones. Burch (1968b) has discussed these results in detail, and shown them to be consistent with the Lockheed experiment, when due account is taken of instrument capabilities.

Burch (1968b) has shown that, on individual passes, while the hard and soft proton zones more or less coincide with the hard and soft electron zones respectively, they do not coincide exactly. For example, during one magnetically quiet (K ~ 0) pass at night the hard electron zone was not even detected, while the hard proton zone stretched some 100 km on both sides of the low-latitude edge of the soft electron zones.

Such differences are important, because, for example, it is well known from ground-based photometric studies that there occur clearly identifiable proton-induced auroras with little flux of electrons. On other occasions, electron and proton precipitation occurs at the same location - a phenomenon that raises a rather large barrier to those who revel in speculations that low-altitude electric fields cause auroral-particle precipitation!

The proton energy spectra found by Aurora 1 are generally consistent with other rocket and satellite data, with a wide range consistent with E_o of some 10 keV to 20 keV. Distinction into two zones of "hard" and "soft" radiation is therefore somewhat difficult, but it is judged to be real (Burch, 1968b). There is no evidence for a continuum energy spectrum with a superimposed narrow regime of energies, as was the case with the electron fluxes. No accurate statements can be made for the proton spectrum below 1 keV because of the large corrections necessitated by the inefficient response of channel multipliers to ions of such low energies.

It is relevant here to note that alpha particles have been observed together with precipitating protons in auroras. The ratio of protons with about 100 keV to alpha particles with about 200 keV (i. e., with the same energy per unit charge) was found to be about 50 to 1 (Reasoner et al., 1968). This finding is consistent with the origin of the ions being the solar wind (or thermalised solar wind) but not magnetospheric ions.

5. DEFINITION OF A POLAR CAP

The concept of a polar "cap" around the north and south magnetic poles has, of course, received a great deal of its strength from the manner in which solar cosmic ray events cover each polar region to produce the well known and clearly defined Polar Cap Absorption (PCA) events.

From another point of view, one may think of the polar cap as being that region bounded at low latitudes by the auroral oval. (Indeed one of the vital experiments that is yet to be thoroughly carried out with a satellite is the determination of the relative locations at the same time of the low-latitude edge of a PCA event - more specifically, where the magnetic cut-off rigidity is zero - and the aurora.)

Again, from one more point of view, one may regard the polar cap as being the region threaded by geomagnetic field lines that are "open" i. e., swept back into the extended geomagnetic tail (Ness, 1965).

A decade or more ago, one might have defined the polar cap as the region above say, magnetic latitude 70°. We now know, of course, that such factors as the local-time and storm-time distortions of the magnetosphere render such a simple definition immaterial and totally inadequate.

In a similar manner, until 1961 the "inner" and "outer" radiation belts were treated as separate entities, rather than as derived from instrumental effects and a gradual variation of particle characteristics with L (cf. reviews by O'Brien, 1967; Hess, 1968). We have already discussed above our rationale in not treating the "hard" and "soft" electron zones nor the "auroral" and "soft" zones (Sharp and Johnson, 1968; Burch, 1968a, respectively) as generically separate zones.

Therefore, to act as a devil's advocate in these proceedings about the polar cap, we now proceed to examine the very definition of Polar Cap, and, in particular, the question as to whether there is such a generically distinct entity as a Polar Cap. As a beginning, we can test the concept against our three relatively sophisticated definitions related above to

Definition 1: distribution of solar cosmic rays
Definition 2: location of auroras
Definition 3: "open" magnetic field lines.

(Other possible definitions based on ionospheric characteristics are left to others.)

Definition 1. Since relativistic solar cosmic rays can penetrate the atmosphere at latitudes down to the equator, we used the concept of a zero-rigidity cut-off to locate the polar cap edges. From this point of view then, the "polar cap" will generally be well within the PCA zone caused mainly by protons with energies around some 10 MeV. We therefore have an immediate problem of nomenclature, but we will ignore it.

To our knowledge, there exist no satellite surveys of the latitude variations of solar protons with energies below about 1 MeV, even though protons with energies down to ~ 100 keV have been measured by rockets during PCA events. The relevant notion that PCA events theoretically should begin near the auroral region and then spread towards the polar cap (Michel and Dessler, 1965) does not appear validated by satellite observations though, to our knowledge, there has been no indisputable proof to the contrary. Indeed, recent results on "screening" of protons even with E ~ 15 MeV (Kane et al., 1968) indicate the great complexity of the problem and its deviation from any simple magnetospheric models. However, McDiarmid and Wilson (1968) do provide a very important preliminary study of this concept, but with 1 to 8 MeV solar protons whose intensity began to increase a few degrees below the abrupt high-latitude boundary of electrons with E > 35 keV.

As far as Definition 1 is concerned, then, we are forced to conclude that available experimental data do not permit us to define a polar cap, nor even to verify its existence as an entity.

Definition 2. In using the location of auroras to delineate the lower edge of a polar cap, we are immediately confronted with the problem that auroras are observed over the magnetic poles (Akasofu, 1964). The auroral arcs are generally oriented towards the sun, which presumably should provide a clue to the topology of the magnetosphere where the causal electrons (presumably) are accelerated. But how then can one even speak of a polar cap when it can be pervaded by auroral formations?

When we also take note of the occurrence of intense high-latitude "spikes" of precipitated particles (McDiarmid and Burrows, 1965), and also of the Aurora 1 finding of fluxes of electrons over the entire high-latitude region we must therefore examine whether there really is a distinct polar cap or whether the region is but a further gradation from the "auroral" and then the "soft" electron zones of Burch (1968a).

Suppose we try to predict (or guess) what the electron fluxes over the polar region should be if it is merely an extension of the lower-latitude regions. It should have a "continuum" electron flux, and indeed the Aurora 1 data often show a precipitation some ten to one hundred times lower than in the "soft" zone, but still measurable at lower energies below 10 keV. For example, on one pass (Burch, 1968b) the flux of electrons with energy between 50 eV and a few keV was of order 10^7 eV cm^{-2} sec^{-1} sr^{-1} and very soft, with the flux of 6 to 25 keV electrons less than 20 cm^{-2} sec^{-1} sr^{-1}. Burch (1968b) states that, whereas the boundary between the "soft" zone and the "polar cap" is not always as sharp as that in Figure 4, "it is a typical feature, especially near the midnight meridian". Thus these data support our devil's advocate position that there is a gradation from the soft zone to the lower fluxes over the polar cap.

Just as the "soft" zone often was invaded by auroral fluxes, so we might guess that they could invade the polar cap itself, although less frequently, and so they do, in the "spikes" of McDiarmid and Burrows (1965) for example. Figure 6 of McDiarmid and Burrows (1965) indicates dramatically that the frequency of occurrence of the spikes is some two hundred times larger at $\Lambda \sim 70^{\circ}$ than at $\Lambda \sim 82^{\circ}$ (some care needs to be taken in weighting the sampling of different values of Λ by the satellite before an exact ratio can be stated). Also the Aurora 1 data show occasional (but smaller) "spikes" after the UV-contamination cases are eliminated.

We are in a quandary as to guessing what the pitch angles of electrons over the polar cap might be, and can only guess that they would show the "mixture as before" viz generally peaked normal to the magnetic field or isotropic over the upper hemisphere during disturbed conditions, with perhaps a rare example of peak fluxes parallel to the field. To quote Hoffman and Evans (1968) "the dusk polar region was usually rather quiet, with activity picking up as the satellite moved poleward. This activity was in the form of bursts of electrons, from less than 1 to 10 seconds in duration, of highly variable spectra and pitch-angle distributions. The activity reached a maximum in the region of highest latitude..." Geomagnetic activity was quite low in the samples presented.

Thus, to all intents and purposes, neither the satellite observations by Alouette, Aurora 1 or OGO-4 nor the visible auroral studies indicate any very clear delineation of a "polar cap".

One might argue against this point by stating that Maehlum (1968) has found a very clear UT control of the boundary between Burch's "soft zone" and "polar cap". This, it might be suggested, indicates a special role for the polar cap. But, to counter this, we point out that McDiarmid and Wilson (1968) have found evidence of a UT control (which they correctly call a control by the orientation of the geomagnetic axis) on the high-latitude boundary of trapping of electrons with $E > 35$ keV. Admittedly, the effect seen by McDiarmid and Wilson is small, being only 2° for an invariant latitude of some 73° (on the average), while Maehlum finds a variation of some 10° at the higher average latitude of Λ of some 78°. But surely this is to be expected viz that the higher the latitude the more susceptable it is to control by the solar-wind angle of impact on the geomagnetic field.

Thus, once again, we find the "polar cap" and the "soft" zone different quantitatively but with similar qualitative characteristics. In other words, generically they are apparently the same.

Definition 3. We then proceed to examine Definition 3, which would have the polar cap as the region threaded by "open" magnetic field lines. We might be pedantic, and invoke Maxwell to invalidate any notion of "open" or "non-closed" magnetic field "lines". But it is generally accepted, following the work of Ness (1965), that the term "open" is given to those magnetic field lines, including those from the dayside, that are swept back in an antisolar direction to form the geomagnetic tail or wake at a radial distance of some 15 R_E and beyond, where there can be recognised a neutral sheet that divides, i. e., keeps open, the field lines from the northern and southern polar regions. Since the geomagnetic field is discussed extensively by Fairfield in these proceedings, only a brief treatment is given here.

Here, at last, in the magnetic field configuration, do we have an opportunity to distinguish a real generic difference. There are, at small values of Λ magnetic field lines that are definitely closed, while at higher values of Λ there are open field lines. The dominant question is: where is the transition from one type to the other, and then, of course, there is the question of the stability of that transition. Is the matter simply as shown in Figure 1 and as discussed by O'Brien (1967) and others, that the auroral region is the transition region itself, separating the polar cap from trapped radiation?

In order to relate the magnetospheric wake to low-altitude polar events, one must be able to "map" the magnetic field lines from beyond say 15 R_E down into the ionosphere. However, such mapping is difficult enough even at much smaller radial distances, say 5 R_E, and so we examine various approaches.

Although statistically there has been a tendency to relate the high-latitude low-altitude "spikes" of electrons to the magnetospheric wake "islands" (McDiarmid and Burrows, 1965) there has been no attempt to use them as tracers of field lines. This is because, at best, energetic particle fluxes are a very unreliable guide to magnetospheric magnetic field configurations. Such particles are subject to such effects as "L-shell splitting", to drift and acceleration by electric fields and so on. (This also necessitates very great caution in relating high-latitude ionospheric effects to localised high-altitude radiation.)

Conceptually, one might use the principle of conservation of magnetic flux to deduce the polar cap. The low-altitude magnetic field is known and the instantaneous magnetic field in a cross-section of the tail, together with knowledge of the dimensions of the tail, could lead one to calculate the size of the polar cap as defined above. For example, using a tail diameter of some 40 R_E and an average tail field of some 15 γ (e.g., Ness, 1965) one might calculate a polar cap that extends down to about $\Lambda \sim 80^o$.

But such simple calculations are of little value for several reasons, ignoring as they do such very important facets as the universal time and local time and seasonal distortions of the magnetosphere, the diamagnetic effects of plasma and particles in the tail region (e.g., Anderson and Ness, 1966) and the resultant corrections to measured field strengths, and so on. Admittedly one could make more sophisticated calculations taking account of such effects, the rotation of the earth etc., but basically one must inevitable be balked also by the difficulty in measurement of B all across the tail region. For instance, a 50 percent uncertainty in B in the quoted example will move the "polar cap" by about 2°.

One might then attempt to determine if magnetic field lines are "open" or "closed" by examining the conjugacy of various phenomena. Since entire symposia have been held on this topic alone, we mention here only the difficulty in deciding whether a field line is really closed even if it passes through conjugate aurora, because we do not know either the manner or the altitude of acceleration of the incident auroral electrons. One must treat with similar caution conjugacy of micro-pulsations.

Therefore, to summarize our attempts to define a "polar cap" we have reached the following situation:

Definition 1 is of no present use because there have been inadequate measurements of low-energy solar cosmic rays.

Definition 3 is of no present use because there has been no tracing or "tagging" of the field lines.

Definition 2 leads to the conclusion that there is no special distinct entity called a "polar cap".

6. DISCUSSIONS

It is becoming reasonably common to divide the region at $\Lambda > 60^{\circ}$ into three zones. For example, Burch (1968a) has done so using satellite data on precipitated electrons. Again, Piddington (1968) has done so in his theoretical discussions on plasma flow from and into the geomagnetic tail.

By contrast, here we have adopted an approach which regards segregation of the zones as being artificial, arising as it does from a too-finely-focussed preoccupation with the differences between the zones and too little attention to their similarities and the occasional detection in each zone of the very same phenomena as are used to classify another zone as unique. We have shown that the very concept of a unique polar cap is inconsistent with auroral data and with low-altitude measurements of precipitated particles.

Let us also briefly examine high-altitude measurements of such particles (an extensive treatment is given elsewhere at this symposium).

Bame et al. (1967) found a plasma sheet at 17 R_E that was 4 to 6 R_E thick at the midnight meridian, flaring out to twice that thickness near dawn and dusk. The sheet was characterized by electrons with energies between ~ 200 eV and some 12 keV, with fluxes up to as high as 10^9 particles cm^{-2} sec^{-1}. Occasional increases in the average energies of the particles gave rise to the so-called "islands".

Low-energy (~ 10 keV) protons were also observed within the plasma sheet, with the same number density as the electrons and hence in considerably smaller fluxes.

Now, outside this plasma sheet the counting rate of the detectors fell below background, i.e., to less than ~ 10^7 particles cm^{-2} sec^{-1} in the above low-energy range. Bame et al. comment that the transition was generally abrupt. However, Bame et al. also note that "there are a few examples of enhanced plasma seen at considerably higher than normal latitudes" (cf. also Vasyliunas, 1968a). Thus, once again, a region is distinguished only on the basis of the probability of a phenomenon, and not on the basis of its presence or total absence.

The greater sensitivity of the Aurora 1 instrument did permit detection of particle fluxes over the poles, in intensities that would not be detected by either the Vela or OGO sensors.

A further very interesting high-altitude effect is the radial variation in the plasma sheet studied by Vasyliunas (1968a) (The Vela satellites in near circular orbits could not survey this effect). There is essentially a dearth of low-energy (~ 2 keV) electrons at dusk and midnight all the way from the inner edge of the plasma sheet at say 11 ± 1) R_E into the outer edge of the plasmasphere at some 4 R_E. During periods of magnetic bay activity, the plasma sheet may move in to ~ 6 to 8 R_E. Since significant fluxes of electrons of such energies are seen at least down to L ~ 4 by low-altitude satellites (e.g., Burch, 1968b) this finding poses a very real problem. Since Frank (1967) has reported the existence of detectable fluxes in this region at high altitudes, using OGO-3 data as did Vasyliunas, we merely note here that some inconsistencies apparently exist; these may arise simply from the extreme variation of this "cavity" with local time, since it is known (Vasyliunas, 1968b) that the plasma sheet extends into the plasmasphere at midday and dawn. However, significant fluxes are observed at low altitudes down to L ~ 4 both at midday and midnight (Burch, 1968b) so the matter certainly warrants urgent attention. In this problem of the radial distribution, unlike the problem of the latitude distribution discussed above, it is not possible to reconcile high-altitude undetectable fluxes with low-altitude detectable fluxes on the basis of instrumental sensitivities, since the fluxes seen at low altitudes are well within the OGO range of detection.

Apart then, from this matter of the radial distribution of the plasma sheet, which pertains to low values of L (~ 4 to 6) at local midnight, there appears to be consistency between low-altitude and high-altitude data. The high-altitude plasma sheet thus crudely corresponds with the "auroral" plus the "soft" zone of Burch (1968a), while the lower fluxes of the "polar cap" would appear to occupy the high-altitude region between the plasma sheet and the magnetopause. When comparison is made of either the energy spectrum or of the occurrence of spikes at high and low altitudes, this association is seen to be acceptable.

7. CONCLUSIONS

From the foregoing, we have questioned the very concept of a "polar cap" per se having a generic uniqueness. We question it on many grounds, such as the fact that auroras occur over the magnetic poles, and that fluxes of electrons with energies between 50 eV and a few keV are precipitated at midnight and midday across the entire region with an input of order 10^7 eV cm^{-2} sec^{-1} sr^{-1}, while occasional very intense spikes of higher-energy electrons are also observed up to at least $\Lambda \sim 83^o$. Proton fluxes over the entire region are undetectable and hence less than ~ 10^6 eV cm^{-2} sec^{-1} sr^{-1}.

On the basis of comparison of high-altitude and low-altitude data, we conclude that, if one wishes to retain a concept of a polar cap, then it should be defined at all altitudes in the manner adopted by Burch (1968a), in that it is the region where the low-energy (~ keV) electron flux drops rapidly by an order of magnitude or more. Then, at high altitudes, the "polar cap" would lie between the plasma sheet and the magnetopause, and be subject to both systematic and sporadic intrusions by the plasma sheet. At low altitudes, the boundary will vary between $\Lambda \sim 74^o$ and 84° during local day and $\Lambda \sim 71^o$ and 83° during local night, where the wide variations are caused by changes in the tilt of the geomagnetic axis with respect to the impinging solar wind (Maehlum, 1968).

To the extent then, that we neglect impact of the solar wind other than normally to the geomagnetic axis, and other than in the noon-midnight plane (two extremely wide-sweeping factors!) we may continue to use a sketch such as Figure 1, with perhaps a somewhat thicker "plasma sheet".

It is interesting to note that the plasma sheet streams in the anti-solar direction, and yet its presumed boundary at low latitudes moves by some ten degrees or more as the earth rotates. It seems not unreasonable to believe that there should be significant seasonal changes in the magnetotail field strength and in the plasma sheet thickness because of this very pronounced effect.

Acknowledgements. It is a pleasure to thank Professor H. Messel for his continued encouragement and support.

REFERENCES

Akasofu, S.-I., The development of the auroral substorm, Planetary Space Sci., 12, 273-282, 1964.

Albert, R. B., Nearly monoenergetic electrons fluxes detected during a visible aurora, Phys. Rev. Letters, 18, 369, 1967.

Anderson, K. A. and N. F. Ness, Correlation of magnetic fields and energetic electrons on the IMP 1 satellite, J. Geophys. Res., 71, 3705, 1966.

Bame, S. J., J. R. Asbridge, E. H. Felthouser, E. W. Hones, and I. B. Strong, Characteristic of the plasma sheet in the earth's magnetotail, J. Geophys. Res., 72, 113, 1967.

Bolyunova, A. D., O. L. Vaisberg, Yu. I. Galperin, B. P. Potapov, V. A. Temny, and F. K. Shuiskaya, Investigation of corpuscles on the Electron 1a and Electron 3 satellites, Space Res., VI, 1966.

Burch, James L., Low-energy electron fluxes at latitudes above the auroral zone, J. Geophys. Res., 73, 3585, 1968a.

Burch, James L., High-latitude satellite observations of electrons and protons at 4000 km, Ph. D. Thesis, Rice University, Houston, Texas, 1968b.

Davis, T. N., The aurora, in Introduction to Space Science, edited by W. N. Hess, Gordon and Breech, N. Y., 1965.

Evans, D. S., On auroral electrons, NASA-GSFC Report No. X-611-67-515, 1967.

Frank, L. A., On the extraterrestrial ring current during geomagnetic storms, J. Geophys. Res., 72, 3753, 1967.

Fritz, T. A. and D. A. Gurnett, Diurnal and latitudinal effects observed for 10 keV electrons at low satellite altitudes, J. Geophys. Res., 70, 2485, 1965.

Hess, Wilmot N., The Radiation Belt and Magnetosphere, Blaisdell Publishing Company, Waltham, Massachusetts, 1968.

Hoffman, R. A. and D. S. Evans, Field-aligned electron bursts at high latitudes observed by OGO-4, J. Geophys. Res., 73, 6201, 1968.

Hones, E. W., S. J. Bame, S. Singer, and R. R. Brown, Observed correlations between electron bursts in the magnetotail and electron precipitation in the auroral zone, J. Geophys. Res., 73, 6189, 1968.

Johnson, R. G., R. D. Sharp, M. F. Shea, and G. B. Shook, Satellite observations of two distinct day-side zones of auroral electron precipitation, Trans. Am. Geophys. Union, 47, 64, 1966.

Kane, S. R., J. R. Winckler, and D. J. Hofman, Observations of the screening of solar cosmic rays by the outer magnetosphere, Planetary Space Sci., 16, 1381, 1968.

McDiarmid, I. B. and J. R. Burrows, Electron flux at 1000 km associated with the tail of the magnetosphere, J. Geophys. Res., 70, 3031, 1965.

McDiarmid, I. B. and J. R. Burrows, Local time asymmetries in the high-latitude boundary of the outer radiation zone for the different electron energies, Can. J. Phys., 46, 49, 1968.

McDiarmid, I. B. and M. D. Wilson, Dependence of the high latitude electron ($E > 35$ keV) boundary on the orientation of the geomagnetic axis, J. Geophys. Res., 73, 7237, 1968.

McIlwain, C. E., Direct measurements of particles producing visible auroras, J. Geophys. Res., 65, 2727, 1960.

Maehlum, B. and B. J. O'Brien, Study of energetic electrons and their relationship to auroral absorption of radio waves, J. Geophys. Res., 68, 997-1010, 1963.

Maehlum, B., Universal time control of the low-energy electron fluxes over the polar caps, J. Geophys. Res., 73, 3459, 1968.

Michel, F. C. and A. J. Dessler, Physical significance of inhomogeneities in polar cap absorption events, J. Geophys. Res., 70, 4305, 1965.

Ness, N. F., The earth's magnetic tail, J. Geophys. Res., 70, 2989, 1965.

O'Brien, B. J., A. Large diurnal variation of the geomagnetically-trapped radiation, J. Geophys. Res., 68, 989-996, 1963.

O'Brien, B. J., High-latitude geophysical studies with satellite Injun III, Part 3, Precipitation of electrons into the atmosphere, J. Geophys. Res., 69, 13-44, 1964.

O'Brien, B. J. and H. Taylor, High-latitude geophysical studies with satellite Injun III, Part 4, Auroras and their excitation, J. Geophys. Res., 69, 45-63, 1964.

O'Brien, B. J., Interrelations of energetic charged particles in the magnetosphere, in Solar-Terrestrial Physics, edited by Newman/King, Academic Press Inc. Ltd., London, p. 169-211, 1967.

Piddington, J. H., The magnetospheric radiation belt and tail plasma sheet, Planetary Space Sci., 16, 703, 1968.

Reasoner, D. L., R. H. Eather, and B. J. O'Brien, Detection of α-particles in auroral phenomena, J. Geophys. Res., 73, 4185-4198, 1968.

Sharp, R. D., R. G. Johnson, M. F. Shea, and G. B. Shook, Satellite measurements of precipitating protons in the auroral zone, J. Geophys. Res., 72, 227, 1967.

Sharp, R. D. and R. G. Johnson, Some average properties of auroral electron precipitation as determined from satellite observations, J. Geophys. Res., 73, 769, 1968.

Taylor, H. E. and E. W. Hones, Adiabatic motion of auroral particles in a model of the electric and magnetic fields surrounding the earth, J. Geophys. Res., 70, 3605-3628, 1965.

Van Allen, J. A., G. H. Ludwig, E. C. Ray, and C. E. McIlwain, Observation of high intensity radiation by satellites 1958 Alpha and Gamma, Jet Propulsion, 588, 1958.

Vasyliunas, V. M., A survey of low-energy electrons in the evening sector of the magnetosphere with OGO 1 and OGO 3, J. Geophys. Res., 73, 2839, 1968a.

Vasyliunas, V. M., Low energy electrons on the day-side of the magnetosphere, J. Geophys. Res., 73, 7519, 1968b.

Westerlund, L. H., Rocket-borne observations of the auroral electron energy spectra and their pitch-angle distribution, Ph. D. Thesis, Rice University, Houston, Texas, 1968.

SATELLITE MEASUREMENTS OF LOW ENERGY ELECTRONS PRECIPITATED AT HIGH LATITUDES

James L. Burch

Department of Space Science
Rice University, Houston, Texas 77001, USA

Abstract. Some of the typical and average properties of the high-latitude zone of low energy electron precipitation as observed by the Aurora 1 satellite are presented. Measurements of electrons with pitch angles less than 30° were made at 4000 km altitude and at magnetic local times within two hours of noon and midnight. The position of the low-latitude boundary of the "soft" electron zone for $K_p \leq 4$ is approximated by $\Lambda_{min}(\text{night}) = 69.5^o - .8\,K_p$ and $\Lambda_{min}(\text{day}) = 78.3^o - 1.3\,K_p$. Average spectra for $.1\text{ keV} \leq E \leq 1.3\text{ keV}$ are represented by the power laws, $J(E) = 2.7 \cdot 10^7 E^{-1.48}$ (night) and $J(E) = 3.0 \cdot 10^7 E^{-2.26}$ (day). No clear dependence of the magnitude of electron fluxes in this region on magnetic activity has been noted. Poleward of this zone, the electron flux is estimated to be less than $.001\text{ ergs cm}^{-2}\text{ sr}^{-1}\text{ sec}^{-1}$.

The average energy influx in the "soft" electron zone is of the approximate order of magnitude to explain the anomalously high levels of F-region ionization observed near the dark pole. It is also noted that the dependence of the properties of this zone on K_p are similar to certain K_p variations of high-latitude F-region ionization and 6300 Å emission.

1. INTRODUCTION

High-latitude measurements of the energy flux of low energy (< 1 keV) electrons were first made aboard polar orbiting satellites by the Lockheed research group (Evans et al., 1967; Sharp and Johnson, 1968). An interesting feature of their observations was the separation of the day side region of electron precipitation into two zones - a lower latitude ($\Lambda \sim 69^o - 75^o$) zone in which both the 80 eV and the 21 keV threshold detectors responded above background and a higher latitude ($\Lambda \sim 75^o - 79^o$) zone in which only the 80 eV threshold detector so responded. A single zone of electron precipitation (both detectors responding) was noted on the night side in the invariant latitude range $66^o - 72^o$.

Recent electron flux measurements in six energy bands between 100 eV and 100 keV aboard the Rice University/Office of Naval Research satellite Aurora 1 have confirmed the existence of the two day-side electron zones and in addition have indicated a similar but more complicated behavior during local night. In an earlier report (Burch, 1968a), three-point electron spectra obtained at 4000 km altitude for magnetic local times within two hours of noon and midnight were plotted for several passes through high latitudes. These passes occurred during a period of slight to moderate geomagnetic activity ($K_p < 4$). On both the day and night sides the lower latitude zone was characterized by considerable fluxes in the 6200 eV - 25,000 eV energy channel which remained nearly constant through a wide range of pitch angles except near 180° (northern hemisphere) where an intensity decrease was often observed. Poleward of this zone much softer electron spectra and rapid temporal and/or spatial variations which precluded determination of any pitch angle dependence were observed over a region which was typically about 5° wide in invariant latitude. At still higher latitudes only very weak electron fluxes were noted.

It is the purpose of this paper to describe in more detail the characteristics of the high-latitude "soft zone" of electron precipitation and its possible connection with phenomena in the polar F2-region. In this vein it has already been noted (Maehlum, 1968; 1969) that certain features of the diurnal variation of high-latitude f_oF_2 can be explained by the universal time dependence of the poleward extent of these low energy electrons and that the low level of F2-region ionization and of 6300 Å intensity near the L-pole are consistent with their high-latitude termination. It will now be shown: (i) that the average electron energy fluxes in the soft electron region are of the right order of magnitude to explain the "auroral cliff" in the topside F2-ionization seen by Thomas et al. (1966); (ii) that the variation of the low-latitude boundary of this region with K_p corresponds approximately to that of the night-side "auroral cliff"; and (iii) that this K_p variation and the lack of an apparent K_p dependence of the high-latitude boundary of this zone or of the average intensity of low energy electrons in its interior are in qualitative agreement with the 6300 Å measurements of Sandford (1968).

2. TYPICAL CHARACTERISTICS OF HIGH-LATITUDE ELECTRON PRECIPITATION

The data presented in this section were obtained during mid-July, 1967, after the satellite had magnetically oriented to such an extent that only those electrons with pitch angles less than about 30° (i.e., precipitated electrons) were counted. The electron spectrometer has been described elsewhere (O'Brien et al., 1967; Burch, 1968b), but a diagram of the arrangement of the channel multipliers and deflection plates is shown in Figure 1 for reference. The six-point energy spectra which will be discussed in this section are based on 0.1-second average count rates in the two 270° channel multipliers closest to the detector axis. The deflection plate voltage is stepped through three levels, yielding energy passbands of 96 eV - 163 eV, 150 eV - 820 eV, 1075 eV - 1910 eV, 1.7 keV - 8.6 keV, 10.8 keV - 19.5 keV, and 17.5 keV - 100 keV. Each passband is sampled once every five seconds, with the sampling scheme requiring 1.5 seconds for the generation of a six-point spectrum. The differential spectra which will be presented are actually zero-order approximations, since the above passbands are defined by 10% response to a "flat" laboratory spectrum. The absolute flux levels are based on a relative laboratory calibration and the channel multiplier efficiency measurements of Frank (1965). His values are somewhat at variance with those of Arnoldy (cf. Hoffman and Evans, 1967). The discrepancies introduce no significant differences in spectral shape for the spectra considered herein, but do produce an error of about a factor of two in the total energy between 100 eV and 1 keV with the higher value resulting from Frank's measurements.

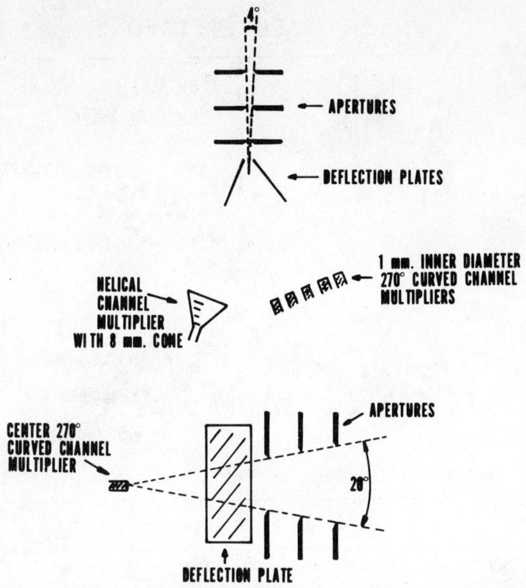

Figure 1. Sketch of the electron spectrometer geometry, showing approximate locations of apertures, deflection plates and channel multipliers.

In Figure 2 we have plotted the response in four of the energy bands sampled during a daytime northern hemisphere pass when K_p and K_c (the local College, Alaska, K index, which we will use as a measure of high-latitude magnetic activity) both had a value of 1. Also shown are six-point spectra taken at two positions along the satellite trajectory. Observations of precipitated electrons during two night-side passes are presented in Figures 3 and 4. These passes occurred during moderately disturbed magnetic conditions ($K_p, K_c = 2 - 3$). We note in these three examples that the spectra in the regions L ~ 5 - 8 during local night and L ~ 8 - 20 during local day are relatively flat or slightly peaked at energies of several keV. They are also characterized by a rising intensity below a few hundred eV and a sharp decrease in intensity above about 10 keV. Such spectra are similar to those which have been observed from sounding rockets directly above visual auroras (Evans, 1967; Chase, 1968; Westerlund, 1969). In each of these cases, as in most cases investigated, the high-latitude decrease of fluxes in the 1.7 - 8.6 keV channel is accompanied by a significant increase in fluxes in the two lowest energy channels, resulting in a rapid spectral softening. This higher latitude "soft zone" is characterized by large flux variations along the satellite's spatial and temporal path and by the sporadic occurrence of narrow "spikes" with somewhat harder spectra. These spikes may be of the same type as those observed at energies above 40 keV by McDiarmid and Burrows (1965), and others. In the day time pass (Figure 2) the satellite did not approach the L-pole sufficiently to cross the high-latitude boundary of these electron fluxes. However, in the two night-side passes (Figures 3 and 4) this termination is evident, occurring at a somewhat higher latitude in Figure 4, which corresponds to a time when the earth's dipole was tilted toward the sun, than in Figure 3, which occurred near the opposite configuration (cf. Maehlum, 1968). The response in all channels is below threshold across the pole except for occasional very

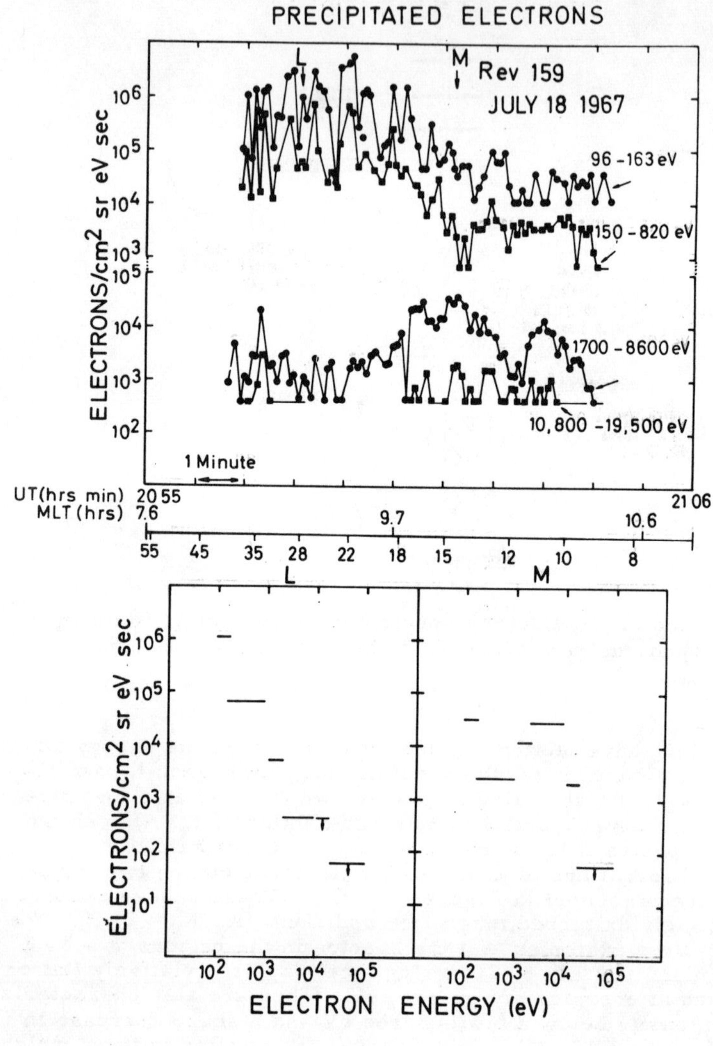

Figure 2. Differential flux in several energy channels during a pass through the polar regions, based on 0.1-sec. count rate averages taken once every 5 sec. Six-point differential spectra are plotted for a few selected points, with passbands defined by 10% response to a flat isotropic laboratory electron spectrum. (Day side; K_p, $K_{\bar{c}}$ ~ 1).

low count rates in the lower energy channels, and this behavior is typical of every pass analyzed.

To support the contention that the "auroral" type energy spectra observed near the low-latitude boundary of electron precipitation are connected with auroral zone phenomena and to illustrate the lack of a clear magnetic activity dependence of the intensity of low-energy electrons in the soft electron zone, the data in Figure 5 are presented. This pass occurred when K_p = 0+ - 1 and K_c = 0. Note that, whereas fluxes in the 1.7 - 8.6 keV energy band are very low except for one sample period, the "soft zone" electron fluxes are still quite intense. This behavior has generally been noted on the night side when K_p ~ 0.

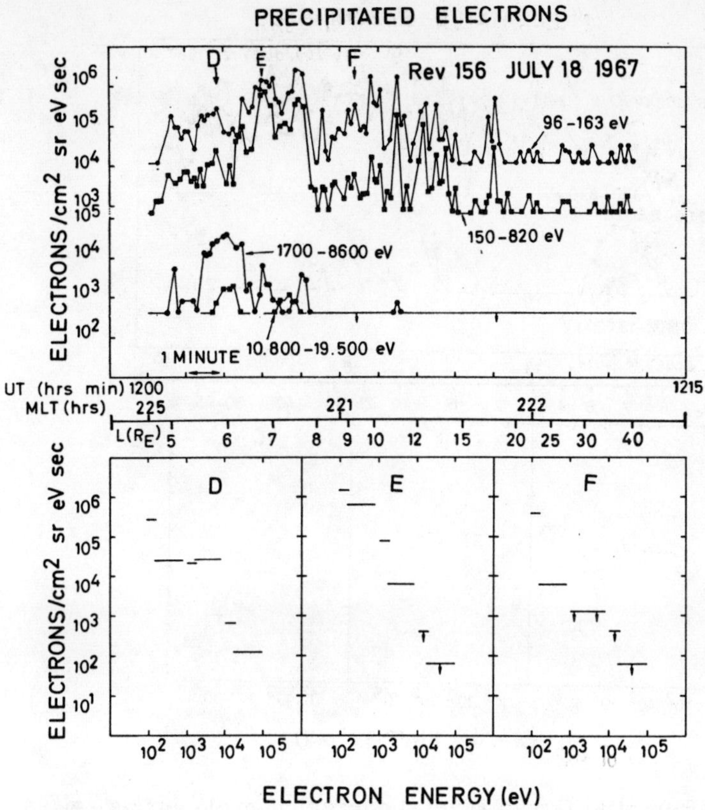

Figure 3. Differential flux in several energy channels during a pass through the polar regions. (Night side; K_p, K_c ~ 3).

The low-energy intensity in the day side high-latitude zone also appears to be independent of K_p variations. However, all the data analyzed to date were obtained during times of relative magnetic quiet ($K_p \leq 4$).

The zones of low-energy electron precipitation shown in Figures 2, 3, 4, and 5 are typical of 36 of the 40 passes analyzed to date. In the other four passes, all on the night side, the intense fluxes of low-energy electrons were accompanied by very intense localized fluxes of electrons of several keV energy. Each of these cases occurred during a period of high-latitude magnetic bay activity. Data from revolution 157 are shown in Figure 6. During this period a negative bay of about 100 γ was evident on the magnetogram from College (about 75° of longitude east of the subsatellite point). Although the total energy influx throughout this region is much higher than that during quiet times, the total number of electrons with energies of a few hundred eV is not significantly greater. It is interesting to note that the high-latitude cutoff of electron precipitation occurs near the latitude which one would expect for this universal time (cf. Maehlum, 1968). If these intense fluxes of "auroral" energy electrons are connected with the expansive phase of a polar substorm, then perhaps the poleward extent of the substorm is controlled in some way by the pre-existing boundary and as such should be under universal time

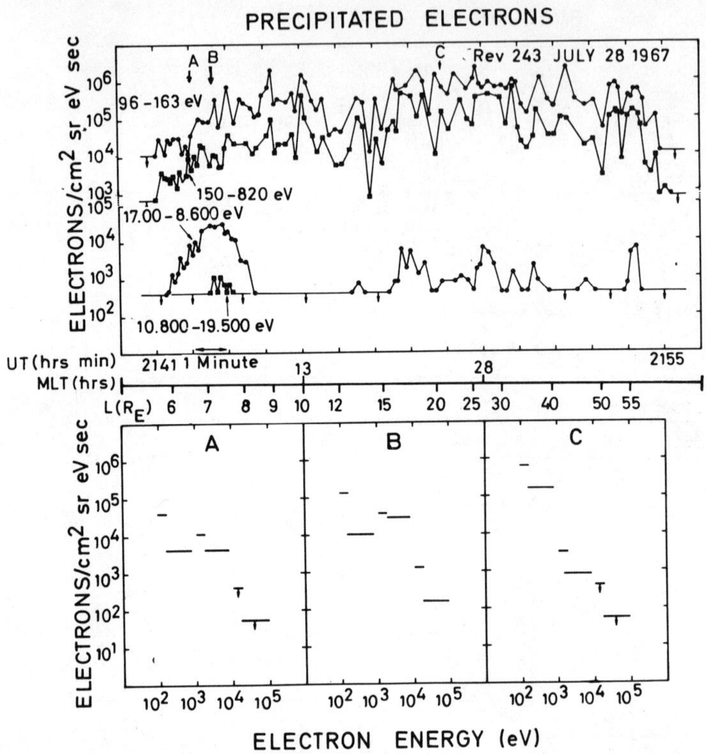

Figure 4. Differential flux in several energy channels during a pass through the polar regions. (Night side; K_p, K_c ~ 3).

control. To our knowledge such a study has not been conducted as yet. The presence of more energetic electrons at higher latitudes during a polar substorm is consistent with the observations of Lin et al. (1968), that electrons of energies greater than 40 keV are present during such disturbances at much higher latitudes than usual, generally filling the region in which high-latitude "spikes" are observed during normal passes.

The morphology of the visual auroral displays, strong polar magnetic disturbances and other phenomena which originate chiefly in the E-region has been studied in some detail and has shown relatively good correlation with the precipitation of higher energy electrons ($E > 1$ keV) (e.g., Brice, 1967). Therefore, in the following sections we will examine some of the average properties of the high-latitude soft electron zone and relate these properties at least coincidentally to some of the unexplained characteristics of the polar F-region.

3. SOME AVERAGE PROPERTIES OF THE HIGH-LATITUDE LOW-ENERGY ELECTRON PRECIPITATION

An average of 428 electron spectra measured during eight passes through the night-side low energy electron zone when $K_p \leq 3+$ and with no polar substorm effects noted can be fit approximately by two power law relations: $J(E) = 2.7 \cdot 10^7 E^{-1.48}$ cm^{-2} sr^{-1} keV^{-1} sec^{-1} for $.1$ keV $\leq E \leq 1.3$ keV, and $J(E) = 3.4 \cdot 10^7 E^{-2.26}$ cm^{-2} sr^{-1} keV^{-1} sec^{-1}

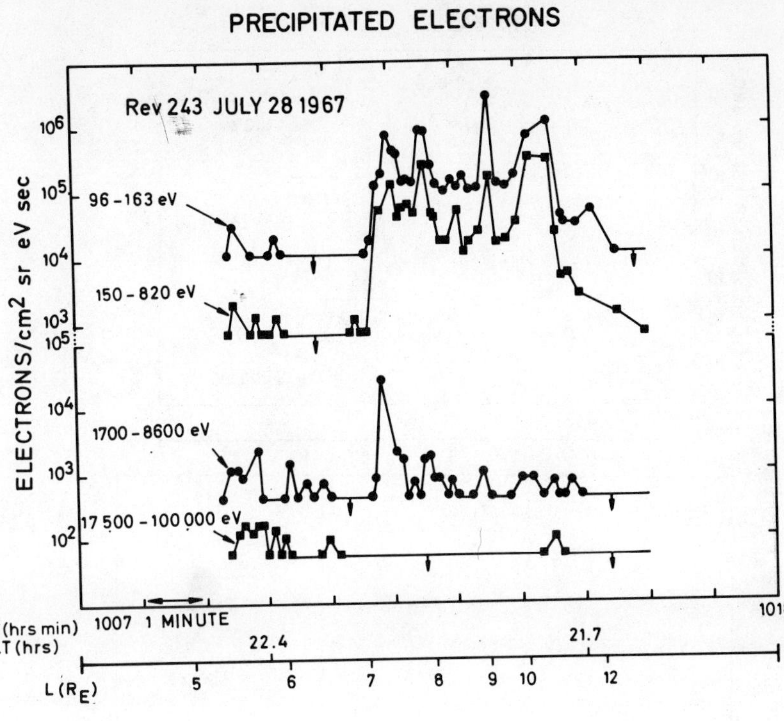

Figure 5. Differential flux in several energy channels during a pass through the polar regions. (Night side; $K_p \sim 0+ - 1$, $K_c = 0$).

for $1.3 \text{ keV} \leq E \leq 100 \text{ keV}$. This spectrum gives an average electron energy of 0.15 ergs cm^{-2} sr^{-1} sec^{-1} with about half this energy being carried by electrons with $E < 1.3$ keV. The calculations of Rees (1964) assuming an isotropic distribution in pitch angle, indicate that the greater part of this low-energy portion will be absorbed in the F-region.

A similar average of 270 spectra obtained during eleven day-side passes when $K_p \leq 3$ yielded a soft zone electron spectrum which is somewhat steeper and more intense than the average night-side spectrum. That is, $J(E) = 3.0 \cdot 10^7 E^{-2.26}$ cm^{-2} sr^{-1} keV^{-1} sec^{-1} for $.1 \text{ keV} \leq E \leq 1.3 \text{ keV}$. This spectrum results in a total energy influx in this energy range of .16 ergs cm^{-2} sr^{-1} sec^{-1}, or about twice the night-side value.

As mentioned above, electron fluxes above the high-latitude termination of the intense low energy electron fluxes were nearly always below threshold. If we assume an energy spectrum similar to that observed in the day and night side soft zones we can set an upper limit of about .001 ergs cm^{-2} sr^{-1} sec^{-1} on the electron fluxes over the poles.

Concerning the location of the soft electron region, it has already been shown by Maehlum (1968) that the high-latitude boundary lies in the ranges $\Lambda = 73^{\circ} - 85^{\circ}$ (day side) and $\Lambda = 70^{\circ} - 79^{\circ}$ (night side) and that the boundary occurs at higher invariant latitudes as the "tilt angle" between the geomagnetic axis and the earth-sun line decreases. Furthermore, no correlation of the position of this boundary with

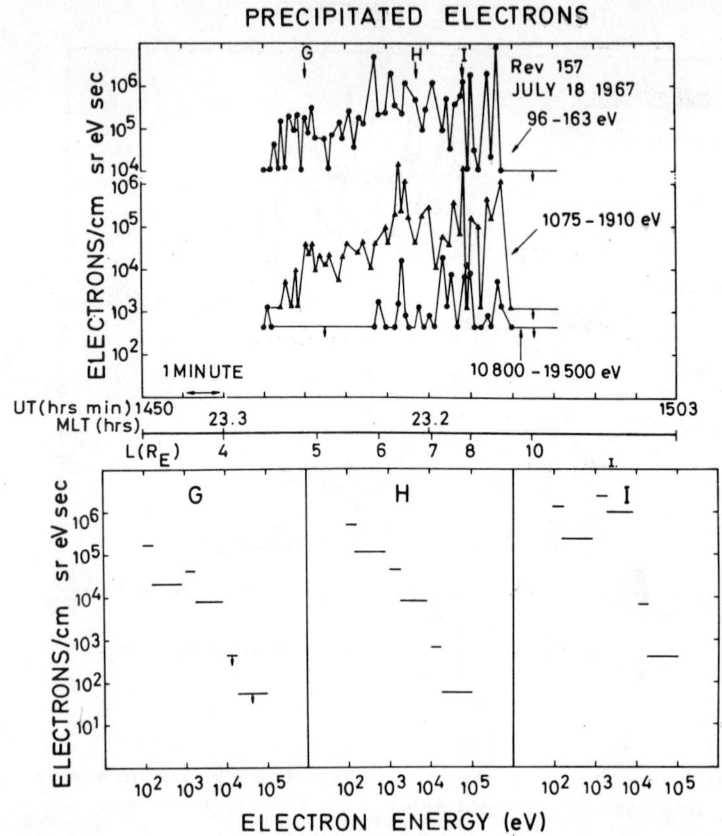

Figure 6. Differential flux in several energy channels during a pass through the polar regions. (Night side; K_p, $K_c \sim 3$).

magnetic activity has been noted. An investigation of the low-latitude boundary of this region has revealed no such universal time control, but does show a clear K_p correlation for $K_p \leq 4$. A regression line $\Lambda_{min} = 69.5^o - .8\,K_p$ has been fitted to the fifteen available boundary crossings on the night side with a correlation coefficient of .41. The twenty values of Λ_{min} sampled on the day side are represented by the regression line $\Lambda_{min} = 78.3^o - 1.3\,K_p$, with a correlation coefficient of .74. It is evident then that when the geomagnetic axis is tilted away from the sun and K_p is very low the soft zone can be quite narrow, and this has been observed. We have not found any cases in which the zone was non-existent although this is possible especially on the day side.

A study of possible systematic variations of spectral hardness and total energy flux with universal time and K_p revealed no apparent correlation for the data analyzed to date.

4. DISCUSSION

We noted in the previous section that, in the absence of polar substorms, the average precipitated energy carried by electrons with energies less than 1.3 keV is roughly twice as great near local noon as near local midnight. The fact that the Lockheed experiments did not show a soft electron zone on the night side could possibly be explained in terms of this diurnal variation. Also, we should mention that the latitude range ($\Lambda = 66^{\circ} - 72^{\circ}$) in which night-side electron precipitation was reported by Sharp and Johnson (1968) is not incompatible with the existence of fluxes at higher latitudes, since their values indicated "centers of gravity" of the total energy across the zone. It is evident in Figures 2, 3, and 4 that for typical cases the total energy flux is greatest near the low-latitude edge of the complete zone of precipitation.

It has been shown by Maehlum (1968; 1969) that universal time variations in f_oF2 observed in the winter polar regions can be explained for observing stations which observe the F-region at invariant latitudes between 70° and 85° by their diurnal motion in and out of the soft electron zone. It was also shown (Maehlum, 1969) that the depletion of ionization (Thomas et al., 1966) and of 6300 Å emission (Sandford, 1968; Eather, 1969) near the poles during periods of low sunspot number is compatible with the lack of significant electron fluxes in that region.

An attempt to estimate the F-region ionization produced by the average electron fluxes described in the previous section is difficult due to the present lack of detailed knowledge of the total ionization cross-sections and recombination coefficients pertinent to that region. Uncertainties in the electron pitch angle distribution complicate this matter further. However, a rough estimate can be made utilizing the experimentally determined energy dissipation function discussed by Rees (1963), and the approximate recombination coefficients given by Donahue (1966) for a temperature of 1000 °K at 220 km altitude. For energies less than 1.3 keV the average night-side soft zone electron energy is about 200 eV. The average flux, J, assuming an isotropic distribution is $7.5 \cdot 10^{8}$ cm^{-2} sec^{-1}. According to Rees (1964), 200 eV electrons which are isotropic in pitch angle at 1000 km will produce a maximum ionization rate at about 250 km and a rate amounting to 10% of the maximum at 200 km. Assuming an average energy loss per ion pair produced of 35 eV, the ionization rate at 250 km is approximated by

$$q_{max} = J \frac{200 \text{ eV}}{35 \text{ eV}} \frac{\rho}{R} \lambda_{max} \frac{n(M)_{250 \text{ km}}}{n(M)_{200 \text{ km}}} \quad (1)$$

where, ρ is the mass density (g/cm^3) at 200 km, R is the atmospheric penetration depth for 200 eV electrons in g/cm^2, λ_{max} is the peak value of the energy dissipation function ($\lambda_{max} \sim 1.5 - 1.9$) and $n(M)_h$ is the number density of ionizable atoms at altitude h. Using the model atmosphere of Rees (1963) we obtain $q_{max} \sim 10^{3}$ ions cm^{-3} sec^{-1}. Using typical coefficients for recombination of O^{+}, the dominant ion in this region, with molecular oxygen and nitrogen given by Donahue (1966), that is, $k_{O_2} = 3.4 \cdot 10^{-11}$ cm^3 sec^{-1} and $k_{N_2} = 1.8 \cdot 10^{-12}$ cm^3 sec^{-1}, the maximum electron density is approximated by $n_e(F2_{max}) \sim 7 \cdot 10^{5}$ ions cm^{-3}. Nominal values of $3 \cdot 10^{7}$ cm^{-3} and $3 \cdot 10^{8}$ cm^{-3} were taken for $n(O_2)$ and $n(N_2)$ respectively (Johnson, 1965). There are, of course, many uncertainties connected with this estimate. However, it is of the right order of magnitude to explain the "auroral cliff" in n_eF2 observed by Thomas et al. (1966). Also, the average position of the low-latitude boundary of the soft electron region agrees very well (for $K_p \sim 0$) with the average position of this auroral ionization cliff. Moreover, the average latitude of the ionization cliff was shown by Thomas et al. to decrease with increasing K_p, although this decrease was somewhat steeper than that observed for the soft electron boundary.

At present the question of particle-excited F-region 6300 Å emission is very much an open one (e.g., Cole, 1965; Dalgarno, 1964; Dalgarno and Degges, 1968). However, some of the observed features of the high-latitude oxygen emission near sunspot minimum indicate a possible relationship with the zones of low-energy electron fluxes discussed herein. As mentioned above, it has already been noted (Maehlum, 1968) that the lack of 6300 Å emission close to the L-pole during low sunspot activity and the tendency for this emission to peak at higher latitudes than the 4278 Å (Eather, 1969) and 3914 Å (Sandford, 1968) N_2^+ emissions are consistent with the high-latitude zone of electrons less energetic than about 1 keV.

We can now make the following additional comparisons. For $K_p \leq 4$, the observations of Sandford (1968) indicate that during 1963 the average high-latitude boundary of detectable 6300 Å emission did not vary appreciably with changing K_p while the peak intensity tended to occur at lower latitudes for higher K_p values. Further the peak 6300 Å intensity increased only slightly with K_p, remaining constant within about a factor of two as K_p varied from 0 to 4. Over this same range of activity the 5577 Å and 3914 Å intensities varied by factors of about 20. These observations are in agreement with the observed K_p variation of the low-latitude boundary of the soft electron zone and the lack of a clear K_p dependence of the position of the high-latitude boundary and the total precipitated energy in the zone.

5. CONCLUSIONS

Some of the typical and average properties of high-latitude low energy electron precipitation observed by the Aurora 1 satellite have been presented. It was shown that the average energy influx is of the right order of magnitude to explain the anomalously high levels of F-region ionization observed near the dark pole. It was also shown that the dependence of the properties of this zone on K_p are similar to certain variations in high-latitude F-region ionization and 6300 Å emission. Based on such comparisons it is suggested that both of these phenomena may be explained by the observed electron fluxes.

Acknowledgements. I am grateful to Dr. B. J. O'Brien for his continuing guidance and advice and to Dr. B. Maehlum for many helpful discussions. The encouragement and support of Cdr. C. W. Causey of ONR is also greatly appreciated.

This research was supported in part by the Office of Naval Research under contract Nonr-4964 (1). Particular thanks are due to the Payload Manager, Mr. Ramon Trachta and to the personnel of the Navy Space Systems Activity, Space Systems Division, of the U.S. Naval Space Surveillance Systems, of Goddard Space Flight Center for STADAN support, the NASA Manned Spacecraft Center and members of the Space Science Facilities at Rice University for their help and support.

REFERENCES

Brice, N. M., Bulk motion of the magnetosphere, J. Geophys. Res., 72, 5193, 1967.

Burch, J. L., Low-energy electron fluxes at latitudes above the auroral zone, J. Geophys. Res., 73, 3585, 1968a.

Burch, J. L., High-latitude satellite observations of electrons and protons at 4000 km, PhD thesis, Rice University, Houston, Texas, May, 1968b.

Chase, L. M., Spectral measurements of auroral-zone particles, J. Geophys. Res., 73, 3469, 1968.

Cole, K. D., Stable auroral red arcs, sinks for energy of D_{st} main phase, J. Geophys. Res., 70, 1689, 1965.

Dalgarno, A., Corpuscular radiation in the upper atmosphere, Ann. Geophys., 20, 65, 1964.

Dalgarno, A. and T. C. Degges, Electron cooling in the upper atmosphere, Planetary Space Sci., 16, 125, 1968.

Donahue, T. M., Ionospheric reaction rates in the light of recent measurements in the ionosphere and the laboratory, Planetary Space Sci., 14, 33, 1966.

Eather, R. H., Latitudinal distribution of auroral and airglow emissions: the "soft" auroral zone, J. Geophys. Res., 74, 153, 1969.

Evans, D. S., Rocket observations of low energy auroral electrons, in Aurora and Airglow, edited by B. M. McCormac, Reinhold Publ. Corp., New York, p. 191, 1967.

Evans, J. E., R. G. Johnson, R. D. Sharp, and J. B. Reagan, Recent results from satellite measurements of low-energy particles precipitated at high latitudes, Space Sci. Rev., 7, 263, 1967.

Frank, L. A., Low-energy proton and electron experiment for the orbiting geophysical observatories B and E, University of Iowa Rep., 65-22, 1965.

Hoffman, R. A. and D. S. Evans, OGO-4 auroral particles experiment and calibration, NASA-GSFC preprint X-611-67-632, 1967.

Johnson, F. S., Structure of the upper atmosphere, in Satellite Environment Handbook, 2nd ed., edited by F. S. Johnson, Stanford Univ. Press, p. 3, 1965.

Lin, W. C., I. B. McDiarmid, and J. R. Burrows, Electron fluxes at 1000-km associated with auroral substorms, Can. J. Phys., 46, 80, 1968.

Maehlum, B., Universal-time control of the low energy electron fluxes in the polar regions, J. Geophys. Res., 73, 3459, 1968.

Maehlum, B., On the high latitude, Universal Time controlled F-layer, J. Atmospheric Terrest. Phys., 31, 531, 1969.

McDiarmid, I. B. and J. R. Burrows, Electron fluxes at 1000 km associated with the tail of the magnetosphere, J. Geophys. Res., 70, 3031, 1965.

O'Brien, B. J., F. Abney, J. Burch, R. Harrison, R. LaQuey, and T. Winiecki, SPECS, a versatile space-qualified detector of charged particles, Rev. Sci. Instr., 38, 1058, 1967.

Rees, M. H., Auroral ionization and excitation by incident energetic electrons, Planetary Space Sci., 4, 1209, 1963.

Rees, M. H., Note on the penetration of energetic electrons into the earth's atmosphere, Planetary Space Sci., 12, 722, 1964.

Sandford, B. P., Variations of auroral emissions with time, magnetic activity, and the solar cycle, J. Atmospheric Terrest. Phys., 30, 1921, 1968.

Sharp, R. D. and R. G. Johnson, Satellite measurements of auroral particle precipitation, in Earth's Particles and Fields, edited by B. M. McCormac, Reinhold Book Corp, Chicago, p. 113, 1968.

Thomas, J. O., M. J. Rycroft, L. Colin, and K. L. Chan, The topside ionosphere, II., Experimental results from the Alouette 1 satellite, in Electron Density Profiles in Ionosphere and Exosphere, edited by J. Frihagen, North-Holland Publishing Co., Amsterdam, p. 322, 1966.

Westerlund, L. H., The auroral electron energy spectrum extended to 45 eV, J. Geophys. Res., 74, 351, 1969.

OGO 4 SATELLITE MEASUREMENTS
OF LOW ENERGY - HIGH LATITUDE ELECTRON PRECIPITATION

R. A. Hoffman

NASA-Goddard Space Flight Center
Greenbelt, Maryland 20771, USA

Abstract. Data from the Auroral Particles Experiment aboard OGO-4 have indicated four regions of low-energy electron precipitation at high latitudes. These have been designated by the terms band region, burst region, polar cavity, and the pre-midnight region. The band region is characterized by a relatively structureless, moderately hard precipitation pattern which generally follows the auroral oval, except extends to lower latitudes at noon. It is especially prevalent in the dawn hemisphere. The burst region begins within the band region and extends polewards, reaching up to 85° at noon, and is characterized by usually a very soft spectrum and large structure in the precipitation pattern. The pitch angle distributions of 2.3 keV electrons within the bursts tend towards anisotropic, with the maximum flux directed along the magnetic field into the atmosphere. The polar cavity is an area of no detectable fluxes, while the pre-midnight region displays structured electron fluxes with widely varying spectra. This paper presents a preliminary summary of all the phenomena observed in the high-latitude region and analyzed to date.

1. INTRODUCTION

Data from charged particle detectors capable of measuring fluxes of electrons whose energies are below a few keV indicate that there exists a region of precipitation of these low-energy electrons poleward of the auroral oval (Sharp and Johnson, 1967; Burch, 1968; Hoffman and Evans, 1968; Hoffman, 1969). The influx of these electrons into the polar ionosphere can be an important energy source for various phenomena in this region (Maehlum, 1969; Burch, 1969).

The Auroral Particles Experiment aboard the Orbiting Geophysical Observatory (OGO) - 4 satellite has made measurements of these low-energy electrons in the high-latitude region for a period of $1\frac{1}{2}$ years. This paper will present a preliminary summary of the phenomena observed in this region and analyzed to date.

2. EXPERIMENT

The satellite was launched on July 28, 1967, into a low-altitude polar orbit having an apogee of 908 km, a perigee of 412 km, and an inclination of 86°. Orbital injection was into a dawn-dusk meridian plane. Precession of the orbit due to this inclination together with the motion of the earth around the sun cause a change in local time of the orbital plane of about $1\frac{1}{2}^{\circ}$ per day.

The satellite was controlled in attitude such that one side of the spacecraft was always pointed towards the earth, while another axis was constrained to the earth, satellite, sun plane.

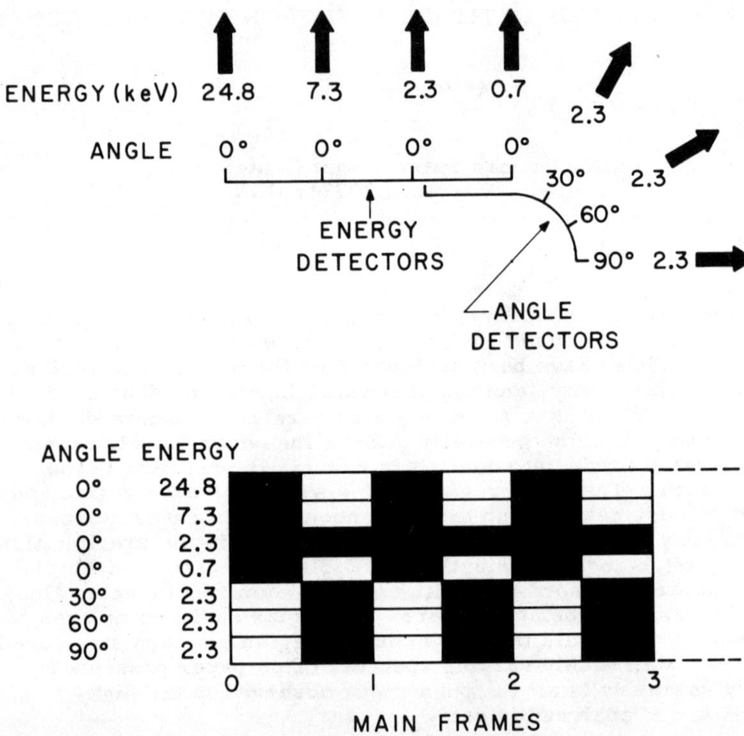

Figure 1. Schematic diagram of the experiment detector system indicating the look directions of the detectors with respect to the earth radius vector, the center energies measured by each detector, and the sampling scheme of the detector outputs, where the blackened area indicates the accumulation period.

The Auroral Particles Experiment contains an array of eight detectors, each comprised of an electrostatic analyzer for species and energy selection and a Bendix channel electron multiplier as the particle detector (Hoffman and Evans, 1967). Four of the detectors always point radially away from the earth (0°), while three others are positioned 30°, 60° and 90° to the earth-spacecraft radius vector (Figure 1). Since the inclination of the earth's magnetic field is nearly 90° in the high-latitude region the look directions sampled are very nearly the pitch angles of the particles detected. The eighth detector is a background detector.

The bandpass of each detector is +19%, -13% of the center energies indicated in Figure 1, while the geometric factor is about $6 \cdot 10^{-5}$ cm^2-sr at the peak of the energy bandpass.

The sampling scheme of the detectors is also illustrated in Figure 1.

The four energy detectors at 0° have their outputs accumulated simultaneously by four logarithmic digital accumulators for a time period of half the main telemetry frame before being read out, while during the second half of the main frame the four angle detectors, all measuring particles at 2.3 keV, have their pulse outputs accumulated simultaneously by the same accumulators before being read.

Center Energy of Detector	Conversion Factor*
(keV)	(to $\frac{\text{electrons}}{\text{cm}^2\text{-sec-sr-keV}}$)
0.7	$6.70 \cdot 10^5$
2.3	$2.03 \cdot 10^5$
7.3	$8.13 \cdot 10^4$
24.8	$3.60 \cdot 10^4$

*to convert from counts/readout for the 4 kilobit/sec tape storage mode of the spacecraft. For 16 kilobits/sec. telemetry rate these factors must be multiplied by 4.

Table 1

Since some of the data will be displayed in the figures as counting rates from the detectors, the conversion factors to fluxes are listed in Table 1. Additional details about the experiment are shown by Hoffman and Evans (1967).

3. DATA EXAMPLES

The data from the experiment at high latitudes indicate four distinct regions of low-energy electron precipitation, each with its own characteristics as observed by the satellite detectors: (i) band region; (ii) burst region; (iii) polar cavity (Maehlum, 1969); (iv) pre-midnight region. These terms are chosen as merely qualitative descriptions of the appearance of the data in the frame of reference of the moving satellite and are not meant to necessarily imply any true physical behavior on the part of the particles.

The first two regions of precipitation are shown in Figure 2, which contains data from a dusk to dawn pass which crossed magnetic local noon at a very high latitude. The relatively structureless, moderately hard precipitation at latitudes below 76° on the dawn side, especially noticeable in the 7.3 keV detector counting rate, is an example of what we term a band. Quite typically for the very low K_p period, of which this pass is an example ($K_p = 0+$), there is no detection of a dusk side band.

Well separated spatially on this pass and polewards of the band is the burst region, where usually the spectrum is very soft and the structure in the precipitation pattern in the frame of reference of the moving satellite is especially apparent at the lowest energy detected. This example of bursts is somewhat atypical in the sense that the 7.3 keV flux is larger than usual, but the entire pass is an excellent example of a complete polar crossing.

The precipitation in the pre-midnight region usually does not display this "two region" nature, as shown in Figure 3. Here, again

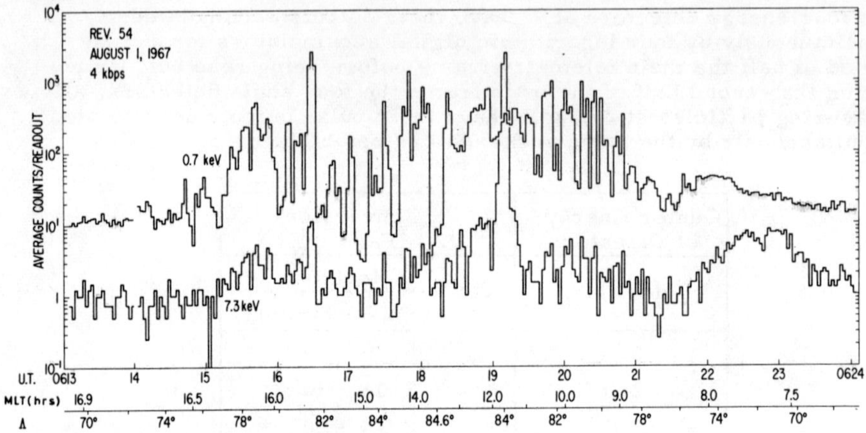

Figure 2. Responses of the 0.7 keV and 7.3 keV detectors to precipitating electrons in the northern hemisphere during a dusk to dawn polar crossing. (K_p = 0+).

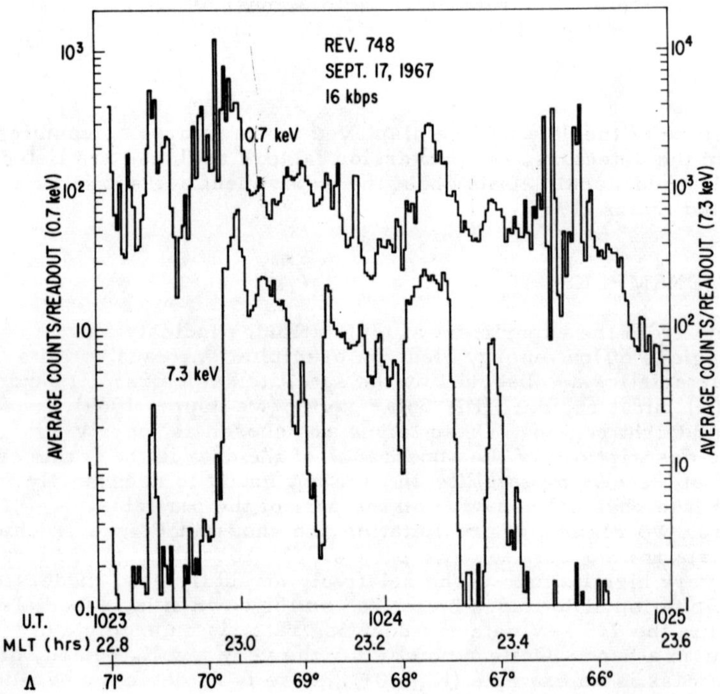

Figure 3. Responses of the 0.7 keV and 7.3 keV detectors to precipitating electrons in the northern hemisphere near midnight. (K_p = 1-).

for a low K_p pass (K_p = 1-), the precipitation is very intense and structured even at 7.3 keV, and the spectrum is widely varying. Early and late in this segment of data the 0.7 keV flux is three orders of magnitude larger than the 7.3 keV flux, while in some of the bursts this ratio is only a factor of 10.

The polar cavity appears as a region of no measureable precipitation by the experiment's detectors. For the first three weeks after launch the detector outputs were almost entirely free from noise, so an upper limit of a few counts/readout could be established over this region. During this period the upper limits on the flux in the energy range of the experiment are given as a few times the conversion factors listed in Table 1. (After three weeks in orbit a noise problem suddenly developed in the outputs of several of the detectors, so that very small fluxes could not be detected. This fairly steady noise output has since been diagnosed as a breakdown in the high voltage coupling capacitor. Unless otherwise indicated the noise has been subtracted from the counting rates plotted in the figures, and in no case was this noise level of any significance in the analyses discussed.)

4. BURST REGION

Because of the recent special interest in the trans-auroral zone region, the characteristics of the precipitating electrons in the burst region will be primarily discussed.

4.1 Structure

Figure 4 shows in great detail the variations in the counting rates of two of the detectors while passing through the burst region. Since the abscissa marks occur every 10 seconds, the structure here appears to have time widths varying from less than one to a few seconds in the satellite frame of reference. The satellite is moving at about 8 km/sec. Of course, there is no way from the particle data alone to distinguish whether this precipitation is spatial or temporal in nature.

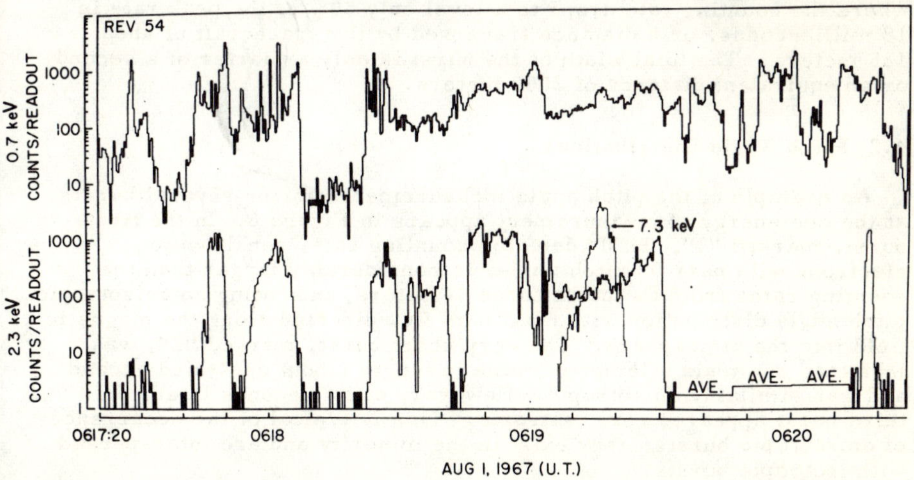

Figure 4. An example of burst activity at invariant latitudes above 80° near local noon in magnetic local time. The telemetry rate was 4 kbps.

A very unusual seemingly monoenergetic burst is apparent in the 7.3 keV detector at 0619:20 U.T. Such higher energy bursts, though rare, always seem to occur while the lower energy detectors are measuring fairly high levels of radiation for some time or spatial interval around the higher energy burst. Assuming reasonable isotropy over the upper hemisphere, an assumption which will be shown later to be not necessarily tenable, this burst contained some 30 ergs/cm^2-sec of energy influx in the bandwidth of the 7.3 keV detector alone.

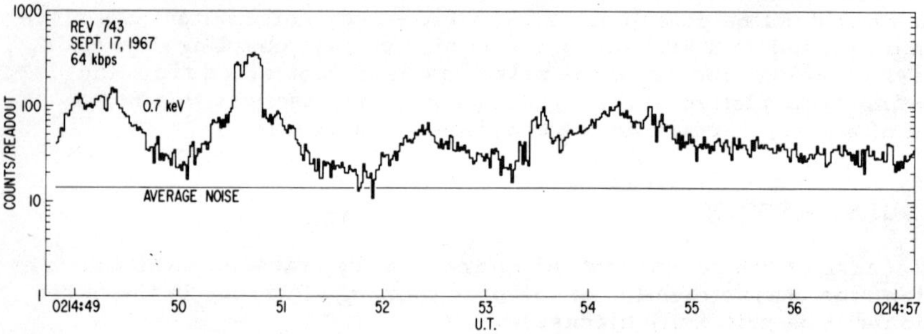

Figure 5. Highest resolution data available from the spacecraft. These bursts occurred at about 03 hours magnetic local time at a latitude of 69°.

An even more detailed example of the distinctness which the bursts can attain is shown in Figure 5. This is the highest resolution data available from the spacecraft, and provides 55 sets of energy measurements and pitch angle distributions each second. The eight seconds of data in the figure begins with several bursts and ends in the region of a quiet band. Note the sharpness of each edge of the burst occurring at 0214:50.7 U.T., especially the trailing edge, where the counting rate drops to a level only 30% of the peak rate in 18 milliseconds, or a distance traversed by the spacecraft of about 135 meters. The total width of the burst is only a quarter of a second, or an equivalent distance of 2000 meters.

4.2 Pitch Angle Distributions

An example of the pitch angle measurements during several bursts at the one energy of measurement appears in Figure 6. In the first burst, marked "1", the 0° detector counting rate, sensitive to electrons with near 0° pitch angle, is considerably larger than the counting rates from the other three detectors, indicating an anisotropic pitch angle distribution with maximum flux directed along the magnetic field into the atmosphere. The very short burst, marked "2", was less than one main telemetry frame in length (about one-third second), and was similarly anisotropic. However, a few seconds later, the third burst appeared very isotropic. This is typical of the occurrence of anisotropic bursts: they exist in the minority and are interspersed with isotropic bursts.

The actual pitch angle distributions for the bursts marked "1" and "2" are shown in Figure 7.

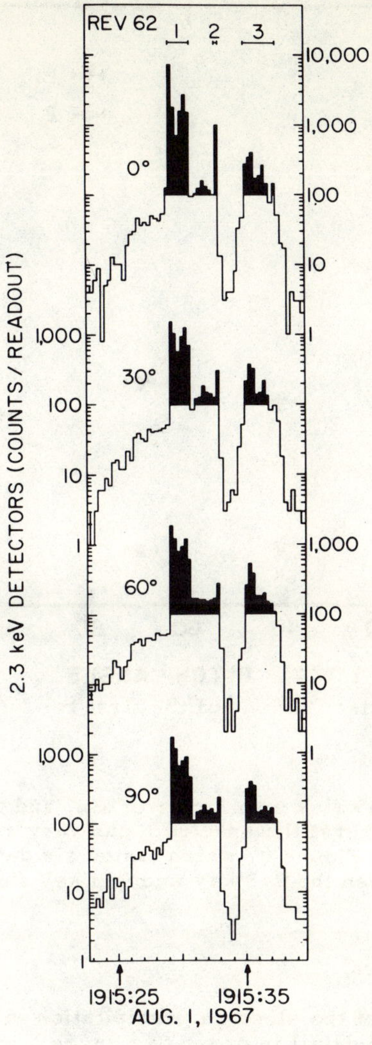

Figure 6. Counting rates of the four detectors with common center energy of 2.3 keV and angle with respect to the radius vector from the earth of 0°, 30°, 60° and 90°. These data were taken at about 21 hours magnetic local time at an invariant latitude of 70° at a telemetry rate of 4 kbps.

4.3 Spectral Characteristics

Energy spectra during bursts are impossible to classify. Usually the spectra are very soft so that the counting rate variations are most apparent at 0.7 keV, and the detector at that energy appears to be measuring the high-energy tail of the precipitating electrons. The steepness of the spectra is indicated by the ratio of fluxes between 0.7 keV and 7.3 keV, which can reach as high as 10^4. At times, however, the spectrum of a burst hardens, and may even appear to be monoenergetic in terms of the energy resolution of the detector array, as illustrated by the 7.3 keV burst in Figure 4. In this case the 7.3 keV

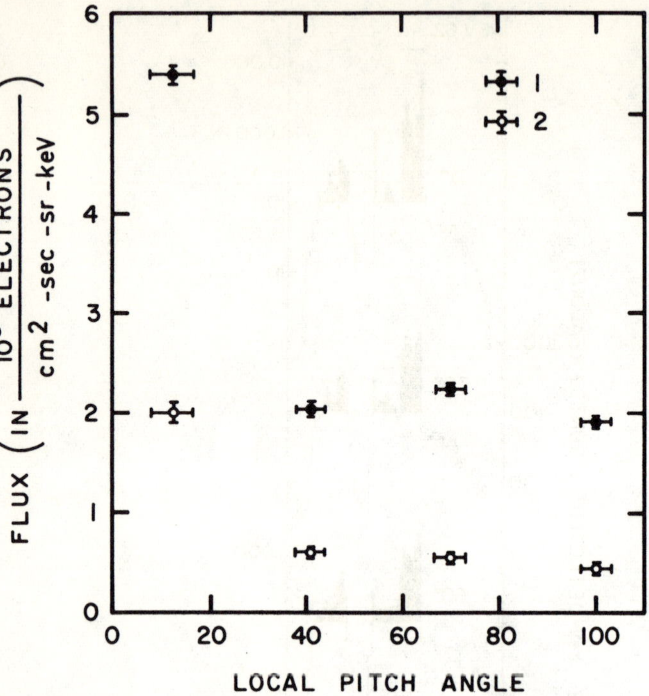

Figure 7. Pitch angle distributions of the first two bursts in Figure 6.

flux exceeded the 2.3 keV flux by a factor of six, and the 0.7 keV flux by 30%. Even within a burst the spectrum can vary widely. During the burst marked "1" in Figure 6, which lasted six data accumulation periods, the ratio between the 2.3 keV and 7.3 keV fluxes varied from 86:1 to 1:1.

4.4 Synoptic Parameters

Several parameters of the electron precipitation in the burst region are now studied in more detail in order to acquire a more comprehensive view of its properties.

Region and probability of occurrence. The first study is to define the region of occurrence of the electron bursts and determine the probability of occurrence within this region. For this purpose all data from July 30 through December 3, 1967, were analyzed. The data were divided into cells of invariant latitude and magnetic local time, $2\frac{1}{2}^o$ by one hour respectively, except for above 85°, where the cells were 5° by two hours. A cell was considered to be sampled if the path length of an orbit through the cell was at least one-half the dimension of the cell in the direction of travel. A cell was considered to have bursts if bursts existed over half of this path length. Thus there was no bias towards the existence of bursts. The criteria for bursts were rapid fluctuations in the counting rate from the 0.7 keV detector, with peak counting rates of at least 10^3 counts/sec (peak particle fluxes above about 10^8 electrons/cm^2-sec-sr-keV or peak energy fluxes above 10^{-1} ergs/cm^2-sec-sr-keV at 0.7 keV). The probability of occurrence of bursts in each cell was then calculated from the cumulative data. A cell was considered to be adequately sampled if

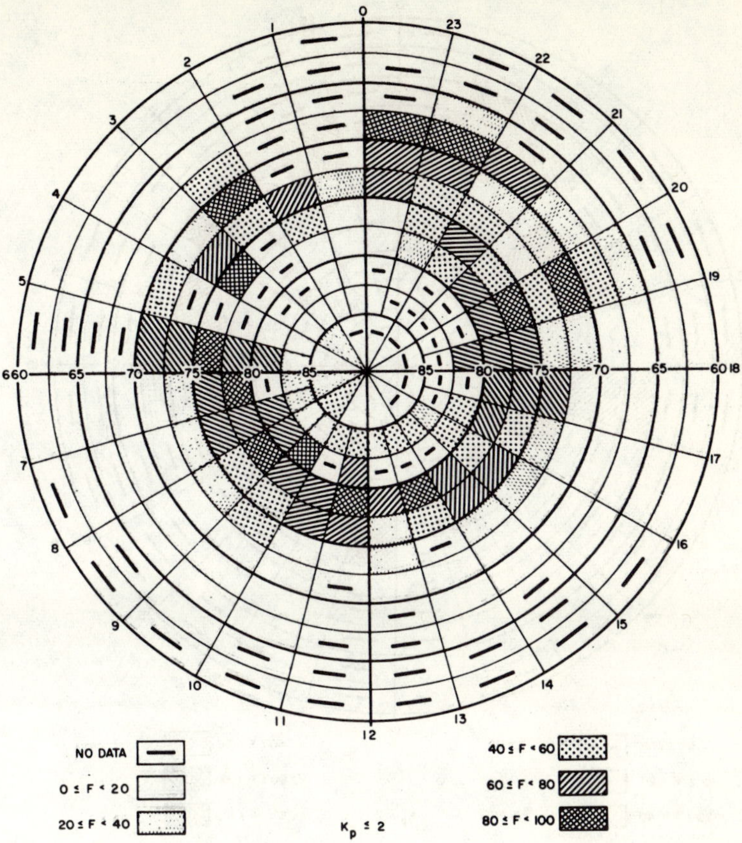

Figure 8. A map of the probability of occurrence of bursts (F) for magnetic activity having $K_p \leq 2$. (Cf. text for the criteria used in defining bursts.)

four or more passes were made through it. The data were also separated into two levels of magnetic activity, divided at the value of $K_p = 2$.

The map of the region and probability of occurrence for low magnetic activity is shown in Figure 8. Note the following features:

(i) The bursts occur at all local times with no indication that the probability of occurrence is maximum at any local time.

(ii) At noon the region extends from 75° to 85°, whereas the northern boundary of the auroral oval lies at about 77° for low magnetic activity (Felstein et al., 1967).

(iii) Before 23 hours the average northern boundary, as distinguished by the jump from less than 20% occurrence to greater than 20%, occurs at about 80°, and there is a sudden displacement in the boundary 5° to the south after 23 hours.

(iv) In general the bursts begin within the auroral oval and extend well to the north of it.

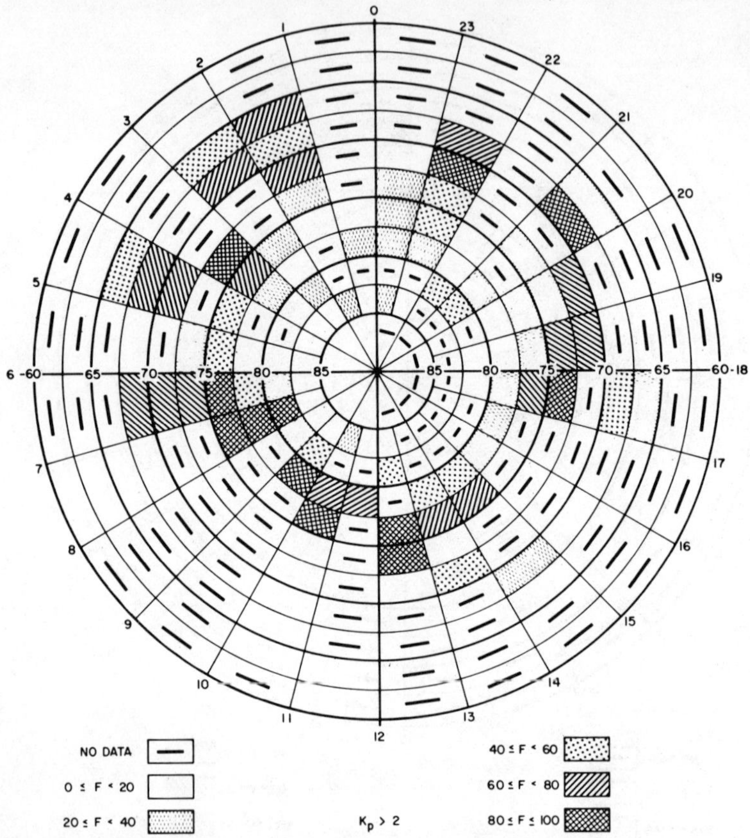

Figure 9. Same as Figure 8 for $K_p > 2$.

Data from more magnetically disturbed periods are shown in Figure 9. The probabilities of occurrence in the heart of the burst region do not change appreciably, but the region of bursts shifts.

(i) In the noon hemisphere the lower latitude boundary lies at least 5° to the equator side of the quiet time boundary.

(ii) The high-latitude boundary in the noon hemisphere moves similarly by about $2\frac{1}{2}^{\circ}$.

(iii) There are indications that in the midnight regions there is an expansion in both directions of the latitude extent of the burst region.

Pitch angle distributions. Examples of pitch angle distributions within individual bursts have shown the existence of both isotropic and field-aligned distributions at the single energy at which such measurements are made. In order to study the extent to which the bursts are observed to be field-aligned, a scatter plot was made (Figure 10) of the total counts within the burst as measured by the 0° detector versus the total counts measured by the 60° detector.

Since the 0° detector is constrained to always point radially away from the earth, while the magnetic field vector is not inclined quite

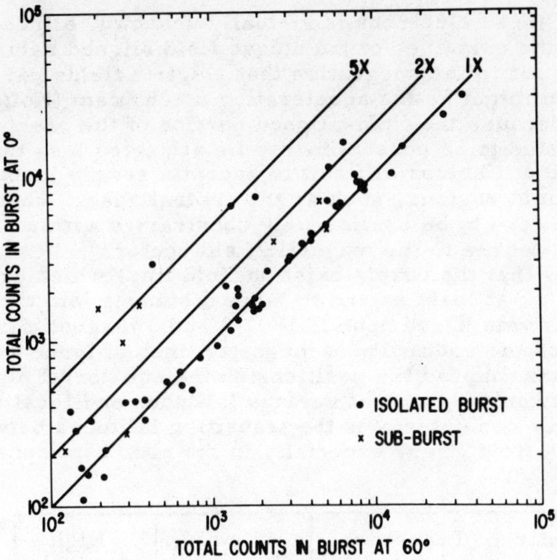

Figure 10. A scatter plot of the total counts within a burst as measured by the 0° detector versus the total counts measured by the 60° detector.

90°, this detector usually measures particles with central pitch angles in the range from a few degrees to 10°, as is apparent from Figure 7. The detector itself accepts particles with pitch angles about $\pm 4^{o}$ around the central look angle.

Isotropic bursts have their data points lying near the line at 45° angle to the axes labeled "1X" in Figure 10. Note that all bursts lie either along this line or above it. Those lying above it show the field-aligned character, with ratios between the fluxes at 0° and 60° of 2 and 5 indicated by the lines labeled "2X" and "5X" respectively. The maximum anisotropy shows a flux ratio of about eight.

The pitch angle distributions can be analyzed into two separate distributions (Figure 7), an isotropic portion which always exists in all four angle detectors, and an anisotropic distribution which, when it appears, is manifest only in the 0° detector.

5. DISCUSSION

Some of the properties of the high-latitude low-energy electron precipitation have been presented in a preliminary summary form. One can expect that these properties probably and in some cases certainly do depend upon various parameters of the magnetosphere, even without knowing the particle source mechanism. It has been shown that the high-latitude boundary of the burst region is universal time controlled (Maehlum, 1968) (i. e., dependent upon the angle of the magnetic axis with the ecliptic plane), and is, therefore, also dependent upon the season of the year. While sequences of passes through the same local times may show similar precipitation patterns, there are also wide variations in the patterns and intensities between passes which are at least partially due to differences in magnetic activity. Indeed, it should be stressed that none of the statistical patterns of electron precipitation are necessarily applicable during polar substorms.

The source of these electrons is virtually unknown, although a consideration of the existence of the unique field aligned nature of some of the bursts has led to the suggestion that electric fields parallel to the magnetic field might be the acceleration mechanism (Hoffman and Evans, 1968). Because the field-aligned portion of the precipitating fluxes always seems to be constrained to lie at angles less than about 30° to the field line it seems difficult to accept a source mechanism located in weak field regions, such as the neutral sheet, where this anisotropy would have to be continuously constrained within pitch angles less than about $\frac{1}{2}$ degree to the magnetic field vector.

It also appears that the bursts exist on field lines which are both closed and open, or at least extend to large distances into the tail region. In an analysis based upon IMP 1, 2 and 3 magnetic field measurements and the continuity of magnetic lines of force ($\nabla \cdot \underline{B} = 0$), Fairfield (1968) has mapped the positions on the equatorial plane of field lines emanating from the earth at various latitudes and local times. From this map one can determine the transition latitudes between closed and "open" field lines, especially in the noon hemisphere.

Magnetic Local Time	Transition Latitude	Low Latitude Boundary	High Latitude Boundary
12	77	75	85
08	77	$72\frac{1}{2}$	≥ 85
16	77	$72\frac{1}{2}$	$\geq 82\frac{1}{2}$
04	73	70	$\geq 77\frac{1}{2}$
20	73	$67\frac{1}{2}$	≥ 80
24	69	$\leq 67\frac{1}{2}$	75

Table 2

In Table 2 we show the latitude of this transition for various local times and the high and low latitude boundaries of the burst region, obtained from Figure 8. In all cases the bursts exist well to each side of the transition latitudes.

As previously mentioned it is not possible to determine from the particle data alone whether the bursts are spatial or temporal in nature. Observations from some other experiment are required to distinguish between the two possibilities, but to date only the photometric data of short lived optical emissions observed from a jet aircraft by Eather (1969) may be relevant. The duration of these "spike precipitation events" was typically 75 to 100 seconds, which if they are interpreted to be spatially stable with the aircraft flying under the forms would correspond to widths of 20 to 25 km, or 3 to 4 seconds in the frame of reference of the satellite. Thus if there is a correspondence between these optical events and our burst type precipitation, this observation of Eather suggests a spacial rather than temporal nature of the bursts.

6. CONCLUDING REMARKS

This paper has presented a preliminary summary of observations of low-energy electrons in the high-latitude region by the Auroral Particles Experiment aboard the OGO 4 satellite. It was based on the first four months of data acquired from the experiment, which has collected data for $1\frac{1}{2}$ years. Each of the items discussed will be studied more extensively, especially in conjunction with other experiments aboard the Observatory, which provides the capability for the interdisciplinary study of geophysical phenomena. Ground level measurements made simultaneously with satellite passes could also be useful in the analysis, particularly to assist in the removal of the spatial/temporal ambiguity in the satellite data, but also to provide measurements of phenomena associated with particle influxes, and to expose the relationships between satellite and ground based synoptic studies.

Acknowledgements. I would like to thank Drs. D. S. Evans and D. J. Williams for beneficial discussions, and Mr. Fred Berko for very significant data analysis and programming assistance.

REFERENCES

Burch, J. L., Low-energy electron fluxes at latitudes above the auroral zone, J. Geophys. Res., 73, 3585, 1968.

Burch, J. L., Satellite measurements of low-energy electrons precipitated at high latitudes, in The Polar Ionosphere and Magnetospheric Processes, edited by G. Skovli, Gordon and Breach, New York, 1969.

Eather, R. H., Latitudinal distribution of auroral and airglow emissions: The "soft" auroral zone, J. Geophys. Res., 74, 153, 1969.

Fairfield, D. H., The average magnetic field configuration of the outer magnetosphere, J. Geophys. Res., 73, 7329, 1968.

Feldstein, Y. I., A. D. Schevnin, and G. V. Starkov, Auroral oval and magnetic field in the tail of the magnetosphere, in The Birkeland Symposium on Aurora and Magnetic Storms, edited by A. Egeland and J. Holtet, Centre National de la Recherche Scientifique, 1967.

Hoffman, R. A., Low-energy electron precipitation at high latitudes, J. Geophys. Res., 74, 2425, 1969.

Hoffman, R. A., and D. S. Evans, OGO 4 auroral particles experiment and calibrations, NASA-GSFC Preprint, X-611-67-632, 1967.

Hoffman, R. A., and D. S. Evans, Field aligned electron bursts at high latitudes observed by OGO 4, J. Geophys. Res., 73, 6201, 1968.

Maehlum, B., Universal-time control of the low energy electron fluxes in the polar regions, J. Geophys. Res., 73, 3459, 1968.

Maehlum, B., On the high latitude, universal-time controlled F-layer, J. Atmospheric Terrest. Phys., 31, 531, 1969.

Sharp, R. D., and R. G. Johnson, Satellite measurements of auroral particle precipitation, in Earth's Particles and Fields, edited by B. M. McCormac, 113, Reinhold Book Corp., Chicago, 1963.

TRAPPED PARTICLE MOTION IN A DISTORTED GEOMAGNETIC FIELD

Juan G. Roederer

Department of Physics
University of Denver, Colorado 80210, USA

Foreword. External geophysical phenomena may be divided into four categories: (i) controlled by the atmosphere; (ii) controlled by the atmosphere, but influenced by the geomagnetic field; (iii) controlled by the geomagnetic field, influenced by the atmosphere, and (iv) controlled by the geomagnetic field. The polar ionosphere belongs right into the middle, between (ii) and (iii).

It is, at present, not yet clear in what proportion the polar ionosphere is produced and maintained by particles from the magnetosphere. And as for those particles which *do* contribute, it is not clear in what proportions they are precipitated out of a trapped population, or come in freshly ejected from the plasma sheet. One thing, however, *is* clear: "polar ionosphericists" and "outer magnetosphericists" must start interacting as strongly as their own subjects of study do. The following sections are written having such an interaction in mind. These sections cover the *adiabatic* behaviour of *trapped* particles only, and do not include such key topics as particle diffusion and precipitation, magnetospheric plasma and instabilities, etc. Their scope, therefore, is extremely restricted. Everybody should be warned that all that follows is nothing but the opening game.

1. INTRODUCTION

The Adiabatic Theory has been developed to simplify the study of charged particles in magnetic fields under certain circumstances. As the name indicates, the Adiabatic Theory is not an exact theory - its results are only approximations which, however, become exactly true when certain "smallness parameters" tend towards zero. Fortunately, this is the situation in many problems dealing with energetic particles trapped in the geomagnetic field.

Numerical use of Adiabatic Theory is particularly important to obtain a physical understanding and/or a quantitative description of the following:

(i) The "ground state" of the magnetosphere, in particular regarding the configuration of trapped particle fluxes during quiet times.

(ii) Corrections that have to be applied to observations during perturbed conditions, in order to eliminate all "trivial" effects of the geomagnetic field caused by its time variations and spatial asymmetries.

(iii) The proper frame in terms of geometry, characteristic times, coordinates and parameters, to be used for the study of trapped particle diffusion, acceleration and precipitation.

The most general equation that describes the motion of a particle of charge q and relativistic mass m in a given magnetic field $\underline{B}$, under the action of an electric field $\underline{E}$ (external and induced) and an external non-electromagnetic force $\underline{F}$, is given by

$$\frac{d}{dt}\left(m\frac{d\underline{r}_p}{dt}\right) = q\left(\frac{d\underline{r}_p}{dt} \times \underline{B} + \underline{E}\right) + \underline{F} \quad (1)$$

The solution $\underline{r}_p = \underline{r}_p(t, \underline{r}_o, \underline{v}_o)$ represents the position of the particle as a function of time and initial conditions $\underline{r}_o, \underline{v}_o$. (1) is described from a given, inertial frame of reference.

The solution of (1) may be arbitrarily complex. For instance, energetic cosmic ray particle trajectories in the geomagnetic field usually are of such nature. But under certain conditions of field geometry, external force intensity and particle energy, periodicities appear in the solution of (1) which then allow a simplified description - especially if one is not interested in the instantaneous, particular value of the phase of the periodic motion in question. There can be as many as three distinct types of periodicities: (i) the cyclotron motion, a periodicity in the particle's motion component perpendicular to the magnetic field; (ii) the bounce motion, a periodic motion up and down the magnetic field line; and (iii) the drift motion, a periodic motion around a closed surface made up of field lines. Periodicity (i) is always the first to appear, and has the highest frequency. It may exist even in absence of (ii) and (iii). For instance, when a not-too-high energy cosmic ray particle approaches close to the earth, it may have a clear-cut cyclotron motion, but no other periodicity. Bounce frequencies are usually several orders of magnitude lower than cyclotron frequencies; drift frequencies are orders of magnitude lower than bounce frequencies. Any magnetic field in which particles have the capability of bounce motion is called a trapping field. If (iii) can occur, we say that it has a configuration of stable trapping.

In the next three sections we shall discuss each one of the three possible periodicities of equation (1), and their application to the study of geomagnetically trapped particles. This is only a review, and only some of the key physical ideas will be touched upon. The reader is referred to publications elsewhere for details (e.g., "The Adiabatic Motion of Charged Particles", by Theodore G. Northrop, Interscience Publishers, 1963; "The Radiation Belt and Magnetosphere", by Wilmot N. Hess, Blaisdell Publishing Co., 1968).

2. CYCLOTRON MOTION AND THE FIRST ADIABATIC INVARIANT

2.1 The Guiding Center Approximation

Cyclotron motion qualitatively represents a helical motion of the particle around a field line. More specifically, cyclotron motion exists if at any instant of time we can find a (moving) frame of reference in which an observer sees the particle in almost circular orbit, at least during a couple of turns. For instance, if the magnetic

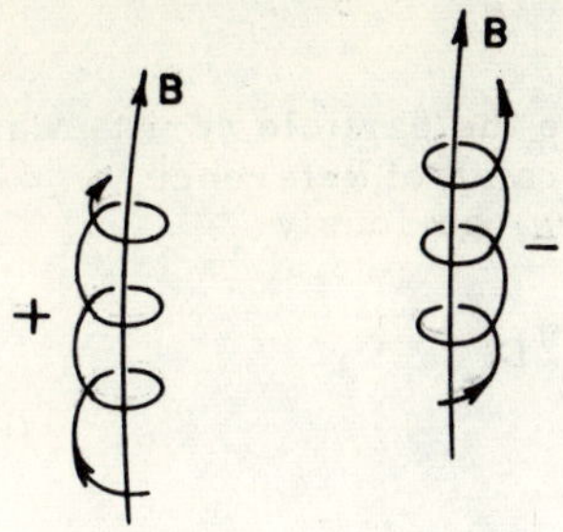

field is homogeneous, and no external forces act on the particle, a frame of reference which moves with the (constant) velocity of the particle parallel to the field $\underline{v}_{\parallel}$, will see the particle in an exact circular motion (opposite charges turning in mutually opposite senses).

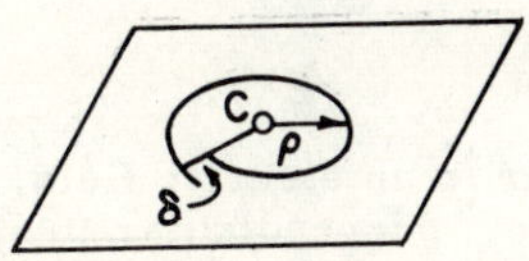

In a general case, however, the orbit will not be exactly circular, but only approximately. We do require, however, that the "non-circularity" δ be extremely small as compared to the (average) radius ρ. If this condition is met, we say that the "guiding center approximation" holds, and we call that particular frame of reference the "guiding center system" (GCS). Point C is the guiding center and the radius ρ_c, called Larmor radius or cyclotron radius, is given by

$$\rho_c = \frac{m^* v_\perp^*}{q B^*} = \frac{p_\perp^*}{q B^*} \tag{2}$$

The associated cyclotron period is

$$\tau_c = \frac{q B^*}{m^*} \tag{3}$$

Starred quantities are values in the GCS, and not those of the original frame, in which (1) is defined. The condition for the guiding center approximation is

$$\delta \ll \rho_c \tag{4}$$

If no system can be found in which (4) holds, there is no cyclotron motion. This happens with energetic cosmic rays in the earth's field.

The instantaneous velocity of the GCS can be divided into two components with respect to the magnetic field

$$\underline{V} = \underline{V}_{\parallel} + \underline{V}_{\perp}$$

The parallel velocity is always equal to the parallel velocity of the particle

$$\underline{V}_{\parallel} = \underline{v}_{\parallel} \tag{5}$$

The perpendicular velocity is called the particle's drift velocity $\underline{V}_D$

$$\underline{V}_\perp = \underline{V}_D$$

because it is the velocity with which we see the particle "drift away" from the initial field line, in the original frame of reference.

The starred quantities in (2) and (3), are, obviously

$$v_\perp^* = |\underline{v}_\perp - \underline{V}_D|$$
$$B^* = B \tag{6}$$
$$m^* = m$$

if $V_D \ll c$, as usually happens. In addition, in most radiation belt situations one has $V_D \ll v_\perp$, and $v_\perp^*$ becomes identical to $v_\perp$.

There are different types of drifts:

2.2 Zeroth Order Drifts

When the particle is subject to a force, or to an electric field, perpendicular to $\underline{B}$, we have the well known force drift or electric field drift, respectively

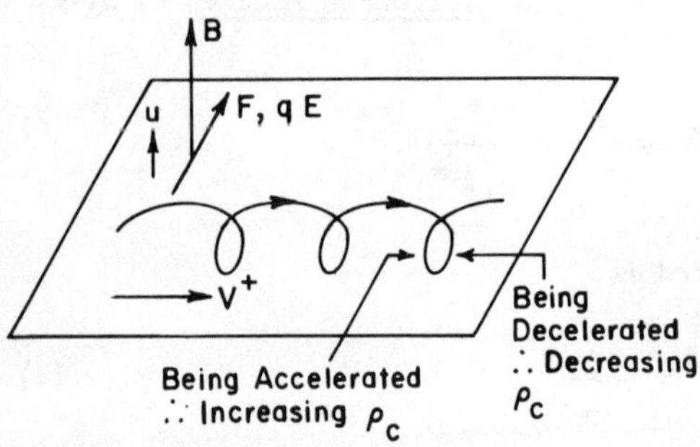

$$\underline{V}_F = \frac{\underline{F}_\perp \times \underline{u}}{qB} \tag{7a}$$

$$\underline{V}_E = \frac{\underline{E}_\perp \times \underline{u}}{B} \tag{7b}$$

$\underline{u}$ is a unit vector parallel to $\underline{B}$.

The condition for (7) is that

$$\frac{F_\perp}{qB}, \frac{E_\perp}{B} < c \tag{7c}$$

This is a very "liberal" condition. Notice in addition that (7a, b) do not depend on the particle's energy. This is why they are called "zeroth" order drifts. Both conserve the particle's energy on the average (the energy does change during a cyclotron turn). In the GCS, the orbit is exactly circular. In that system, an induced electric field $E^* = \underline{V}_{F,E} \times \underline{B}$ appears, which exactly cancels the action of the external force, leaving the particle in pure circular motion.

2.3 First Order Drifts

These are due to an inhomogeneous magnetic field. If there is a transverse gradient in the intensity of the magnetic field ($\underline{\nabla}_\perp B$), and a curvature of the field line (both always come together in curl-free regions), we have gradient drift $$\underline{V}_G = -\frac{m v_\perp^2}{2 q B^2} \underline{\nabla}_\perp B \times \underline{u} \tag{8a}$$

curvature drift $$\underline{V}_C = \frac{m v_\parallel^2}{q R_c B} \underline{n} \times \underline{u} \tag{8b}$$

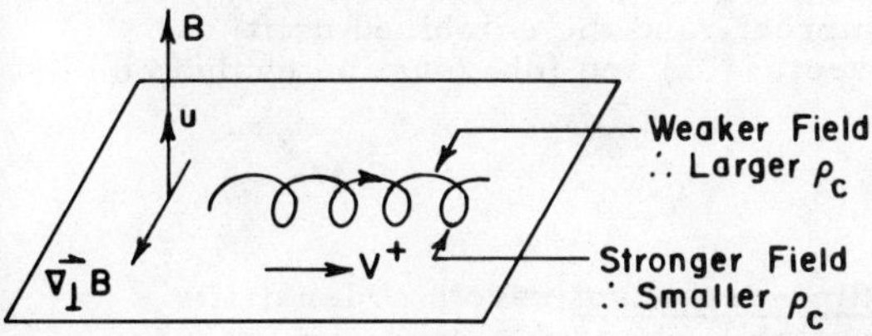

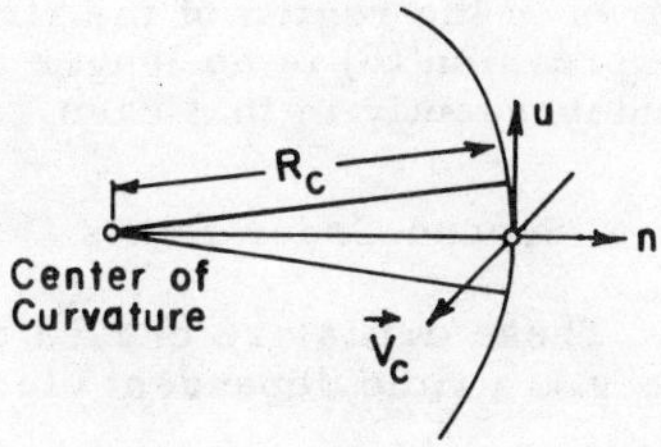

In a current-free region one has

$$\underline{\nabla}_\perp B = -\frac{B}{R_c} \underline{n} \tag{9}$$

and (8a) and (8b) combine to give the total first order drift

$$\underline{V}_T = \frac{m v^2}{2 q B^2} (1 + \cos^2 \alpha) \underline{\nabla}_\perp B \times \underline{u} \tag{8}$$

We have introduced the local <u>pitch angle</u> α, given by

$$\sin \alpha = \frac{v_\perp}{v} \qquad \cos \alpha = \frac{v_\parallel}{v} \tag{9}$$

The condition for (8) is

$$\frac{B}{|\nabla_\perp B|} \gg \rho_c \tag{10}$$

This condition is much more restrictive than (7c). Furthermore, the drift (8) is energy dependent. All this justifies the name "first order drift". If (8) acts together with (7), energy is <u>not</u> conserved [(8) may drive the particle <u>along</u> the external force]. In the case of a pure gradient drift ($\alpha = 90^\circ$), the orbit is again circular in the GCS, which follows the particle in its drift motion with a velocity (8a). Condition (10) leads to condition (4). This time, the induced electric field E^* slightly accelerates and decelerates the particle just the right amount to maintain a circular orbit in spite of the inhomogeneous field.

The curvature drift (8b) is physically caused by an inertial force (the centrifugal force) that is seen by an observer in a system that follows the particle with its direction-changing parallel velocity along the curved field line. This is why this system has to move additionally "sidewise" with velocity (8b), so that an additional electric field can be induced, to cancel the action of the centrifugal force, and again put the orbit in nicely circular shape in the GCS.

Condition (10) may be satisfied at one portion of a field line, but may break down at another one, for the same particle. For instance, if field lines curve through the neutral sheet of the geomagnetic tail very abruptly, a particle spiralling along that field line towards the neutral sheet may suddenly be "untrapped" or jump to another field line, when reaching the sharply curved portion.

Notice that during times of a strong ring current, relation (9) breaks down in the region of the ring current, and the combined drift expression (8) is no longer correct. (8a) and (8b) must be evaluated independently in that case.

2.4 Second Order Drifts

These drifts are caused by time-dependent zeroth order drifts (e.g., a time dependent electric field). In such a case, an observer who moves with the time-dependent zeroth order drift velocity, is really located in an accelerated frame of reference, and therefore experiences an inertial force. This inertial force causes an additional, second order drift. There are many types of these drifts. The most important one is the so-called polarization drift, caused by a time-dependent transverse electric field

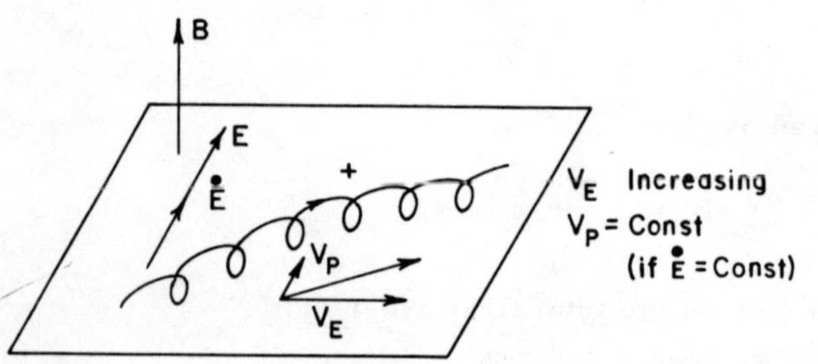

$$\underline{V}_p = \frac{m\dot{\underline{E}}_\perp}{q B^2} \tag{11}$$

This drift, which is charge dependent, is responsible for the dielectric properties of a plasma. The condition for the guiding center approximation to hold under the action of $\dot{E}_\perp$ is

$$\frac{E_\perp}{\dot{E}_\perp} >> \tau_c \tag{12}$$

This means that $V_p << V_E$, which explains the name "second order drift". The energy of a particle subject to a polarization drift changes, but only to the extent that it is needed to accelerate its zeroth order drift motion. The energy in the GCS remains constant (in absence of a gradient drift).

2.5 The First Adiabatic Invariant

If the guiding center approximation holds at all times for a given particle, drifts and time variations of B conserve the following quantity

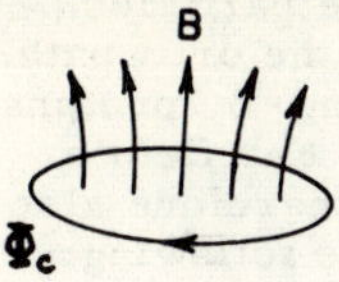

$$M = \frac{p_\perp^{*2}}{2 m_o B} = \frac{q^2}{2\pi m_o} \Phi_c = \text{const} \qquad (13)$$

This quantity is defined in the GCS and is called the relativistic magnetic moment, or first adiabatic invariant. m_o is the rest mass, Φ_c is the magnetic flux through one cyclotron orbit turn. The "ordinary" magnetic moment is given by

$$\mu = \frac{M}{\gamma^*} \qquad \gamma^* = \frac{m^*}{m_o} \qquad (14)$$

In the non-relativistic case, $\mu = M$. μ is equal to the product i A, where i is the equivalent current of a gyrating particle, and A the area of the cyclotron orbit.

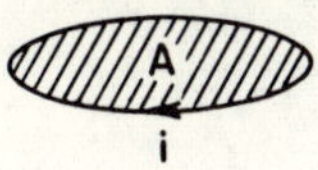

In the non-relativistic case, we have

$$M = \mu = \frac{T_\perp^*}{B} \qquad (15)$$

$T_\perp^*$ is the kinetic energy of the perpendicular motion. Since usually the drift velocity is much smaller than the particle velocity, we have $v_\perp^* = v_\perp$, and

$$\mu = \frac{T_\perp}{B} = \tfrac{1}{2} m v^2 \frac{\sin^2\alpha}{B} \qquad (16)$$

If there is no acceleration of the particle along the field line, which means that the field line is an equipotential in the external force field, v is constant along the field line and so is

$$\frac{\sin^2\alpha}{B} = \text{const} \qquad (17)$$

The condition for the validity of the conservation of M is given by

$$\frac{B}{\dot{B}}, \frac{F}{\dot{F}}, \frac{E}{\dot{E}} \gg \tau_c \qquad (18)$$

Time derivatives in (18) have to be interpreted as variations as seen by the particle. The first terms are thus characteristic time scales of change.

2.6 Application to Equatorial Particles in the Magnetosphere

We are now in position to discuss the motion of particles in the equatorial (minimum - B) surface of the geomagnetic field (i.e., particles with 90° equatorial pitch angles).

Let us first assume that no external forces act on the particles. The constancy of M then implies constancy of B along the drift path. This means that particles will follow contours of constant B, protons westwards, electrons eastwards. Figure 1 reproduces Fairfield's (1968) measured average constant - B contours, which therefore also represent drift paths of equatorial particles. Notice the following:

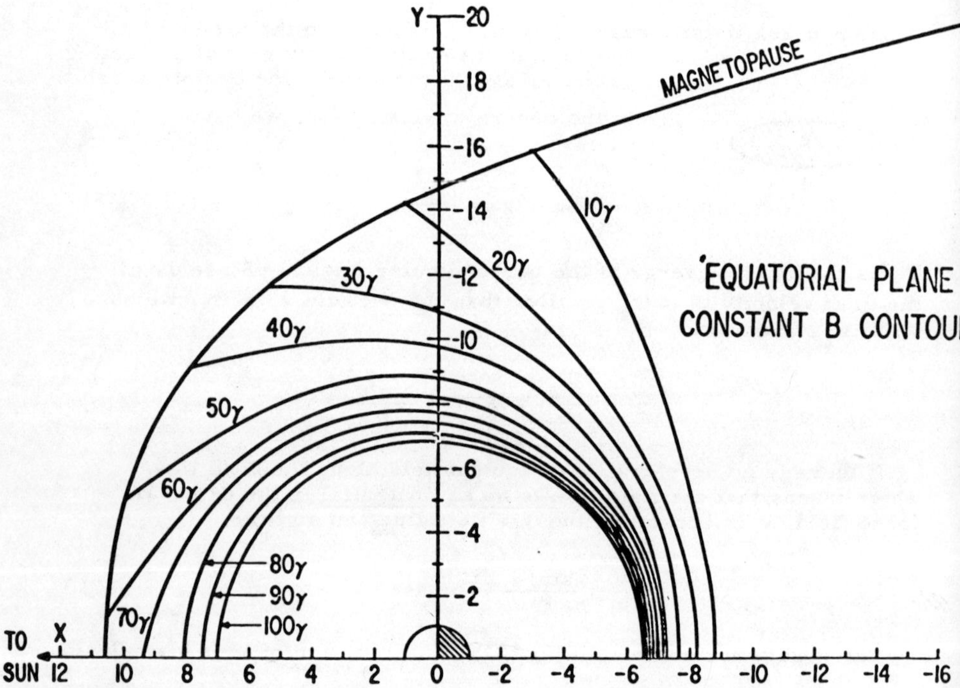

Figure 1. Contours of constant magnetic field magnitude in the equatorial surface (Fairfield, 1968).

(i) The contours' asymmetry increases as one moves away from the earth. Within about 4 R_E they should be circles, as prescribed by a dipole field.

(ii) The asymmetry is such that a given drift path has its closest approach at magnetic midnight. Magnetic field lines through a given drift path thus intersect the earth at a lower latitude at night than on the day side, this difference being more pronounced as the drift path is located further out. If these equatorial particles are suddenly scattered and dumped into the atmosphere, their precipitation pattern would thus have a shape similar to the auroral oval (if they come from constant B-rings near to the magnetopause on the day side).

(iii) Contours are closer to each other at midnight. This means that the transverse gradient of the field is larger there. Thus, according to (8), a particle's drift velocity will be larger at night than at noon; each particle spends more time on the dayside than on the night side during its drift. In other words, there are always more particles on the day side than on the night side of a given contour.

(iv) Constant - B contours going through the midnight meridian at distances greater than about 7 R_E are not closed around the earth. Particles drifting along them remain in the magnetosphere only a limited time, running into the boundary at the flanks of the magnetosphere. These are "pseudotrapped" particles. Particles on closed contours, instead, are stably trapped. The "last" closed contour, or "equatorial boundary of stable trapping" maps fairly well into the region of maximum precipitation of soft auroral particles (Hartz and Brice, 1967), or the equator-ward edge of the auroral oval.

(v) The orbit of a synchronous satellite (6.6 R_E) cuts through different drift contours as a function of local time. In particular, at midnight it samples the outermost B-ring, at noon the innermost one. Along each ring, the directional, equatorial particle flux j_o is constant (Liouville's theorem), therefore j_o can be only a function of B. If j_o decreases outwards (i.e., with decreasing B), a directional detector, looking into the equatorial plane, would reveal a diurnal variation of its counting rate with maxima occurring at local noon. This is indeed seen with a detector on the satellite ATS-1, Figure 2 (Brown, 1968). Higher energy particles have a steeper gradient, hence a greater diurnal amplitude of counting rate. Day to day variations seen in Figure 2 are probably due to both, diffusion and long term changes in the magnetospheric configuration.

We now briefly discuss the case of equatorial particles subject to an external electric field. We shall assume this field to be everywhere perpendicular to the field lines and of "Brice type", with two major constituents: a corotational field ($\underline{E}_{cor} = -\underline{\nabla}(-\frac{K}{r})$; K = 102 kV, r in earth radii), and a dawn-dusk "cross-tail" field ($\underline{E}_{cross} = -\underline{\nabla}(Cr \sin\phi)$; ϕ : longitude east of midnight, C $\simeq$ 2 kV during substorm times, and maybe 2 times smaller during quiet times).

In presence of an electric field, an equatorial particle is simultaneously exposed to both drifts (7b) and (8a). The first one neither depends on energy nor on the electric charge, the second one does depend on both. If the particle is of very low energy (say, a few keV), the electric drift will dominate, and both, electrons and protons will drift nearly along equipotentials. Their instantaneous energy

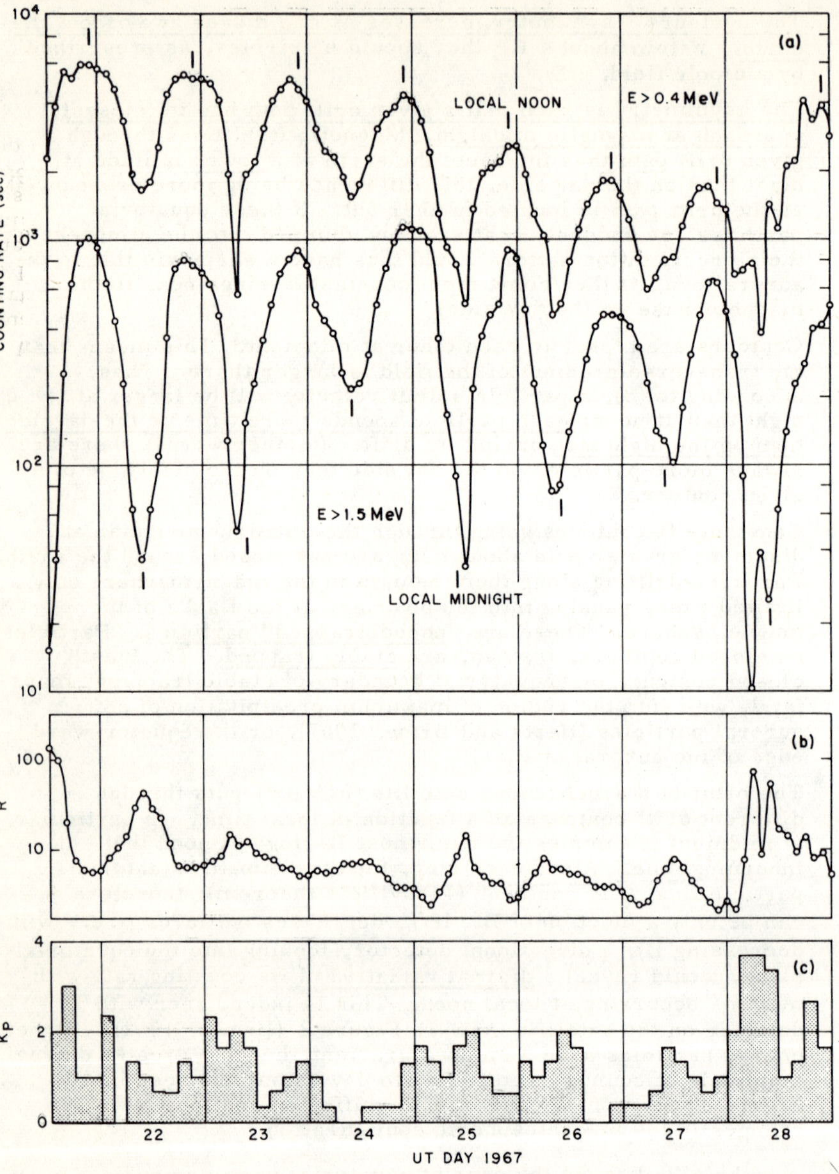

Figure 2. (a) Diurnal variations of electron counting rates at 6.6 R_E caused by the asymmetry of the magnetospheric field (Brown, 1968). Tick marks above the upper curve indicate local noon; similar marks below the lower curve indicate local midnight. (b) The ratio $R = CR_{E > 0.4} / CR_{E > 1.5}$, an index of the electron energy spectrum. Notice the effect of a steeper radial gradient for the higher energy particles. (c) 3-hourly average Kp.

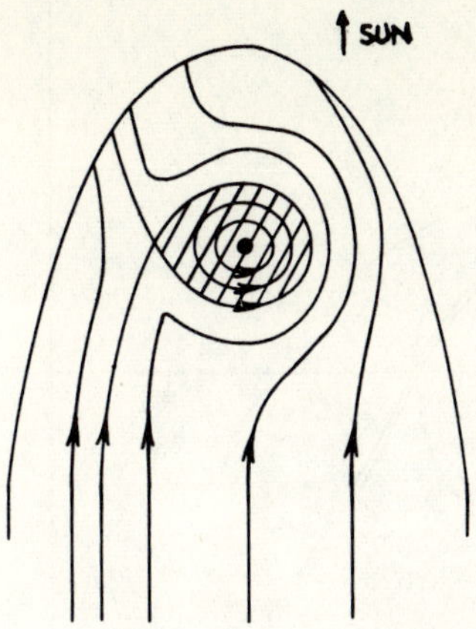

$(T = T_\perp)$ will be given by the conservation of μ (15 or 16), i.e., it will be proportional to the local B value along the given equipotential. Notice in the figure the well defined corotational region. If the particle's energy is high ($\gtrsim$ 100 keV), the electric drift can be neglected, and electrons and protons drift in opposite senses along constant B lines, as discussed above.

In the intermediate range of energies, the drift motion is more complicated. Electrons in general behave rather "normally" because the corotational electric drift and the gradient drift both point in roughly the same direction. For protons, however, both drifts point in roughly opposite directions, and depending on which is stronger, the proton will drift eastwards or westwards. Figures 3 and 4 show computed drift paths of 1 keV electrons and protons, respectively, injected at four positions of the synchronous orbit (using a magnetospheric model, Section 3.4). Notice that closed proton paths are seen only in the evening sector. Notice furthermore that in this case closed proton paths do not encircle the earth.

In general, the effect of a "Brice-type" electric field on stably trapped electrons is to displace the whole drift path in the dawn-dusk direction. Mapping these paths along field lines onto the earth yields contours with dawn-dusk asymmetries. This asymmetry should be greatly enhanced during the action of a substorm-type cross-tail electric field. It should be further noted that electrons on closed drift paths are always accelerated and decelerated in such a way as to have maximum energy near the dawn meridian and minimum energy in the dusk side.

The general "recipes" to compute equatorial particle drift paths are as follows:

Method I (constant electric field): (i) give initial point P_o and initial T_o, B_o, and electrostatic potential V_o. Compute the two constants: $\mu = \frac{T_o}{B_o}$, and total energy $W = T_o + q V_o$. (ii) go to any other longitude and instruct computer to find a point P' such that the quantity $\mu B' + q V'$ is equal to W ± error (use iterative method). Sometimes no such point can be found; this means that the particle has turned around before reaching this longitude. Sometimes two solutions are found (cf. protons on dusk side in Figure 4).

Method II (can be used with time dependent fields, and is faster): (i) give initial point, initial T_o, and compute μ and initial total drift velocity. (ii) integrate the drift velocity vector. For each step compute the particle's energy from $T' = \mu B'$. (By the way, it is found that substorm-type time dependent electric fields have a tremendous effect on redistribution of low energy particles in the equatorial plane.)

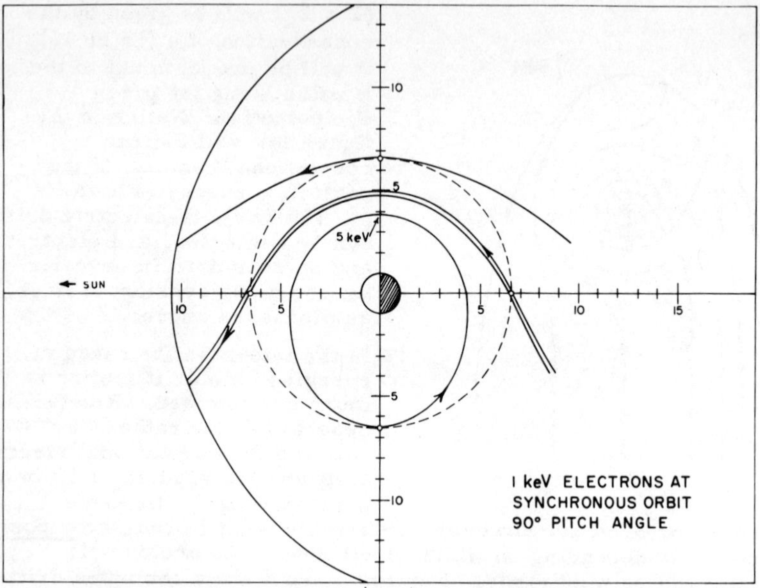

Figure 3. Electron drift paths in the equatorial surface for Mead's model of the magnetosphere, and for a Brice-type electric field. These electrons pass through the synchronous orbit (6.6 R_E) at noon, dusk, midnight and dawn, respectively, with 1 keV. Their energy varies along the drift path (Hones and Roederer, unpublished).

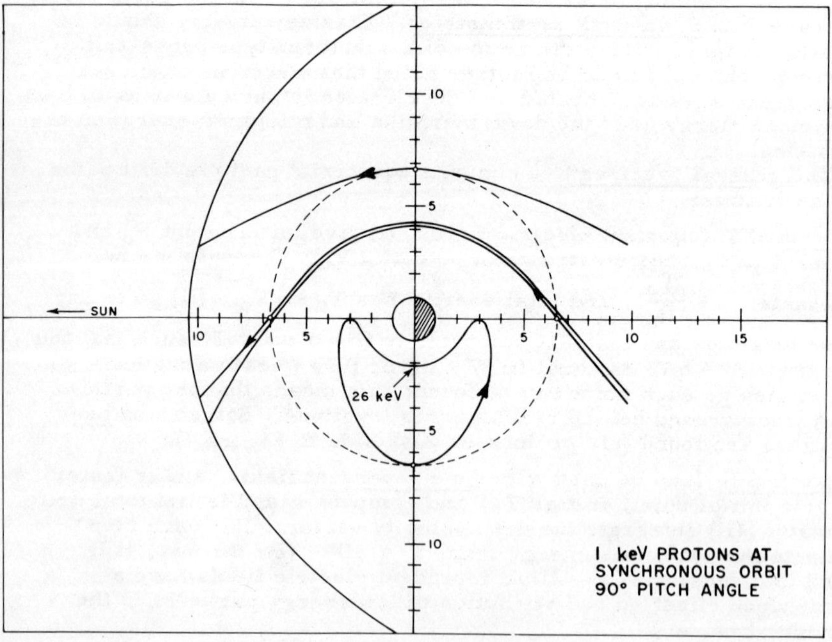

Figure 4. Same as in Figure 3, for protons. Notice the closed drift path on the dusk side.

3. BOUNCE MOTION AND THE SECOND ADIABATIC INVARIANT

3.1 The Second Periodicity

Equation (1) can be rewritten for components parallel to the field line, and converted into

$$\frac{d}{ds}(m v_{\parallel}) = -\frac{\partial}{\partial s}\left(\mu B(s) + U(s)\right) \tag{18}$$

s is the field line arc-length (usually measured from the equatorial minimum - B point), U(s) is the potential of all external forces (including the electrostatic field) along the field line. A particle thus behaves in its parallel motion, as if it were moving in a fictitious potential $\mu B + U$. For the parallel energy T we have

$$T_{\parallel} + \mu B + U = \text{const} = T_{\parallel}^{o} + \mu B^{o} + U^{o} \tag{19}$$

If the field line is equipotential, this converts to

$$T_{\parallel} + \mu B = \text{const}$$

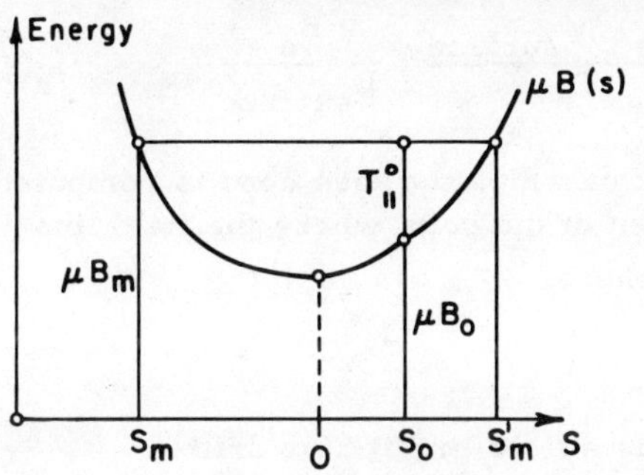

If the field geometry is such that B increases towards both ends of a field line, we shall have particle trapping. This occurs on closed field lines in the magnetosphere. The points where the particle reverses its parallel motion are the mirror points s_m, s_m'. The field intensity there is B_m. From (16), (14), and (13) we have, taking into account that in this case v is constant,

$$\frac{\sin^2\alpha(s)}{B(s)} = \frac{\sin^2\alpha_o}{B_o} = \frac{1}{B_m} \tag{20}$$

and

$$M = \frac{p^2}{2 m_o B_m} \tag{21}$$

The integral along the field line

$$\tau_b = \int_{s_m}^{s_m'} \frac{ds}{v_{\parallel}(s)} = \frac{1}{v}\int_{s_m}^{s_m'} \frac{ds}{\left(1 - \frac{B(s)}{B_m}\right)^{\frac{1}{2}}} \tag{22}$$

is the half-bounce period. It is a measure of the second periodicity of equation (1). Usually $\tau_b \gg \tau_c$.

If we now let U vary along the field line, e.g., if there is a parallel component of the electric field, (20), (21), and the second equality in (22) are no longer valid, and mirror points and bounce period will be different. In particular, we shall have, in view of (19) and (16), for the non-relativistic case

$$\frac{[W - U(s)] \sin^2 \alpha(s)}{B(s)} = \text{const} = \frac{T_o \sin^2 \alpha_o}{B_o} = \frac{(W - U_m)}{B_m} \tag{23}$$

W is the (constant) total energy $T_o + U_o$.

For the bounce half-period, in turn (non-relativistically)

$$\tau_b = \int_{s_m}^{s_m'} \frac{ds}{v \cos \alpha(s)} = \int_{s_m}^{s_m'} \frac{ds}{\{\frac{2}{m} [W - U(s) - \mu B(s)]\}^{\frac{1}{2}}} \tag{24}$$

Notice how an electric field parallel to the magnetic field lines $[U(s) = qV(s)]$ can affect the position of each mirror point (23) and the bounce period (24). Likewise, it will affect the "loss cone" at any point of the field line, defined by

$$\sin^2 \alpha_{loss} = \left(1 + \frac{U_o - U_{surface}}{T_o}\right) \frac{B_o}{B_{surface}} \tag{25}$$

T_o and B_o are taken at the point on which the loss cone is computed, $U_{surface}$ and $B_{surface}$ are taken at the point where the field line intersects the absorbing atmosphere.

3.2 Drift Shells

When a particle bounces along a field line it also drifts perpendicularly to it. We shall assume this drift to be "slow": a particle has to bounce several times before drifting more than one Larmor radius away from the initial field line. We shall call the field line along which a particle is bouncing at a given time, the "guiding field line". If we are not interested in the actual, instantaneous position of the guiding center along a field line, it will be sufficient to specify the guiding field line. In other words, if we average out the bounce phase, we are left with the guiding field line. (Similarly, if we average out the cyclotron phase, we are left with the guiding center.)

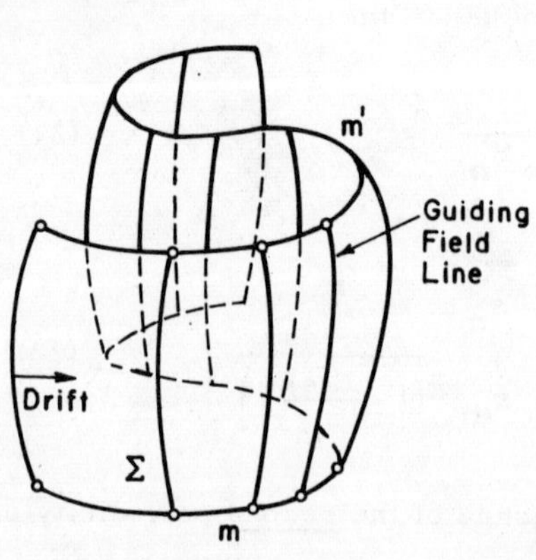

Each guiding field line is "limited" by the particle's mirror points. As the particle drifts, its guiding field line generates a surface called drift shell Σ. The mirror points generate the two limiting curves m, m', called mirror point paths or traces (or rings, if the surface is closed).

In order to define a drift shell, we have to: (i) identify the initial guiding field line; (ii) find its velocity of displacement; and (iii) integrate this velocity, identifying all subsequent guiding field lines.

(i) Guiding field line identification. There are many ways to geometrically or analytically define a field line in a given magnetic field. The most practical one for our purposes is the following: Consider a given, fixed surface Ω orthogonal to all field lines (in usual trapping geometries this is always possible). Each field line is completely specified by the position of its intersection point 0 on the fixed surface Ω. If α, β is a convenient curvilinear coordinate system on Ω, only two values α_i, β_i are necessary to specify a given field line.

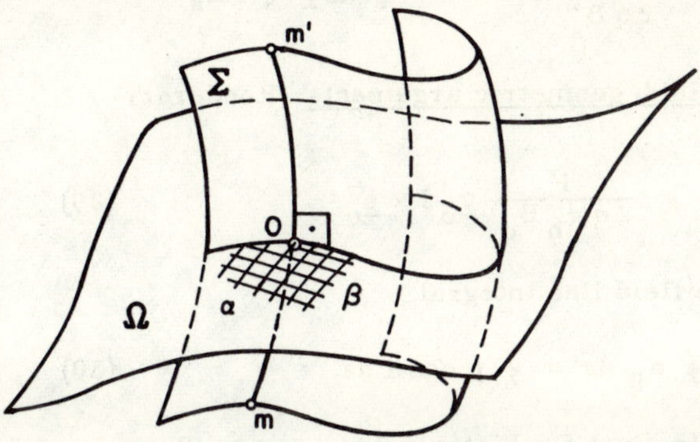

A particularly convenient reference surface Ω in the geomagnetic field is the minimum - B equator. In the case of a pure dipole, this is a plane, and the two most convenient α, β parameters are the polar coordinates of 0

$$\alpha = r_o = L \text{ (in earth radii)}$$
$$\beta = \phi_o \text{ (longitude)} \qquad (26)$$

(ii) Guiding field line drift velocity. The velocity of the reference point 0 will be a sort of bounce-average drift velocity. But one has to be very careful: if $\underline{V}_s$ is the drift velocity of the particle while its guiding center is passing through point P_s, the displacement of point 0 associated to the displacement $\underline{V}_s\,\delta t$ will be given by $\underline{00'}$. 0' is obtained by tracing the field line ℓ' that goes through P_s' down to the reference surface. We call the associated displacement $\underline{00'} = \underline{V}_{os}\,\delta t$. The bounce-average drift velocity of 0 will then be given by the bounce-average of all associated velocities $\underline{V}_{os}$

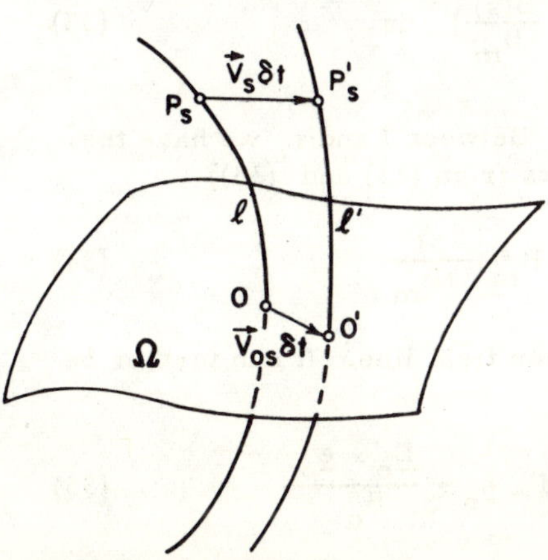

$$<\underline{V}_o> = \frac{1}{\tau_b} \int_{s_m}^{s_m'} \underline{V}_{so} \frac{ds}{v_{\|}} \qquad (27)$$

The associated velocity $\underline{V}_{so}$ is related to the actual drift velocity $\underline{V}_s$ by a field-geometric transformation

$$\underline{V}_{so} = \underline{\underline{T}}_s \underline{V}_s \tag{28}$$

In the case of a dipole, the tensor $\underline{\underline{T}}_s$ reduces to the scalar $\cos^{-3}\lambda$, where λ is the latitude of the point P_s. In the general case, the associated velocities $\underline{V}_{os}$ need not be parallel to each other.

If $\underline{V}_s$ is given by (7b) and (8), and calling $\underline{e}$ the unit vector parallel to $\underline{B}$), that is, if

$$\underline{V}_s = \frac{\underline{E}_s \times \underline{e}_s}{B_s} + \frac{m v_s^2}{2 q B_s^2} (1 + \cos^2 \alpha_s) \, \underline{\nabla}_\perp B_s \times \underline{e}_s$$

it can be shown on purely field-geometric arguments (Roederer, unpublished) that

$$<\underline{V}_o> = \frac{1}{2 q \tau_b B_o} \underline{\nabla}_o J \times \underline{e}_o \tag{29}$$

In this expression, J is the field line integral

$$J = \oint p_{\|} \, ds = \oint p \cos\alpha \, ds \tag{30}$$

taken for a full bounce cycle (just for historical reasons). B_o and $\underline{e}_o$ are defined at the reference point 0 of the guiding field line. The gradient $\underline{\nabla}_o J$ is perpendicular to the field line, computed at 0 and taking into account conservation of total energy W and magnetic moment $\underline{M}$, i.e., taking neighboring field lines and integrating (30) between mirror points given through (21) by

$$B_m = \frac{1}{\mu} \sqrt{2m(W - U_m)} \tag{31}$$

If the field lines are equipotentials (no parallel electric field or forces), (30) can be written in the form

$$J = 2p\, I \tag{32}$$

where

$$I = \int_{s_m}^{s_m'} \left(1 - \frac{B(s)}{B_m}\right)^{\frac{1}{2}} ds \tag{33}$$

is a purely field-geometric integral. Between I and τ_b we have the following useful relation (which follows from (22) and (33))

$$v \, \tau_b = I + 2 B_m \frac{\partial I}{\partial B_m} \tag{34}$$

The derivative is taken along a given field line. It can further be shown that in this case

$$<\underline{V}_o> = \frac{p}{q \tau_b B_o} \underline{\nabla}_o I \times \underline{e}_o + \frac{\underline{E}_o \times \underline{e}_o}{B_o} \tag{35}$$

The first term represents the bounce-average gradient and curvature drift, the second is the bounce-average electric field drift.

(iii) Shell tracing. Given a particle injected along a field line in a given magnetic and electric field, one can find the intersection of all guiding field lines with the reference surface Ω, by integration of (35). The full shell is then determined by tracing the field lines that go through this intersection to the local mirror points (given by (31)). This procedure is quite cumbersome and lengthy. In the next section we shall prove a second conservation theorem, which can be used more conveniently for shell tracing.

3.3 The Second Adiabatic Invariant

The quantity J (30) is an adiabatic invariant which is conserved during the drift of a trapped particle, as long as fields and forces change only very little during an interval of time of the order of a bounce period. There is no particular name for J, except "Jay", "Second adiabatic invariant" or "Longitudinal invariant". The conservation theorem follows at once from (29): since $\langle \underline{V}_o \rangle$ is of the form "scalar times $(\underline{\nabla}_o J \times \underline{e}_o)$", the drift occurs perpendicularly to $\underline{\nabla}_o J$, i.e., parallel to a J = const surface. In other words, J is constant for all field lines traced by the particle's guiding center during its drift. Remember carefully that J must be computed for constant M and W. The mirror point traces m and m' are thus determined through (31).

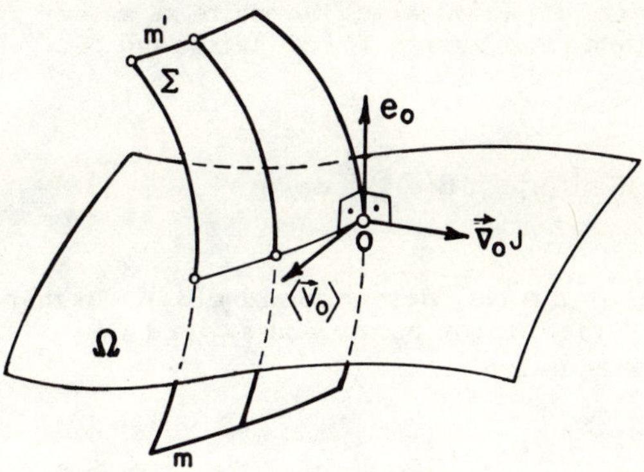

Let us consider the particular case of zero electric field. In that case, conservation of J implies conservation of the geometric integral I (33). Furthermore, since U = 0, (31) leads to conservation of the mirror point field intensity B_m. In summary, we now have

$$\begin{aligned} I &= \text{const} \\ B_m &= \text{const} \end{aligned} \tag{36}$$

(36) tell us that in absence of electric fields, drift shells are independent of particle energy.

When we have an electric field, but with field lines being equipotentials (no parallel electric field), we can combine (31) and (32) to conclude that the field-geometric quantity $I\sqrt{B_m}$ is conserved. In this case, the set of equations to be used in shell tracing is

$$\begin{aligned} I\sqrt{B_m} &= \text{const} \\ \frac{B_m}{\sqrt{W - U}} &= \text{const} \\ W = T + U &= \text{const} \end{aligned} \tag{37a}$$

The equivalent relativistic relations are

$$I\sqrt{B_m} = \text{const}$$

$$\frac{B_m}{\gamma^2 - 1} = \text{const} \tag{37b}$$

$$\gamma + \frac{U}{m_o c^2} = \text{const}$$

$$\left(\gamma = \frac{m}{m_o}\right)$$

If there is a parallel electric field, i. e., if field lines are not equipotentials anymore, the conservation of J still holds (under the condition of "slow" variations), but (32) does not hold. In that case we have to use the general expression of J (30), which we write as a function of magnetic moment M and total energy W (cf. integrand of (24))

$$J = 2\int_{s_m}^{s_m'} \{2m(W - U(s) - \mu B(s))\}^{\frac{1}{2}}\, ds \tag{38}$$

The mirror points s_m, s_m' are, as always, determined by (31). In that expression U_m is the potential at the mirror point, and s_m and s_m' must be found as solutions of the equation

$$B(s_m) = \frac{1}{\mu}\{2m(W - U(s_m))\}^{\frac{1}{2}}$$

It is easy to realize that parallel electric field can have a profound effect on particle shell geometry.

In this lecture we shall not discuss the effect of electric fields on particles mirroring away from the equator.

3.4 Shell Tracing and Magnetospheric Models

We shall first give the "recipe" to find shell field lines by using relations (36). We inject a particle at a given field line and determine B_m and I through (20) and (33), respectively. Then we go to any other longitude, on which we want to find the shell field line. On that longitude, we first determine the B_m = const trace, on which the mirror points must necessarily lie. Then we instruct the computer to find, with an iterative method, the field line along which the value of (33), computed between B_m = const points, is equal to the initial I (± prefixed error). Notice that in this way, we can find shell field lines at arbitrary longitudes, without having to proceed in small steps from the initial one. Of course, if the drift time is wanted, we have to integrate (35) (only the first term), step by step.

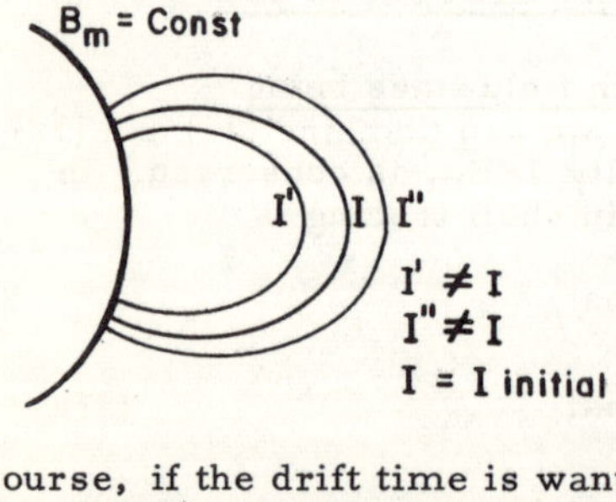

In order to compute drift shells in the magnetosphere, we have to use a magnetic field model. There are two types of magnetospheric models currently used in the literature: the "physical" model of Mead and Williams (Mead, 1964; Williams and Mead, 1965) and the "mathematical" image-dipole model (Hones, 1963).

The Mead-Williams model uses a boundary surface model and places currents on that surface in such a way as to cancel the total field (earth's dipole field plus the field of boundary currents) outside the magnetosphere. The earth's dipole is taken as stationary, and perpendicular to the ecliptic. A current sheet, lying in the geomagnetic equator, is added in the night side to simulate the neutral sheet of the tail. There are four parameters in this model: the stand-off distance R_s to the subsolar point of the magnetopause, the near and far distances to the edges of the tail current sheet, R_n and R_f, respectively, and the tail field intensity B_T (field intensity near the center of the current sheet in absence of other sources). Typical quiet time values are

$$R_s = 10\ R_E\ ,\quad R_n = 10\ R_E\ ,\quad R_f = 200\ R_E\ ,\quad B_T = 15\ \text{gammas}$$

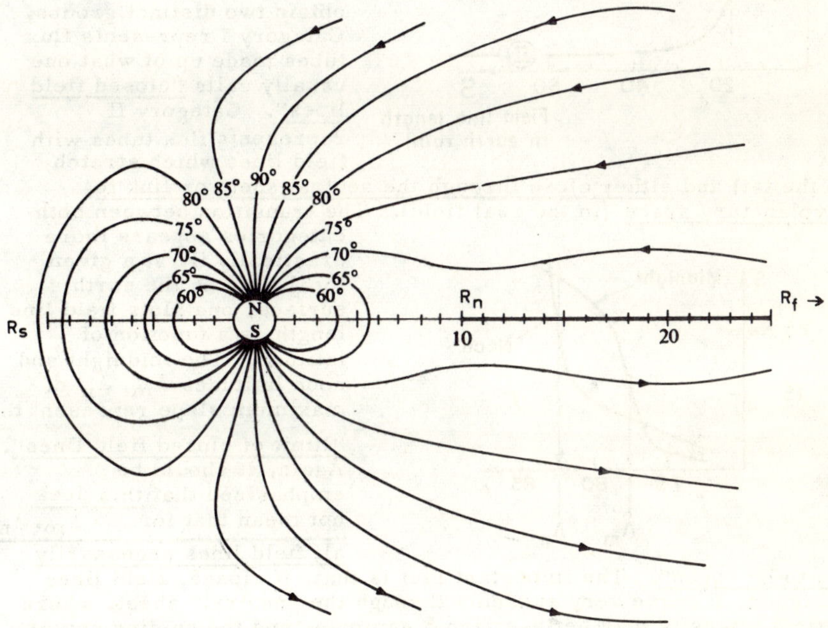

Figure 5. Field lines in the Mead-Williams model of the magnetosphere (Williams and Mead, 1965). R_s: stand-off distance; R_n, R_f: near and far edges of the neutral sheet, respectively. Field lines are labeled with the geomagnetic latitude of their intersection with the earth's surface.

Figure 5 shows the general field configuration of this model in the noon-midnight meridian.

The other model frequently used in the literature is Hones' image-dipole model. In this model the effect of the magnetopause currents is simulated by a large image dipole placed between the earth and the sun. Again, the earth's dipole is taken perpendicular to the ecliptic, and so is the image dipole (same direction as that of the earth's dipole). The parameters governing this model are the ratio r of the image dipole moment to that of the earth's, and the distance S between the two. Typical values for r and S that reproduce Fairfield's results (Figure 1) are

$$r = 40 \qquad S = 38\ R_E$$

To this field, a tail current sheet has to be added. A virtue of the Hones model is that it can be expressed in analytical form.

H. Sauer (unpublished) has recently developed a model which takes into account the dipole tilt.

All models and, as a matter of fact, the real magnetospheric field, have two categories of field lines, according to the distribution of lengths of flux tubes. If we compute the magnetic flux that leaves the earth's surface contained in field line tubes of lengths between S and S + dS, we obtain two distinct groups. Category I represents flux tubes made up of what one usually calls "closed field lines". Category II represents flux tubes with field lines which stretch into the tail and either close through the neutral sheet or link to interplanetary space (in the real field).

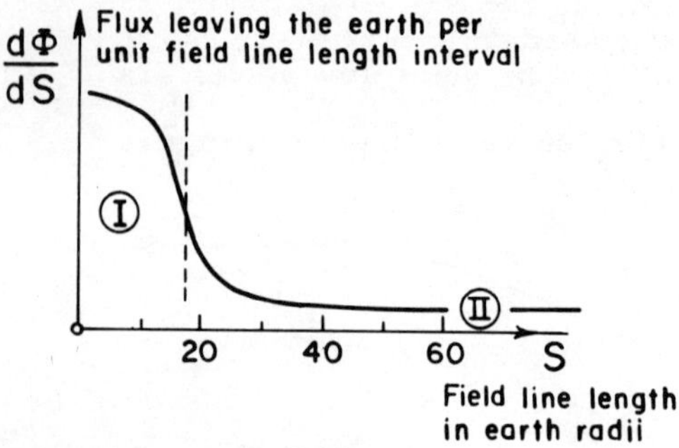

The transition between both categories appears more drastically if, at a given local time on the earth's surface, one plots field line length as a function of latitude. The midnight and noon latitudes Λ_m, Λ_n of maximum slope represent the "limit of closed field lines". Again, it should be emphasized that this does not mean that for $\lambda > \Lambda_m, \Lambda_n$, all field lines necessarily have to be "open". The important fact is that, if closed, field lines of category II curve very suddenly through the "neutral" sheet, where the field intensity may be less than 5 gammas, and the guiding center approximation very likely breaks down, except perhaps for extremely low energies. Figure 6 depicts how these "transition latitudes" depend on the magnetospheric model parameters (using Mead's model). Notice that the boundary is controlled more efficiently by the tail field intensity than by the stand-off distance.

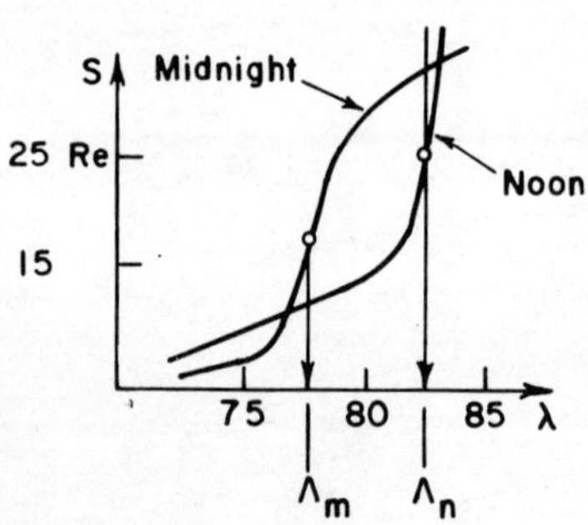

3.5 Shell Splitting

Take a particle that starts at a given longitude ϕ, circling around a given field line and mirroring at a value B_m. The integral (33) computed along the field line between the two mirror points has a value I. This means that when drifting through any other longitude, for example 180° away, this particle will bounce along a field line that

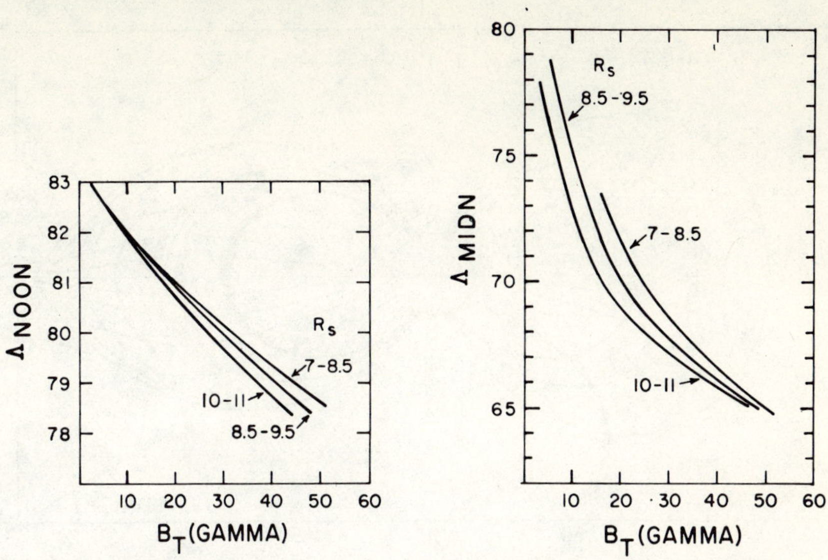

Figure 6. Latitude of the earth intersection of the noon and midnight limit of "closed" field lines, as a function of the tail field intensity B_T and for different values of the stand-off distance R_s, for the Mead-Williams' model (Roederer, 1969).

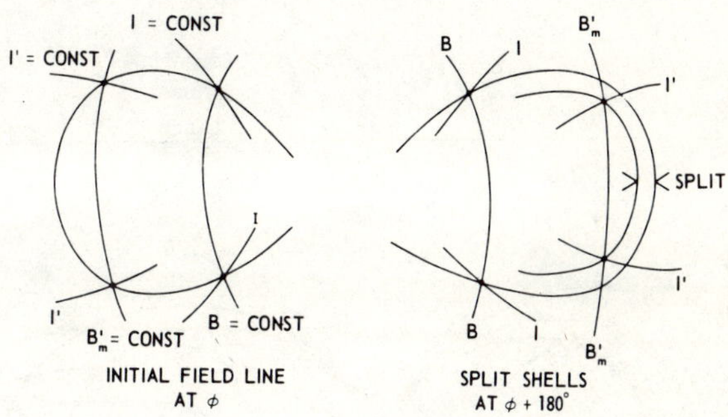

passes through the intersection of the corresponding I = constant and B_m = constant surfaces. Now take a particle that starts on the same initial field line but mirrors at a lower value $B'_m < B_m$. Its integral I' will also be smaller $(I' < I)$.

After 180° longitudinal drift this second particle will be travelling along a field line that passes through the intersection of the surfaces I' = constant and B'_m = constant. Only in case of perfect azimuthal symmetry (as in the pure dipole) will these surfaces intersect exactly on the same line as that of the first particle. In the general case, particles starting on the same field line at a given longitude will populate different shells, according to their initial mirror point fields or, which is equivalent, according to their initial equatorial pitch angles. (Of course, all these different shells would be tangent to each other at the initial field line.)

The Mead model was used to compute magnetic shells (Roederer, 1967) for particles initially mirroring on a common field line and having equatorial pitch angles with cosines 0.2, 0.4, 0.6, 0.8, and nearly 1

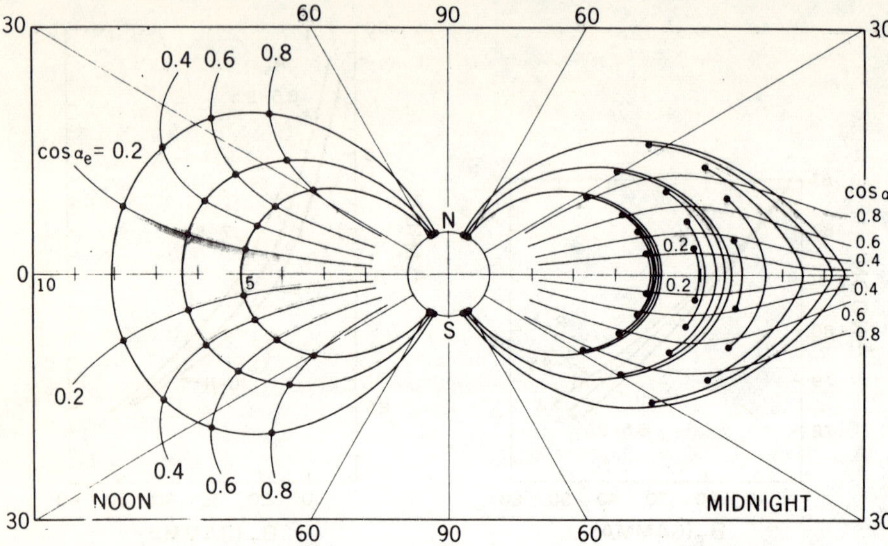

Figure 7. Computed shell splitting for particles starting on common field lines in the noon meridian (Roederer, 1967). Dots represent particles' mirror points. Curves giving the position of mirror points for constant equatorial pitch angle α_o are shown.

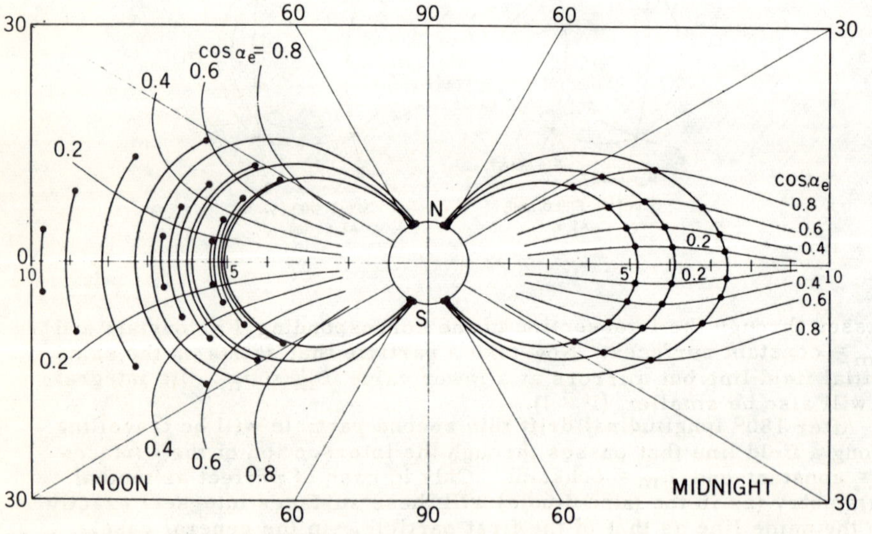

Figure 8. Same as in Figure 7, for particles starting on common field lines in the midnight meridian.

(mirroring close to the earth's surface). Figure 7 shows how particles, starting on a common line in the noon meridian, do indeed drift on different shells which intersect the midnight meridian along the field lines shown in the figure. The dots represent particles' mirror points. Curves giving the position of mirror points for constant equatorial pitch angles are traced for comparison (in a dipole field they would be constant latitude lines). Notice the change (decrease) in equatorial pitch angle for the same particle when it drifts from noon to midnight. Figure 8 depicts the same features for particles starting on a common field line in the midnight meridian. In this case, again, the equatorial pitch angle changes considerably when the particle drifts to the opposite meridian (increasing at noon).

Notice from Figures 7 and 8 that, as equatorial pitch angles increase, shell splitting is directed radially inward for particles starting on the same field line at noon and radially outward for particles starting on a common field line at midnight. Furthermore, shell splitting is maximum for particles mirroring close to the equator; for a given change in pitch angle (in degrees) the radial displacement of the shell will be greater for equatorial particles.

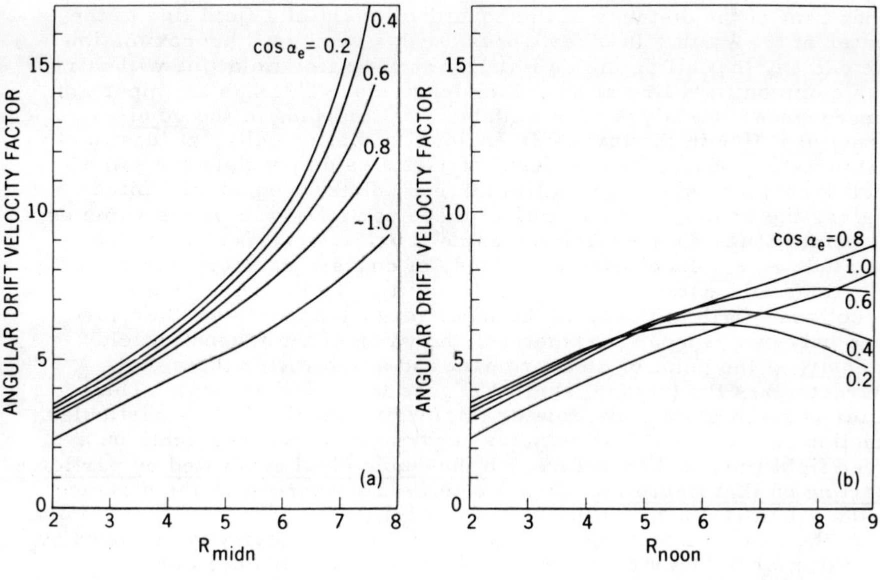

Figure 9. Value of the geometric factors intervening in the angular drift velocity as a function of the equatorial distance of the corresponding field line, for different pitch angles, and for midnight and noon, respectively (Roederer, 1967). To obtain drift velocities in degrees/sec multiply the above values by the factor $2.321 \cdot 10^{-2} (\gamma^2 - 1)$ times (rest mass in electron masses).

Numerical calculations (Roederer, 1967) reveal a considerable local-time dependence of the equatorial drift velocity, as it has been predicted for equatorial particles in Section 2.6. The geometrical factors appearing in the first term of Equation (35) were computed and represented in Figure 9 as a function of the equatorial distance of the corresponding field line for different pitch angles, and for noon and midnight, respectively. For a better understanding angular drift factors are shown. For radial distances $< 3\ R_E$ we observe a dipolar dependence. Beyond $3\ R_E$ there is a considerable departure. Drift velocities on the nightside are indeed appreciably higher than on the dayside. The peculiar inversion of the pitch-angle dependence occurring on the dayside (Figure 9) is due to a shift in relative importance of curvature drift versus gradient drift (particles mirroring at high-latitude experience more of a dipole-type field and drift faster). Calculations are in excellent agreement with drift period observations in the outer magnetosphere (Brown, 1968).

A close inspection of the local-time dependence of the drift velocity (not shown here) again leads to the important conclusion that a given particle trapped in the outer magnetosphere ($\gtrsim 6\ R_E$) spends up to 2/3 to 3/4 of its total lifetime in the dayside. In other words, there is always a higher probability of finding a given particle in the dayside than in the nightside.

3.6 Trapped Particle Flux Mapping

For the case of the internal geomagnetic field in absence of external perturbations - i.e., within about $4\ R_E$ - it can be shown that the distance between split shells is relatively small, say a fraction of 1 per cent of the distance of the equatorial point of a field line to the center of the earth. In other words, with a very good approximation one can say that all particles initially on the same field line will mirror on a common field line at any other longitude. This has an important consequence: it enables a two-dimensional mapping of the omnidirectional flux in the three-dimensional radiation belts, at least up to distances of about $4\ R_E$. Indeed, if particles do populate the same shell irrespective of their initial mirror points or equatorial pitch angles, the omnidirectional flux of these particles will be the same on all points of the shell having the same B value, irrespective of the longitude, i.e., local time (provided, of course, no appreciable injections or losses occur during the drift). To describe omnidirectional particle fluxes in the inner magnetosphere, we therefore need only two "space" parameters: the value of the magnetic field intensity at the point of measurement and a parameter that characterizes the (unique) shell that goes through that point. This latter is McIlwain's L parameter (McIlwain, 1961). L is a particular function of I and B_m that remains constant (within 1 per cent) on a given field line, and therefore, on the whole shell generated by particles starting on that field line. In a dipole field it represents the distance to the equatorial point of the field line (26).

In the outer magnetosphere, the azimuthal symmetry is removed by the external field sources, and shell splitting is considerable. Omnidirectional particle fluxes cannot be ordered in a two-dimensional way. Particles converging from different directions to a given point where the omnidirectional detector is located belong to different shells, and will nowhere else converge to a common point (i.e., to the same detector) except at the initial position (or its conjugate point).

Beyond $5\ R_E$ only the directional flux of particles measured within a narrow interval of pitch angles is physically meaningful and can be ordered in a two-dimensional way, independent of longitude or local time. The two parameters I, B_m or any well-defined functions of both, can be used as coordinates of the two-dimensional space for directional particle flux mapping.

McIlwain's L value is such a well-defined function of I and B_m and therefore, may still be used as one of the coordinates for directional flux mapping in the outer magnetosphere. However, when doing so, the magnetic field in McIlwain's computer code must be replaced by a field model which contains all external sources responsible for the asymmetry in the outer magnetosphere. Computed in this way, L will not be constant along a field line anymore and cannot be interpreted as the radial distance to the equatorial point of the field line. It will be, however, constant along the mirror point trajectories ("B-L" rings) of the particles to be described.

In view of this, care has to be taken at high latitudes with the "invariant latitude" Λ, defined by

$$\cos^2 \Lambda = \frac{1}{L} \tag{39}$$

Beyond $L \gtrsim 5$, Λ = constant contours do not in general represent particle shell field line intersections with the earth's surface. Only for particles mirroring far away from the equator at very low altitudes (at high latitudes) and which therefore spend most of their bouncing time in the internal geomagnetic field, does Λ accurately represent a shell intersection. Intersections of field lines belonging to shells of particles mirroring close to the equator do not follow constant invariant latitude contours, and are thus universal time controlled. Based on this discussion, notice the following:

(i) A particle precipitation pattern in the "closed field line" region that follows a Λ = const contour and, hence, does not show a UT dependence, would indicate that the source population which feeds

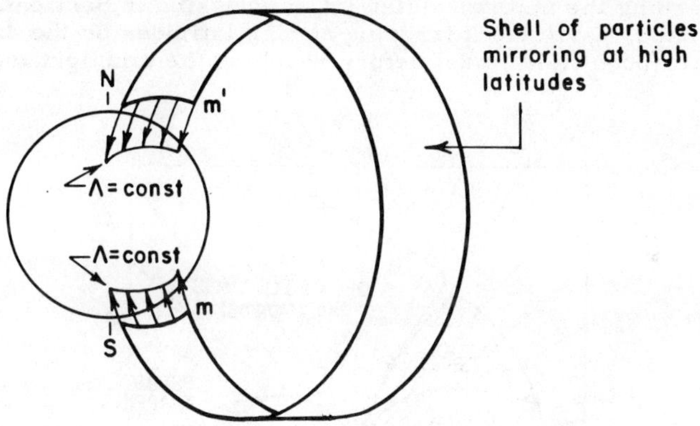

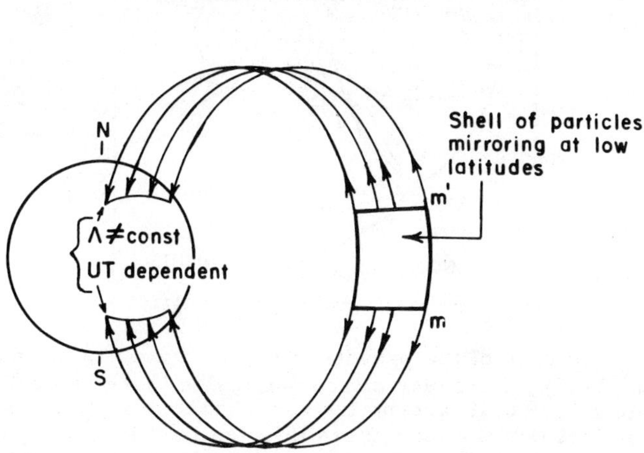

this precipitation consists of particles which on the average are mirroring close to the earth (far away from the equator).

(ii) A precipitation pattern in the closed field line region that does not follow Λ = const contours and which does show a UT dependence may be fed by a source population which on the average is mirroring far from the earth (close to the equator).

(iii) Dawn-dusk asymmetries in precipitation pattern geometry very likely indicate that electric fields are at work.

In spite of losing some of its physical significance for processes involving particles mirroring far out in the magnetosphere, the invariant latitude Λ is still the best surface coordinate for most of the field-controlled phenomena.

No work has so far been published on particle shell geometry in a tilted dipole magnetospheric model. In this case, considerable seasonal variations are to be expected.

3.7 The Trapping Boundary

When a field line has its equatorial point beyond about 8 R_E, a fraction of the particles mirroring on it can only be "pseudo-trapped", being unable to complete a drift around the earth (Roederer, 1967). In particular, particles mirroring at low latitudes in the back side abandon the model magnetosphere through the boundary 30° to 40° before reaching the noon meridian (cf. discussion in Section 2.6). On the other hand, particles mirroring at high latitudes on the dayside run into the tail (open field lines) before reaching the midnight meridian.

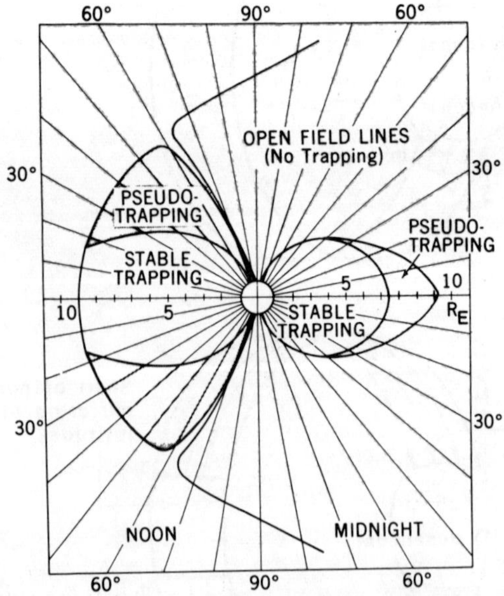

Figure 10. Location of the pseudo-trapping regions in the magnetosphere (Roederer, 1967). Particles mirroring inside those regions are unable to complete a 180° drift around the earth. Those injected into the left side will be lost into the tail; those injected into the right portion will abandon the magnetosphere through the boundary on the dayside. The inner limits of pseudo-trapping are not field lines but cut across them.

Figure 10 shows computed limits between stable trapping and pseudotrapping regions on the noon-midnight meridian. At other local times, both regions approach more closely the boundary; going from noon to midnight one disappears at the expense of the growth of the other.

The theoretical boundary of stable trapping is thus a surface that separates two regions (not field lines) in which lie the mirror points of particles whose shells are either closed around the earth, or open, respectively. Whatever particles one finds mirroring in the region of pseudotrapping, must have been injected only a short time before (minutes or hours, depending on their energy and other factors such as electric fields and diffusion). This is why one should expect low and fluctuating fluxes in the regions of pseudotrapping, against larger, more smoothly varying fluxes in the region of stable trapping. The boundary of stable trapping should be sharp only when "seen" with a directional detector. The regions of pseudotrapping are limited on their "outer" side by the boundary of closed field lines (Section 3.4).

When looking at particles mirroring at very low altitudes, one should expect the trapping boundary at the noon meridian to be situated a few degrees below the boundary of closed field lines. On the midnight meridian, it should practically coincide with it. When looking at particles precipitating from a trapped source into low altitudes, the trapping boundary should be mainly determined by the boundary of the source population, projected along field lines onto the earth's surface.

4. THE THIRD ADIABATIC INVARIANT

If we have a closed shell, the drift motion is periodic, and represents the third periodicity of equation (1). The drift period is given by

$$\tau_d = \oint \frac{dx}{\langle V_o \rangle} \tag{40}$$

with the integral carried out along the intersection of the closed shell with the reference surface (equator). Normally we have

$$\tau_d \gg \tau_b \gg \tau_c$$

If we are not interested in the drift phase, i.e., the actual, instantaneous guiding field line along which the particle is bouncing, it is enough to specify the drift shell, and no more. Time changes of the field and forces will change the particle's drift shell. If those changes have characteristic times that are of the order of, or are smaller than, the drift period, step by step shell tracing (integration of (29) or (35)) has to be applied during the transient state. This may be

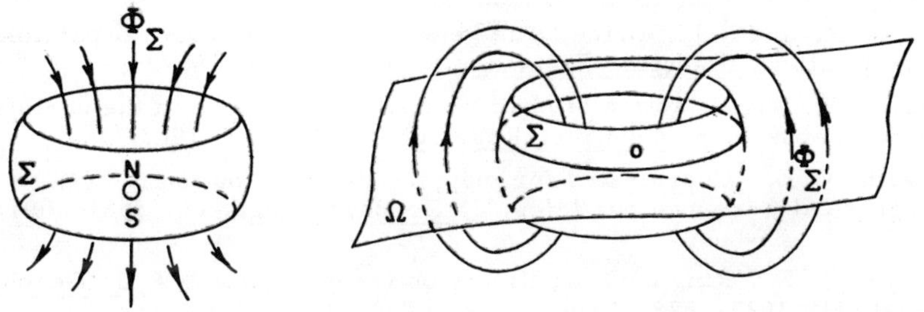

difficult because the induced electric field has to be included in $\underline{E}_o$ (35). If, however, time changes occur slowly, as compared to τ_d, one can conveniently use a third adiabatic conservation theorem. The theorem states that the magnetic flux Φ_Σ encircled by a drift shell remains constant, even if the shell changes (provided the above stated condition holds). Notice that, for the case of the earth, this flux is equal to the flux through the reference (equatorial) plane outside the intersection curve $\underline{o}$. In a slowly time-dependent magnetospheric configuration, the equatorial trace of a shell will thus change in such a way as to maintain the outside flux constant. This will also reflect on changes of precipitation patterns and trapping boundaries.

We shall not describe here the details of numerical procedures for applications of the conservation of the third invariant. Let us just state that slow compressions of the magnetosphere displace drift shells as a whole towards the night side (and viceversa). The actual displacement of the midnight side is strongly influenced by the behaviour of the currents in the neutral sheet and the tail.

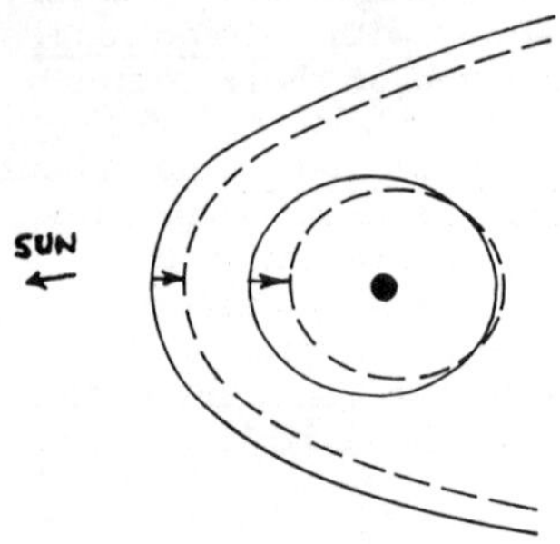

Magnetospheric compressions and neutral sheet current increases are in general positively correlated with increases in Kp. Precipitation patterns of trapped particles should therefore be expected to move equatorwards on the dayside and slightly polewards at night, during Kp increases.

Conservation of the third invariant has been used by McIlwain (1966), and Søraas and Davis (1968), to correct trapped particle intensities for adiabatic effects caused by the ring current, and by Roederer (1969) to correct particle fluxes at the synchronous orbit for adiabatic effects caused by overall changes in the magnetospheric configuration.

The use of the third invariant will come into full display the day one will start to study the effects of a rotating, tilted dipole on particle shells in the outer magnetosphere. Some surprises may be in store, especially in relation to low energy particles whose drift periods are of the order of a day (and for which conservation of the third invariant thus breaks down at all times).

REFERENCES

Brown, W. L., Energetic outer belt electrons at synchronous altitude, in Earth's Particles and Fields, edited by B. McCormac, Reinhold, New York, pp. 33-43, 1968.

Fairfield, D. H., Average magnetic field configuration of the outer magnetosphere, J. Geophys. Res., 73, 7325-7338, 1968.

Hartz, T. R. and N. M. Brice, The general pattern of auroral particle precipitation, Planetary Space Sci., 15, 301-329, 1967.

Hones, E. W., Jr., Motion of charged particles trapped in the earth's magnetosphere, J. Geophys. Res., 68, 1209-1219, 1963.

McIlwain, C. E., Coordinates for mapping the distribution of magnetically trapped particles, J. Geophys. Res., 66, 3681-3691, 1961.

McIlwain, C. E., Ring current effects on trapped particles, J. Geophys. Res., 71, 3623-3628, 1966.

Mead, G. D., Deformation of the geomagnetic field by the solar wind, J. Geophys. Res., 69, 1181-1195, 1964.

Roederer, J. G., On the adiabatic motion of energetic particles in a model magnetosphere, J. Geophys. Res., 72, 981-992, 1967.

Roederer, J. G., Quantitative models of the magnetosphere, Rev. Geophys., 7, 77-95, 1969.

Søraas, F. and L. R. Davis, Temporal variations of the 100 keV to 1700 keV trapped protons observed on satellite Explorer 26 during first half of 1965, NASA-GSFC preprint X-612-68-328, 1968.

Williams, D. J. and G. D. Mead, Night-side magnetospheric configuration as obtained from trapped electrons at 1100 kilometers, J. Geophys. Res., 70, 3017-3029, 1965.

THE INFLUENCE OF THE GEOMAGNETIC TAIL ON ENERGETIC-PARTICLE PRECIPITATION IN THE POLAR IONOSPHERE

George C. Reid

Cooperative Institute for Research in the Environmental Sciences
University of Colorado
and
ESSA Research Laboratories, Boulder, Colorado, 80302, USA

Abstract. Two distinct groups of energetic particles produce substantial amounts of ionization in the polar ionosphere. They are respectively the solar protons responsible for polar-cap absorption (PCA), and the aurorally-associated electrons whose effects can extend deep into the polar cap during polar substorms.

The geomagnetic tail has an important influence on both of these groups, and the observational and theoretical evidence bearing on this point are reviewed here, with an emphasis on ground-based radio observations. The passive role of the tail in acting as a time-varying guide for solar protons, and its dynamic role as an accelerator of plasma during substorms are both discussed.

1. INTRODUCTION

The presence of the geomagnetic tail influences the structure of the polar ionosphere in a variety of ways, as we shall hear in some detail in the papers to follow. The purpose of the present paper is to review the effects of energetic-particle precipitation on the polar ionsphere, and to examine the relationship that exists between these effects and the geomagnetic tail.

In attempting to review the experimental evidence that bears on the polar ionosphere, one is immediately struck by the relative lack of data obtained from the region well within the normal auroral oval, with which this conference is chiefly concerned. This is probably due to a combination of inaccessibility of ground bases in this region, and a preoccupation on the part of most ionospheric physicists with the more dramatic phenomena that take place in the auroral zone. While reasonably large bases do exist at L-values greater than about 20 (e.g., Thule in the northern polar cap, and McMurdo and Vostok in the southern polar cap), they do not yet appear to have been fully exploited from the point of view of ionospheric physics, and the experiments carried out there have employed relatively simple techniques, such as ionospheric sounding and the monitoring of cosmic-noise absorption by riometers. One notable exception has been the operation of a D-region partial-reflection experiment at Resolute Bay ($L \approx 100$) in northern Canada (Belrose et al., 1966).

A considerable amount of information has come from the use of techniques that do not involve placing equipment in the region itself, e.g., by using long-path radio propagation across the polar cap, or by using satellites that cross the polar cap. In the latter category, of course, the Alouette topside-sounder satellites have provided a wealth of data that will be presented in some of the papers to follow. Unfortunately, the long-path propagation techniques tend to provide information about the lowest part of the ionosphere (this is particularly true of observations of LF and VLF propagation and of VHF forward-scatter), while the topside sounders reveal mainly only the region above the peak of the F2-region. There is a very distinct and important gap in between that would best be covered by the use of rocket techniques, and the provision of an adequate rocket facility within the polar cap merits high priority. The application of more sophisticated ground-based techniques, such as Thomson scatter, to the problems of the polar ionosphere is also of the first importance.

2. POLAR-CAP ABSORPTION AND SOLAR PROTONS

The most dramatic manifestation of energetic-particle precipitation in the polar ionosphere is, of course, the now familiar polar-cap absorption event, in which streams of energetic solar protons and alpha particles create enormous enhancements in the ionization of the mesosphere. The influence of the tail on this phenomenon is profound, and in the absence of a tail the geographical distribution of PCA would show quite different characteristics from those observed. PCA is caused chiefly by protons with energies between 1 and 100 MeV, and the effect of a dipole outer geomagnetic field would be to allow these particles free access to the earth only at magnetic latitudes higher than about 77°; below this latitude there would be a steady decrease in absorption reaching very small values at latitudes below about 65°. In actual fact, the absorption is usually observed to be quite uniform at all magnetic latitudes higher than about 65° - 70°, and it was realized at an early stage that the most probable explanation of this fact was the permanent existence of a direct and easy magnetic access route to the earth for protons with energies as low as a few MeV (Reid and Leinbach, 1959). It is now known that the tail does indeed

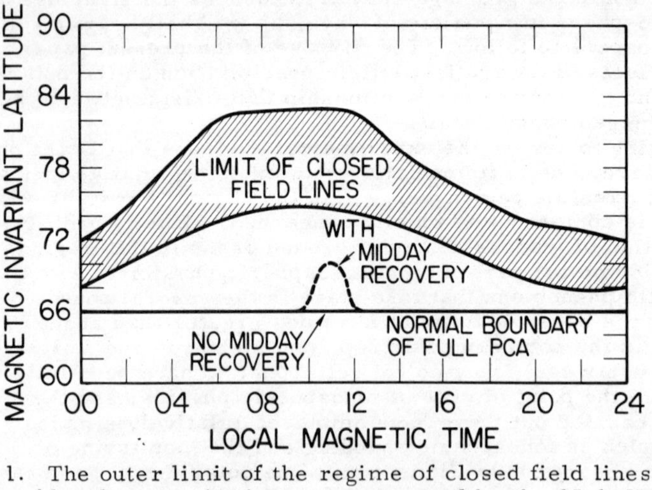

Figure 1. The outer limit of the regime of closed field lines as a function of local magnetic time, as suggested by the 35-keV trapped electron data obtained from Alouette 2 by McDiarmid and Burrows (1968). Also shown is the approximate low-latitude boundary of the region in which polar-cap absorption appears at its full intensity.

provide such an access route, though the detailed paths followed by the protons are still uncertain.

The entry of the protons is not simply a matter of following the open field lines of the tail, however, as Figure 1 illustrates. This shows the range of invariant latitudes corresponding to the outer limit of trapping of 35-keV electrons as observed over a period of about 8 months by the Alouette 2 satellite (McDiarmid and Burrows, 1968). The authors suggest that this boundary marks the true limit of the regime of closed field lines within the magnetosphere, and is quite distinct from the boundary of stable long-term trapping, which occurs at considerably lower latitudes. Also shown in Figure 1 is the normal boundary of the region in which PCA is experienced in its full intensity when the midday recovery phenomenon (Leinbach, 1967) is present and absent. The midday recovery effect is not yet understood completely, but it is more properly an auroral-zone phenomenon than a true polar-cap effect, and will not be discussed here.

The normal PCA boundary is close to the lower limit of the observed closed-field-line boundaries during the nighttime period from 2000 to 0100, as might be expected. Most of the PCA observations refer to periods of moderate geomagnetic disturbance, while most of the trapping-boundary observations correspond to quiet periods, so that a displacement of the PCA boundary towards lower latitudes is not surprising. Furthermore, the relatively large Larmor radii of the protons compared to that of the 35-keV electrons will carry them to greater depths in the magnetosphere. During daytime hours, however, the PCA boundary remains at a more or less constant magnetic latitude, in marked contrast to the behavior of the trapping boundary, which moves up some 7° in going from midnight to noon. The most likely explanation lies in the existence of quasi-trapped orbits for the protons, capable of carrying them from an injection point in the tail around the closed-field-line region of the dayside to an eventual ejection in the tail for those that have not precipitated into the atmosphere. The existence of these quasi-trapped orbits thus extends the direct influence of the tail well into the closed-field-line domain on the dayside of the magnetosphere as far as access for particles from interplanetary space is concerned. The midday recovery effect is almost certainly a phenomenon of this quasi-trapped regime, and needs further investigation.

The available observational evidence thus points to the fact that protons in the 1 to 100 MeV energy range have relatively easy access to the earth via the tail. The details of their entry into the tail, however, are as yet far from clear. If interconnection exists between the interplanetary and geomagnetic fields, then protons of all energies ought to have prompt access to all polar-cap latitudes. If, on the other hand, the tail remains closed to long distances downstream, protons with Larmor radii small compared to the lateral dimensions of the tail would diffuse into the core (corresponding to the region close to the geomagnetic poles) only slowly over a period of some hours (Michel and Dessler, 1965), and PCA would be observed to decrease with increasing latitude above the auroral zone. The observational evidence on this point is somewhat conflicting. Hakura (1967) has analyzed ionosonde data obtained during several PCA events from a variety of high-latitude stations, and has presented evidence that a typical event grows in three distinct stages. First, a slight enhancement in absorption appears near the geomagnetic poles, which is followed by a sharp increase in the auroral zones, followed in turn by a more gradual filling-in of the entire polar cap. This analysis, however, uses the parameter Δf_{min}, defined as the departure of the instantaneous minimum frequency at which an echo is recorded from its monthly median value. This is a qualitative parameter whose usefulness is difficult to gauge, and it would be desirable to verify the result with more quantitative measurements where possible in future events.

There are a number of factors that tend to confuse the interpretation of PCA records, particularly near the beginning of an event, when the absorption levels are low. For example, both riometers and ionosondes do not distinguish between PCA and auroral absorption events, and the weak early PCA is often obscured by the presence of auroral activity, expecially in the vicinity of the auroral zones, where the earliest onset might be expected. If the sun is above the horizon, intense solar radio noise from the proton flare often masks the onset of the event completely, and if the sun is below the horizon the greatly reduced nighttime sensitivity of the ionosphere usually prevents the occurrence of an event being detected until sunrise. Direct satellite measurements of the proton fluxes also suffer from the fact that several minutes usually elapse between the time of crossing the auroral zone and arrival near the geomagnetic pole, and this interval may be critical in attempting to look for variations in onset time as a function of latitude. In one event in which clear riometer recordings were obtained across the southern polar cap, that of February 1965 (Reid and Sauer, 1967), there was no evidence of any differences in onset time of the event, but the point is obviously one that requires further study.

Perhaps the most valid evidence on the question of the ease with which solar protons can penetrate the tail has come from comparisons between observations carried out on pairs of distant satellites, one of which is located within the tail and the other in nearby interplanetary space. Results reported by Krimigis et al. (1967) using data obtained during the event of 7 July 1966 from Explorer 33 and Injun 4 certainly point to rapid access of low-energy (~ 0.5 MeV) protons to the tail. Conflicting evidence, however, has recently been presented by Montgomery and Singer (1969), who compared proton measurements made on two VELA satellites during events in 1967, and found delays ranging between 15 minutes and 2 hours over a wide energy range. It is entirely possible that the detailed structure of the tail is highly variable, and that protons can penetrate relatively directly at some times, and are effectively excluded for appreciable lengths of time at others, a conclusion that is consistent with the observations of Williams and Bostrom (1967). Obviously this is an area in which further observations are needed, and in which ground-based observations of the polar ionosphere can play an important part through their ability to record continuously on one field line of the tail.

In spite of the observations of relatively rapid access of low-energy protons to the tail, it appears that there is a characteristic variation in proton flux across the polar cap, in the sense that fluxes close to the auroral zones are consistently higher than the fluxes well within the polar caps. This feature is difficult to detect on the basis of ionospheric observations alone, due to the above-mentioned frequent contamination of auroral-zone PCA observations by auroral absorption, and to the bias of daytime PCA effects towards protons of energies greater than about 10 MeV, where the effect seems to be less pronounced. Several satellite observations have shown the effect, however, and Figure 2 is an example, taken from the work of Blake et al. (1968). This shows the flux of protons with energies between 1.1 and 1.6 MeV as a function of geomagnetic latitude during the event of August 28, 1966. The observations were carried out on a satellite that crossed the southern polar cap at an altitude close to 3000 km near the noon-midnight meridian. Looking at the nighttime observations, on the left of the diagram, the fluxes in the 80-85° latitude range are lower by a factor of about two than the fluxes near the auroral zone. This effect cannot be attributed to a delay in propagation of the fluxes to the region downstream in which the

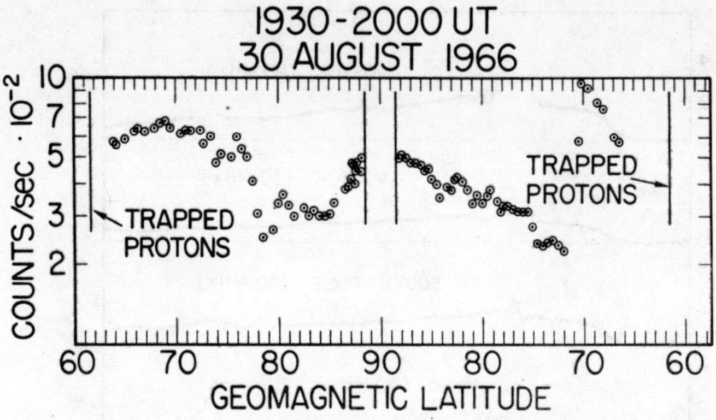

Figure 2. Counting rate of solar protons with energies between 1.1 and 1.6 MeV as a function of geomagnetic latitude over the southern polar cap on 30 August 1966. Data were obtained by Blake et al. (1968) from a solid-state counter telescope on board the satellite 1966-70A.

high-latitude field lines are connected to the interplanetary field, since the high-latitude flux was found to be lower even when the interplanetary flux was decreasing with time. It is difficult to interpret this effect without assuming that the "core" region of the tail (i.e., the region far removed from both the neutral sheet and the magnetopause) is at least partially sheltered from the solar-proton flux in the 1-MeV energy region.

Figure 2 also shows an enhancement in proton flux in the neighborhood of the pole itself, though the maximum latitude reached by the satellite on this particular orbit was about 88°. This is an example of the type of spatial variation across the polar cap that is frequently observed, though with less consistency than the auroral-zone enhancement discussed above. Further examples of polar-cap structure in low-energy proton fluxes have been reported by Paulikas et al. (1966), by Williams and Bostrom (1967) and by Hudson and Anderson (1968), and similar effects have been seen by high-latitude riometers (Reid and Sauer, 1967). Figure 3 shows riometer recordings of the PCA event of February 1965 as observed at Vostok, South Pole, and Byrd, and shows a remarkable increase in PCA at Vostok of about two hours duration, accompanied by much less significant effects at the two lower-latitude stations. Structure in the low-energy proton flux was observed over the northern polar cap during the same period by a polar-orbiting satellite (Paulikas et al., 1966). The origin of this structure remains obscure; it may be connected with the filamentary structure of the interplanetary magnetic field (McCracken and Ness, 1966; Reid and Sauer, 1967), but strong arguments against this point of view have been presented by Hudson and Anderson (1968), who observed persistence of structure of this kind over several hours. It seems likely, however, that the structure is associated with the existence of an anisotropy in the proton flux in interplanetary space, disappearing when the proton flux becomes isotropic (Hudson and Anderson, 1968). The possibility that the auroral-zone enhancement in low-energy proton flux is also closely linked to anisotropy in the flux is a distinct possibility that deserves some attention.

There is one further group of phenomena that is related in some way to changes in the structure of the tail, but that remains unexplained in its details. These are the temporal changes that take

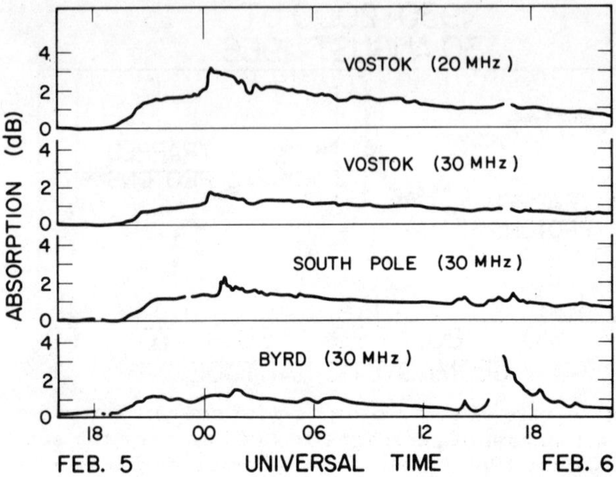

Figure 3. Southern polar-cap riometer recordings of the PCA event of February 1965 (Reid and Sauer, 1967). The L-value at Byrd is about 7, at South Pole about 15, and at Vostok greater than 50.

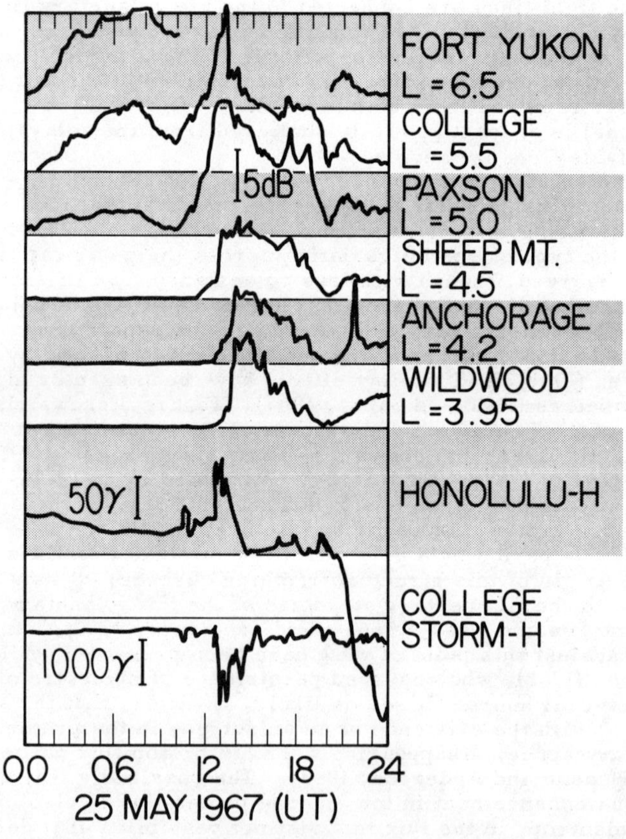

Figure 4. Polar-cap absorption recorded on 25 May 1967 by a latitudinal chain of 30-MHz riometers in Alaska. Also shown are the variations in the horizontal component of the geomagnetic field at Honolulu and at College.

place in the position of the edge of the PCA region in association with magnetic storms. Figure 4 shows the PCA recorded by a latitudinal chain of riometers in Alaska extending from L = 3.95 up to L = 6.5 during the maximum of the event that began on 23 May 1967. Also shown are the variations in the horizontal magnetic field at Honolulu and at College (the latter on a reduced scale). Two sudden commencements occurred in rapid succession during this period, the first one at 1021 UT, the second at 1235 UT. In association with the second of these, PCA began to appear at the lower-latitude stations which had previously experienced no absorption. (The pronounced decrease subsequently experienced at all stations, and centered at about 2000 UT, is a good example of the midday recovery phenomenon mentioned above, and shows the progressive deepening and broadening of the effect with decreasing latitude).

The appearance of solar protons at unusually low magnetic latitudes following the onset of a geomagnetic storm is now a well-established phenomenon, but the details of the onset of the effect are still not understood. In the example shown here, there is some evidence of a progressive motion of the PCA onset to successively lower latitudes with time, but there is no unique feature of the magnetograms that can be unambiguously linked with the cutoff decrease. The closest link with the low-latitude magnetogram is with the onset of the main phase of the storm, following the initial phase in which H is enhanced. Yet at the time of maximum PCA at the low-latitude stations, H has only been reduced to its pre-storm value. At College, the initial phase is apparently obscured by a polar substorm, but this has already passed its maximum before the cutoff decrease occurs at the two lowest-latitude stations. Undoubtedly this phenomenon is associated with the sweeping back of magnetic field into the tail that must occur at the beginning of a magnetic storm, or with the opening of fresh field lines on the nightside of the magnetosphere, but both of these effects must be studied in more detail before the whole process is fully understood.

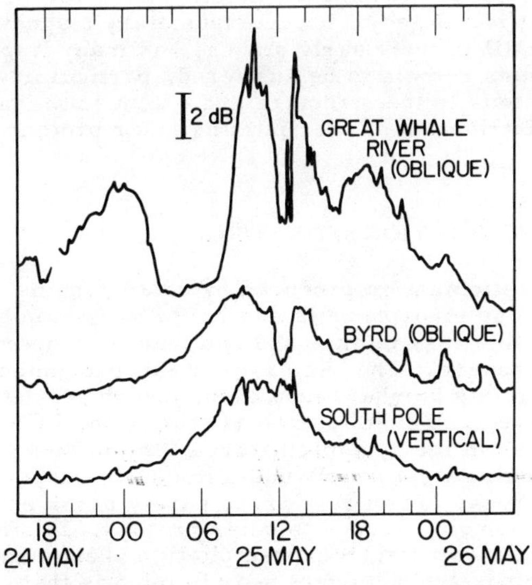

Figure 5. Polar-cap absorption recorded by 30-MHz riometers at the conjugate stations of Great Whale River, Quebec, and Byrd, and at South Pole during 24-26 May, 1967. The antennas at Great Whale River and Byrd are beamed at an angle of 45° to the zenith.

Another storm-associated effect that is somewhat rarer also occurred during this period, and is illustrated in Figure 5. Here we show the riometer recordings from Great Whale River and its conjugate station, Byrd, both at L = 7, and from South Pole, at L = 15, during the same period of May 1967. The lower ionosphere above Great Whale River was in sunlight from about 0900 to 0120 UT, accounting for the larger magnitude of the absorption there as compared to the two Antarctic stations, which were in darkness. A marked decrease in absorption occurred at the two conjugate stations following the first of the two magnetic-storm onsets at 1021 UT; there was no corresponding decrease in the proton flux outside the magnetosphere as observed by satellites, and there is also no corresponding decrease in absorption at South Pole, nor at the Alaskan stations shown in Figure 4, and the only reasonable interpretation of the decrease is that it represents a short-lived increase in the local solar-proton cutoff associated with the storm, that rapidly disappears with the onset of the second storm. This may be an example of the initial-phase storm increase in cutoffs discussed by Ortner et al. (1961), but the initial phase following the sudden commencement at 1021 UT was very moderate in intensity, and much weaker than that following the second sudden commencement at 1235 UT, when the cutoffs were rapidly restored.

These have been presented here as examples of the effects that changes in the outer magnetosphere associated with magnetic storms can have on the polar ionosphere during a solar-proton event. Obviously it is a field in which further study is required, and in which ionospheric observations can once again make an invaluable contribution through their ability to monitor effects taking place along a single geomagnetic field line that extends into the outer magnetosphere.

Summarizing the effects of the tail on PCA events, then, we have seen that the chief importance of the tail lies in its provision of a relatively easy access route for solar protons entering the atmosphere from outside the magnetosphere. This fact in turn suggests that solar protons can be used as a probe of the detailed structure of the tail and the way in which it is related to the interplanetary magnetic field. Such studies are still in their early stages, and many important and challenging questions remain to be answered, particularly with regard to temporal variations in the structure and extent of the tail as revealed by the distribution of precipitating solar protons over the polar caps.

3. POLAR-CAP IONIZATION BY ENERGETIC ELECTRONS

In addition to the ionization produced by solar protons during PCA events, the polar-cap ionosphere is subject to ionization by energetic electrons, though in a less intense and continuous manner than is the auroral-zone ionosphere. The geomagnetic tail undoubtedly exerts a strong influence on this bombardment also, though the details of the mechanisms involved are even less clear than in the PCA case. The chief difference lies in the dynamic nature of the influence of the tail, in which mechanisms as yet poorly understood are responsible for accelerating electrons, in contrast to the purely passive role of the tail in allowing or denying access to solar protons. In attempting to describe these polar-cap electron-precipitation phenomena, perhaps the single most important picture to bear in mind is that of the polar substorm, as developed by Akasofu (1964) and others. As revealed by all-sky camera recordings of visible aurora, a substorm consists of three distinct phases. (i) During the initial quiet phase, before the beginning of the substorm proper, the aurora consists of a series of quiet homogeneous arcs aligned with the auroral oval in the nighttime sector of the oval, with faint bands roughly aligned with the sun-earth direction extending across the polar cap within the oval. (ii) During the second expansive phase of the substorm, the arcs lying in the auroral oval brighten and move rapidly poleward, while the faint polar-cap bands tend to disappear. (iii) During the third recovery phase the poleward motion stops, and is replaced by a slower motion back towards the auroral oval, ending in a re-establishment of the initial quiet condition.

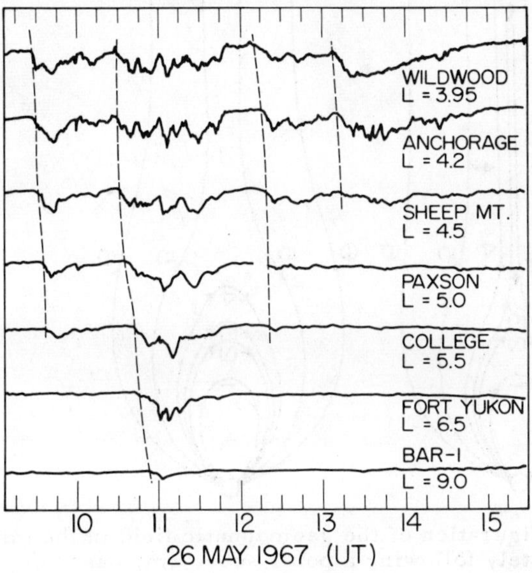

Figure 6. Riometer records of energetic electron precipitation associated with polar substorm activity over Alaska on 26 May 1967.

The expansive phase of the visual auroral substorm has its exact analog in auroral radio-wave absorption, as has been shown, for instance, by Jelly (1968). An example of the poleward motion of substorm auroral absorption is shown in Figure 6, which shows recordings made by the Alaskan riometer chain during the late stages of the magnetic storm of May 25-26, 1967. A series of four distinct events can be seen, each of which had its earliest onset at the lowest latitudes, with a steady progression to higher latitudes. In this particular instance, the worldwide level of magnetic activity was high (K_p about 6), and the most intense activity took place at latitudes considerably lower than that of the auroral zone. Consequently, only one of the events penetrated to the northernmost station at L = 9; more typically, the most intense activity occurs near L = 5, and the northward sweep carries further into the polar cap.

There can be little doubt that the major electron-precipitation events that occur well within the polar cap are a result of the expansion of substorm activity in the direction of the poles. Satellites in the distant geomagnetic tail frequently record rapid-onset bursts, or "islands", of energetic electrons (Anderson, 1965; Bame et al., 1967), and these tend to be accompanied by auroral-zone and polar-cap electron precipitation as revealed by riometers (Parthasarathy and Reid, 1967; Hargreaves et al., 1968) and by magnetometers (Rothwell and Wallington, 1968). Intense "spikes" of energetic-electrons are observed by polar-orbiting satellites over the polar cap (McDiarmid and Burrows, 1965), and they show a good correspondence, at least in a statistical sense, with high-latitude auroral absorption events (Jelly, 1968). It appears that all of these phenomena form part of the overall substorm effect, and that they are all manifestations of a cataclysmic release of magnetic-field energy stored in the tail.

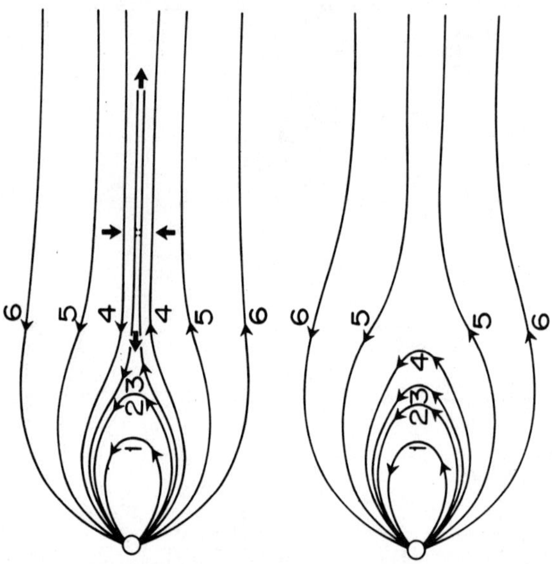

Figure 7. The configuration of the geomagnetic field in the tail during and immediately following a polar substorm, based on the model of Axford (1966). The heavy arrows represent the directions of plasma flow during the substorm, and the numbers indicate consecutive field lines that would form closed loops in the absence of any external field sources.

The details of the substorm mechanism are as yet far from clear, but an intuitively appealing picture has been presented by Axford (1966) (also Atkinson, 1966 and Piddington, 1968), based on the magnetic-field merging mechanism of Sweet (1958) as modified by Petschek (1964). The mechanism is illustrated in Figure 7, which shows two configurations of the magnetic field in the tail, corresponding to the onset and the termination of a substorm. In the upper diagram, representing the onset, some triggering mechanism, as yet unknown, causes open field lines lying close to the neutral sheet (such as that labelled 3 in the diagram) to merge, giving rise to a plasma flow with directions shown by the heavy arrows. The inward flow, carrying with it the freshly closed field line, must give rise to rapid energization of the plasma frozen to the field, through both the Fermi and betatron mechanisms, while the outward flow beyond the merging region will effectively expel plasma from the geomagnetic field. The vertically directed flow toward the neutral sheet carries fresh field lines into the merging region, and the process continues until the flow is choked off, possibly by the electric field set up by longitudinal drift of the trapped energetic plasma in the recently closed flux tubes. At the end of the substorm, represented by the lower diagram, the field lines labelled 3 and 4, which under normal conditions form part of the tail near magnetic midnight, now form closed loops that are probably greatly distended near their outer extension by the diamagnetism of the energetic plasma they contain. The net result of the substorm is that energy has been extracted from the solar wind and turned first into magnetic energy in the tail, and then through field-line merging into plasma energy in the "cusp" region of closed field lines. During the inward flow of the newly reconnected field lines, Fermi acceleration brought about by the shortening of the field lines is likely to result in lowering of mirror-points and consequently in precipitation which will rapidly spread to higher latitudes, accounting for the temporary expansion of the auroral phenomena into the polar cap during the expansive phase of the substorm.

Although the details remain to be studied, this qualitative picture promises to explain most of the phenomena described above, and provides a long-needed mechanism for transforming solar-wind energy into the energy released by auroral and polar-cap electron precipitation and the associated ionospheric currents, and the energy stored in the outer radiation belts and the associated ring current.

The faint quiet-time polar-cap auroras, aligned with the sun-earth direction, remain somewhat mysterious. They presumably represent precipitation linked to some kind of stable structure in the distant tail during quiet conditions, and their disappearance with the onset of a substorm is likely to be related to the onset of the partial collapse of the tail structure as outlined above. Once again, however, the details are not yet understood.

4. CONCLUSION

Summarizing, this paper has attempted to show that the existence of the geomagnetic tail is of major importance with regard to the energetic particle component of the polar ionosphere. The tail operates in two quite distinct ways: (i) its relatively open structure allows energetic particles from outside the magnetosphere, such as solar protons, to enter the ionosphere with little hindrance, and (ii) at the same time the dynamical processes associated with a magnetic structure of this kind are responsible for generating energetic particles, some of which enter the polar ionosphere and create a substantial amount of sporadic ionization. The equally important influence of the tail on the ambient ionospheric plasma in the polar-cap region will be described in the papers to follow.

REFERENCES

Akasofu, S.-I., The development of the auroral substorm, Planetary Space Sci., 12, 273, 1964.

Anderson, K.A., Energetic electron fluxes in the tail of the geomagnetic field, J. Geophys. Res., 70, 4741, 1965.

Atkinson, G., A theory of polar substorms, J. Geophys. Res., 71, 5157, 1966.

Axford, W.I., Magnetic storm effects associated with the tail of the magnetosphere, Proc. ESRO Colloquium, Stockholm, ESRO SP-8, 1966.

Bame, S.J., J.R. Asbridge, H.E. Felthauser, E.W. Hones, and I.B. Strong, Characteristics of the plasma sheet in the earth's magnetotail, J. Geophys. Res., 72, 113, 1967.

Belrose, J.S., L.R. Bode, and L.W. Hewitt, A preliminary investigation of diurnal and seasonal changes in electron number density over Resolute Bay as observed by partial reflections, in Electron Density Profiles in Ionosphere and Exosphere, edited by J. Frihagen, North-Holland Publishing Company, Amsterdam, p. 37, 1966.

Blake, J.B., G.A. Paulikas, and S.C. Freden, Latitude-intensity structure and pitch-angle distributions of low-energy solar cosmic rays at low altitude, J. Geophys. Res., 73, 4927, 1968.

Hakura, Y., Entry of solar cosmic rays into the polar cap atmosphere, J. Geophys. Res., 72, 1461, 1967.

Hargreaves, J.K., E.W. Hones, and S. Singer, Relations between bursts of energetic electrons at 17 earth-radii in the magnetotail and radio absorption events in the ionospheric D region, Planetary Space Sci., 16, 567, 1968.

Hudson, P.D. and H.R. Anderson, Non-uniformity of solar protons over the polar caps on 24 March 1966, Department of Space Science Report, Rice University, Houston, Texas, 1968.

Jelly, D.H., Apparent poleward motion of onsets of auroral absorption events, Can. J. Phys., 46, 33, 1968.

Krimigis, S.M., J.A. Van Allen, and T.P. Armstrong, Simultaneous observations of solar protons inside and outside the magnetosphere, Phys. Rev. Letters, 18, 1204, 1967.

Leinbach, H., Midday recoveries of polar cap absorption, J. Geophys. Res., 72, 5473, 1967.

McCracken, K.G. and N.F. Ness, The collimation of cosmic rays by the interplanetary magnetic field, J. Geophys. Res., 71, 3315, 1966.

McDiarmid, I.B. and J.R. Burrows, Electron fluxes at 1000 kilometers associated with the tail of the magnetosphere, J. Geophys. Res., 70, 3031, 1965.

McDiarmid, I.B. and J.R. Burrows, Local time asymmetries in the high-latitude boundary of the outer radiation zone for the different electron energies, Can. J. Phys., 46, 49, 1968.

Michel, F.C. and A.J. Dessler, Physical significance of inhomogeneities in polar cap absorption events, J. Geophys. Res., 70, 4305, 1965.

Montgomery, M.D. and S. Singer, Penetration of solar energetic protons into the magnetotail, Los Alamos Scientific Laboratory Preprint LA-DC-9271, 1969.

Ortner, J., H. Leinbach, and M. Sugiura, The geomagnetic storm effect on polar cap absorption, Ark. f. Geofys., 3, 429, 1961.

Parthasarathy, R. and G.C. Reid, Magnetospheric activity and its consequences in the auroral zone, Planetary Space Sci., 15, 917, 1967.

Paulikas, G.A., S.C. Freden, and J.B. Blake, Solar proton event of February 1965, J. Geophys. Res., 71, 1795, 1966.

Petschek, H.E., Magnetic field annihilation, in AAS-NASA Symposium on the Physics of Solar Flares, edited by W.N. Hess, NASA SP-50, 425, 1964.

Piddington, J.H., The magnetospheric radiation belt and tail plasma sheet, Planetary Space Sci., 16, 703, 1968.

Reid, G.C. and H. Leinbach, Low-energy cosmic-ray events associated with solar flares, J. Geophys. Res., 64, 1801, 1959.

Reid, G.C. and H.H. Sauer, Evidence for nonuniformity of solar-proton precipitation over the polar caps, J. Geophys. Res., 72, 4383, 1967.

Rothwell, P. and V. Wallington, The polar substorm and electron "islands" in the earth's magnetic tail, Planetary Space Sci., 16, 1441, 1968.

Sweet, P.A., The neutral point theory of solar flares, in Electromagnetic Phenomena in Cosmical Physics, edited by B. Lehnert, Cambridge University Press, p. 123, 1958.

Williams, D.J. and C.O. Bostrom, The February 5, 1965, solar proton event. 2. Low-energy proton observations and their relation to the magnetosphere, J. Geophys. Res., 72, 4497, 1967.

EFFECTS OF LOW ENERGY ELECTRON PRECIPITATION ON THE UPPER ATMOSPHERE

M. H. Rees

Laboratory for Atmospheric and Space Physics
University of Colorado, Boulder, Colorado 80302, USA

Abstract. The physical processes which occur when a primary auroral electron stream penetrates into the atmosphere are examined. A typical soft electron spectrum observed in the polar cap region, and a typical auroral oval electron spectrum are used to derive the associated ionization and secondary electron fluxes. Photon emission rates of several measureable spectral features are computed and it is shown that the optical aurora originates higher in the atmosphere in the polar cap region than it does in the auroral oval.

1. INTRODUCTION

Several papers in these proceedings (O'Brien, Burch, Hoffman) show that the spectral characteristics of precipitated electrons in the region of the auroral oval and in the polar cap are not the same. The electron energy spectra measured over the polar cap are significantly softer than those over the oval, i.e., the former contain a larger fraction of low energy electrons. Differences in temporal characteristics are also observed, precipitation in the polar cap usually being of shorter duration (in the satellite frame of reference) than that observed in the auroral oval. This difference led Hoffman to name the polar cap precipitation "bursts" and the oval precipitation "bands". The observations may reflect the spatial extent of the precipitation rather than a temporal variation. As in most geophysical observations there are exceptions to the average behavior described above. An advantage of satellite measurements is that over a period of months of observation typical patterns generally emerge. We shall use typical electron energy spectra to examine the effect of these bombarding fluxes on the earth's atmosphere.

2. EXPERIMENTAL DATA

The electron energy spectra shown by Hoffman (1969) are adopted. To avoid misleading labels the burst will be called "polar cap" spectrum while the band will be called "auroral oval" spectrum. Four energies were sampled between 0.7 and 25 keV. In reality, the spectra may well extend beyond these experimental cutoffs (as was pointed out by Hoffman). A high energy tail is of little consequence with regard to excitation of optical radiations though it would increase ionization deep in the atmosphere. Phenomena dependent on the electron density, for example cosmic noise absorption, would be affected. Additional flux of electrons with energy less than 0.7 keV would enhance excitation of optical emission features above about 170 km. Spectra of precipitating electrons generally have a peak at some energy but this is a highly variable parameter. It can only be surmised that the peak lies below 0.7 keV in the present case. For lack of information the energy range is arbitrarily limited to the observed values.

The differential number flux may be described, approximately, by a power law

$$N(E)\,dE = N_o\,E^{-\alpha}\,dE \text{ electrons cm}^{-2}\text{ sec}^{-1}\text{ sr}^{-1}\text{ keV}^{-1} \qquad (1)$$

$$0.7 \le E \le 25 \text{ keV}$$

where E is in keV. For the polar cap spectrum $N_o = 10^9$ and $\alpha = 3.0$ while the auroral oval spectrum is satisfied by $N_o = 1.3 \cdot 10^7$ and $\alpha = 1.25$. The omnidirectional flux is

$$\mathcal{F} = \pi N_o \left(\frac{E_{max}^{1-\alpha} - E_{min}^{1-\alpha}}{1-\alpha}\right) \text{ electrons cm}^{-2}\text{ sec}^{-1} \qquad (2)$$

the quantity in parenthesis is the normalization factor.

3. IONIZATION RATE

The principal initial energy loss process involves ionization of the neutral atmosphere; in the energy range considered radiative losses are negligible. For electrons with initial energy greater than 100 eV approximately $\Delta E = 35$ eV are lost for each ion pair formed, independent of the molecular species. To a good approximation the normalized energy dissipation distribution function $\lambda(z/R)$, where z/R is the fractional atmospheric depth of penetration, is independent of the initial electron energy, E. The number of ion pairs produced at any altitude is (Rees, 1963)

$$q(e) = \mathcal{F}\,\frac{1-\alpha}{E_{min}^{1-\alpha} - E_{max}^{1-\alpha}} \cdot \int_{E_{min}}^{E_{max}} \frac{E^{1-\alpha}}{r_o\,\Delta E} \cdot \lambda(z/R) \cdot$$

$$\frac{(O_2)_z + 0.92\,(N_2)_z + 0.56\,(O)_z}{(O_2)_R + 0.92\,(N_2)_R + 0.56\,(O)_R}\,dE \text{ cm}^{-3}\text{ sec}^{-1} \qquad (3)$$

The range for electrons with energy E is r_o (in atm.-cm), while $(X)_z$ and $(X)_R$ are the number densities at atmospheric depth z and R (gm-cm^{-2}) respectively. The coefficients of the (X)'s represent a small correction which takes account of differences in ionization efficiency of the atmospheric constituents. The formulation does not explicitly consider the constraining force of the magnetic field. It has been argued (Rees, 1963), however, that for an extended zone of precipitation the result is unaffected as long as the lines of force are

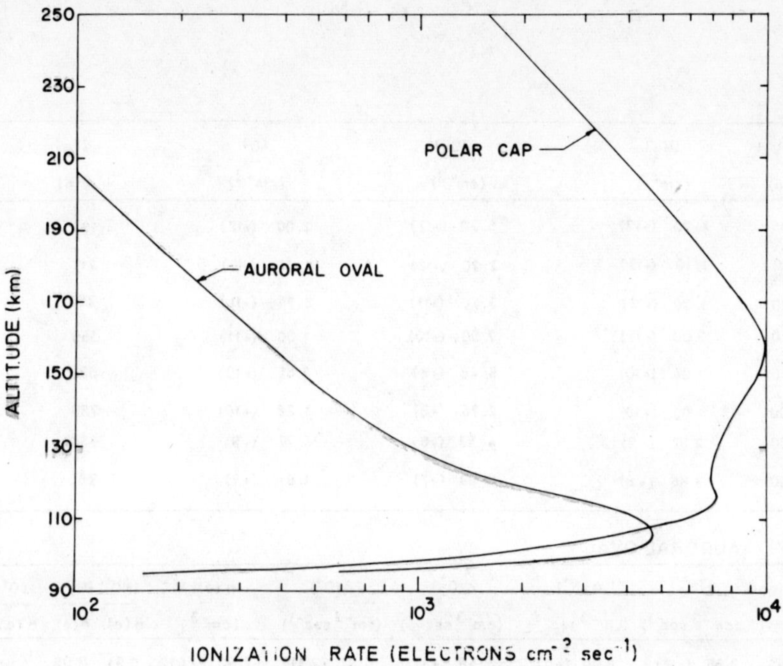

Figure 1. Altitude profiles of the ionization rate corresponding to the auroral oval and polar cap electron precipitation events.

close to vertical which is the case in the polar cap and in the auroral oval ionosphere. The quantity $\lambda(z/R)$ is a function of the angular distribution of the electrons. Hoffman had several detectors which sampled a range of pitch angles at one energy on the same satellite. The majority of the observations gave an isotropic pitch angle distribution over the upward hemisphere, though occasionally interesting anisotropies were recorded. In the present computations an energy dissipation distribution corresponding to an isotropic angular distribution is adopted. The ionization rate profiles corresponding to the polar cap and auroral oval spectra are shown in Figure 1. It is immediately apparent that energy deposition peaks at a greater altitude in the polar cap than in the oval, a direct consequence of the softer spectrum. The auroral oval flux is small, and could be associated with the almost permanent weak precipitation which defines the oval (in contrast to the small scale, short lived, intense precipitation).

The production rates of individual ion species, particularly those which take part in chemical reactions, are of interest. Ionization and dissociative ionization cross-sections compiled by Kieffer and Dunn (1966) are used, and the relevant formulae for N_2^+, N^+, O_2^+ and O^+ are given by Walker and Rees (1968). The rate at which these ions are formed, q(X), is shown in Table 1. The rate of formation of N_2^+ ions in the excited $B^2\Sigma$ state is 0.11 to 0.13 of the total N_2^+ production (Aarts et al., 1968) and the resulting 0-0 band emission at 3914 Å amounts to about one half of the total emission rate from this state to the ground state of the ion. This straightforward relationship provides a convenient method to deduce auroral ionization from N_2^+ photon emission rate.

A fraction of the ions is formed in metastable states. For example, the rate of formation of $O^+(^2P)$ ions is (Walker and Rees, 1968)

Height (km)	$[N_2]$ (cm^{-3})	$[O_2]$ (cm^{-3})	[O] (cm^{-3})	$[T_g]$ (°K)
95	2.20 (+13)	5.30 (+12)	2.00 (+12)	193
100	1.10 (+13)	2.20 (+12)	1.40 (+12)	217
110	1.70 (+12)	3.70 (+11)	2.75 (+11)	282
120	2.00 (+11)	7.50 (+10)	1.00 (+11)	350
140	3.04 (+10)	9.48 (+9)	2.65 (+10)	630
160	1.00 (+10)	2.76 (+9)	1.28 (+10)	789
200	2.02 (+9)	4.53 (+8)	4.77 (+9)	930
250	3.86 (+8)	6.89 (+7)	1.81 (+9)	982

AURORAL OVAL

Height (km)	$q(N_2^+)$ ($cm^{-3} sec^{-1}$)	$q(N^+)$ ($cm^{-3} sec^{-1}$)	$q(O_2^+)$ ($cm^{-3} sec^{-1}$)	$q(O^+)$ ($cm^{-3} sec^{-1}$)	n(e) (cm^{-3})	$[NO^+]/n(e)$	$[O_2^+]/n(e)$	$[O^+]/n(e)$
95	3.56 (+2)	8.90 (+1)	8.58 (+1)	6.31 (+1)	2.90 (+4)	0.91	0.09	
100	2.20 (+3)	5.50 (+2)	4.40 (+2)	3.95 (+2)	7.75 (+4)	0.80	0.20	
110	2.54 (+3)	6.34 (+2)	5.52 (+2)	5.32 (+2)	1.01 (+5)	0.71	0.29	
120	8.41 (+2)	2.10 (+2)	3.15 (+2)	4.20 (+2)	7.50 (+4)	0.72	0.28	
140	2.74 (+2)	6.84 (+1)	8.53 (+1)	1.92 (+2)	5.57 (+4)	0.80	0.19	0.01
160	1.40 (+2)	3.49 (+1)	3.84 (+1)	1.30 (+2)	5.36 (+4)	0.80	0.16	0.03
200	3.86 (+1)	9.65 (+0)	8.65 (+0)	6.12 (+1)	4.82 (+4)	0.74	0.15	0.10
250								

POLAR CAP

Height (km)	$q(N_2^+)$ ($cm^{-3} sec^{-1}$)	$q(N^+)$ ($cm^{-3} sec^{-1}$)	$q(O_2^+)$ ($cm^{-3} sec^{-1}$)	$q(O^+)$ ($cm^{-3} sec^{-1}$)	n(e) (cm^{-3})	$[NO^+]/n(e)$	$[O_2^+]/n(e)$	$[O^+]/n(e)$
95	9.49 (+1)	2.37 (+1)	2.29 (+1)	1.68 (+1)	1.48 (+4)	0.95	0.05	
100	9.63 (+2)	2.41 (+2)	1.93 (+2)	1.73 (+2)	5.04 (+4)	0.85	0.15	
110	3.43 (+3)	8.58 (+2)	7.47 (+2)	7.20 (+2)	1.18 (+5)	0.69	0.31	
120	3.32 (+3)	8.29 (+2)	1.24 (+3)	1.66 (+3)	1.54 (+5)	0.61	0.38	0.01
140	3.69 (+3)	9.22 (+2)	1.15 (+3)	2.59 (+3)	2.16 (+5)	0.64	0.31	0.05
160	4.00 (+3)	1.00 (+3)	1.10 (+3)	3.74 (+3)	3.18 (+5)	0.57	0.27	0.16
200	1.56 (+3)	3.90 (+2)	3.50 (+2)	2.47 (+3)	4.14 (+5)	0.34	0.16	0.50
250	3.52 (+2)	8.80 (+1)	6.28 (+1)	1.06 (+3)	6.39 (+5)	0.05	0.03	0.92

Table 1

$$q(O^{+}\ ^{2}P) = 0.2\ \left(q(O^{+}) - 0.5\ q(O_2^{+})\right) \tag{4}$$

With a 5 second lifetime for the excited state most atoms will radiate at the easily observable wavelengths 7319 and 7330 Å.

4. ION CHEMISTRY

Below about 200 km effects of ion diffusion on the electron density profile are small, and chemical equilibrium may be assumed. Positive atomic ions disappear by ion atom interchange and charge transfer, followed by dissociative recombination. The rate coefficient for ion atom interchange between O^+ and N_2 increases rapidly with vibrational development in N_2, while dissociative recombination rates depend on the electron temperature. The reactions, with their appropriate rate coefficients are described by Walker and Rees (1968). Rate coefficients involving reactions with metastable species may be different from reactions with ground state species; there is an unfortunate lack of relevant cross-sections. The problem is discussed in the above cited reference. The ion composition and electron density profiles are computed corresponding to steady state chemical equilibrium. It was noted earlier that the satellite measurements cannot distinguish between temporal and spatial variations (Recent measurements from an airborne platform suggest that the variations are spatial in origin.) and it seemed unwarranted to attempt solving the time dependent coupled equations since the duration of the precipitation events is unknown. Suffice it to note that the lifetime of some of the molecular ions may be tens of seconds. The resulting electron density height profile and the relative ion composition are listed in Table 1, together with the atmospheric composition and temperature. The same model atmosphere has been used consistently in all the computations presented in this paper.

Absorption of cosmic noise may be computed from the electron density and the collision frequency height profiles (Thrane and Piggott, 1966). The calculations predict absorption of about 0.1 db for the auroral oval precipitation and 0.06 db for the polar cap event (referred to 30 MHz signal). The polar cap precipitation involved an energy flux eight times larger than the auroral oval event but ionization occurred at an altitude where the collision frequency is too small to produce any appreciable absorption. Caution is therefore suggested in the use of absorption data to deduce particle precipitation characteristics.

5. SECONDARY ELECTRON SPECTRUM

Excitation of electronic levels in the neutral gas from which observable emission lines and bands originate is produced principally by secondary electrons. Using 500 eV primary electrons Silverman and Lassettre (1964; 1965) have measured the cross-section for the production of secondary electrons of a given energy in N_2 and O_2. The cross-section may be related to the differential oscillator strength using the Bethe-Born approximation and this formulation is used to extend the experimental results to a wider energy range of primary electrons. Theoretical computations for photoionization of O are used to derive the electron impact differential oscillator strength following the work of Seaton (1959) relating the two processes (Takayanagi and Takahashi, 1966).

The production rate of secondary electrons with energy E_s is (Rees et al., 1969)

$$P(E_s) = \frac{F(N_2)}{2.53}\left(q(N_2^+) + q(N^+)\right) + \frac{F(O_2)}{2.77}\,1.5\,q(O_2^+)$$

$$+ \frac{F(O)}{2.71}\left(q(O^+) - 0.5\,q(O_2^+)\right) \qquad (5)$$

where the F's are shape factors appropriate to each species, the denominators are normalizing factors, and the q's are the ion production rates listed in Table 1. In this formulation the shape factor is independent of the primary electron energy, E_p, an approximation valid in the high energy limit and when $E_s \ll E_p$. In practice, it is sufficient to limit the calculations to secondary electrons with energy less than 100 eV. A small fraction of the secondary electrons have sufficient energy to produce further ionization amounting to an additional 15%. This is neglected. Initial energy spectra of secondary electrons at the height of maximum ionization are shown in Figure 2 for the oval and polar cap spectra. The peak is at about 4 to 5 eV.

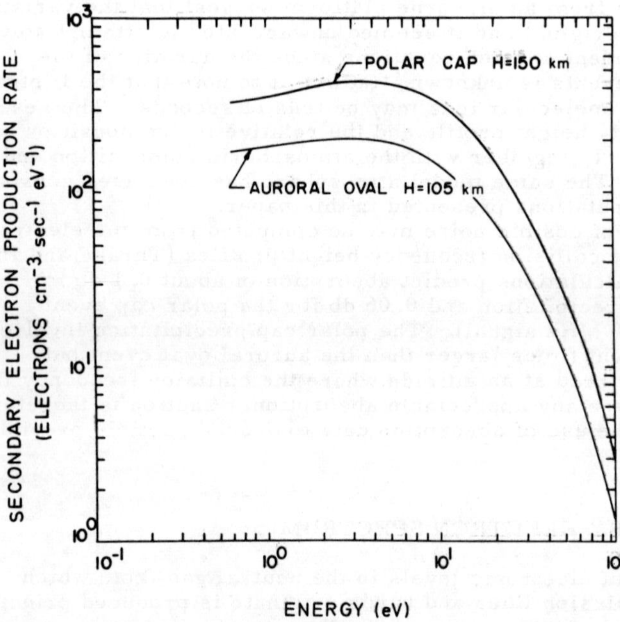

Figure 2. Differential production rates of secondary electrons at the altitude of maximum ionization for auroral oval and polar cap auroras.

Detectors measures the electron flux, not the production rate. The differential flux of secondary electrons at any altitude is

$$\phi(E_s) = \frac{\int_{E_s}^{\infty} P(E'_s)\, dE'_s}{\frac{dE}{dX}\Big|_{elect.} + \frac{dE}{dX}\Big|_{neut.}} \tag{6}$$

where dE/dX is the stopping power. For electrons

$$\frac{dE}{dX}\Big|_{elect.} = \frac{1.95 \cdot 10^{-12}}{E}\, n(e) \tag{7}$$

and for energy loss by collisions with the neutral gas

$$\frac{dE}{dX}\Big|_{neut.} = \sum_j n_j W_j \sigma_j (E) \tag{8}$$

in the continuous slowing down approximation (Stewart, 1969). n_j is the number density of species j and $W_j\, \sigma_j(E)$ is the energy dependent stopping cross-section, where W_j is the threshold energy for the inelastic collision (Dalgarno et al., 1969).

Secondary electron fluxes at 150 km are shown in Figures 3a and b, corresponding to the polar cap and auroral oval cases respectively. The spectral shapes are essentially the same, reflecting only minor differences in the relative ion production rates. The magnitude, of course, is a function of the primary flux. Also shown are the primary spectra, between 0.7 and 25 keV, at 150 km, and at 893 km, the altitude of the satellite at the time the measurements were made. Energy degradation occurs, particularly in the soft component (Rees, 1969). The primary number flux is always considerably less than the secondary flux.

The results of several measurements of secondary electron fluxes have appeared in the literature. Results to 1967 have been presented in a summary figure by Pfister (1967). The energy range between 1 eV and 10 keV is presented. Recent high resolution spectra extending from 45 eV to 100 keV have been obtained by Westerlund (1969). Calculations predict a steeper secondary spectrum above 30 eV than has been observed. In the energy range between about 50 and 300 eV the observations very likely represent a mixture of the primary and secondary fluxes. It is disturbing to note the absence of appreciable structure in the observed spectra below 15 eV. Such structure derives from the large stopping cross-section corresponding to excitation of vibrational levels in nitrogen and several electronic transitions in nitrogen and oxygen.

6. EXCITATION OF OPTICAL EMISSION FEATURES

The production of excited neutral species is largely by secondary electrons. At any altitude it is given by

$$\eta(E_s) = n_j \int_{E_t}^{\infty} \phi(E_s)\, \sigma_j(E)\, dE \tag{9}$$

where $\sigma_j(E)$ is the excitation cross-section and E_t the threshold for the process.

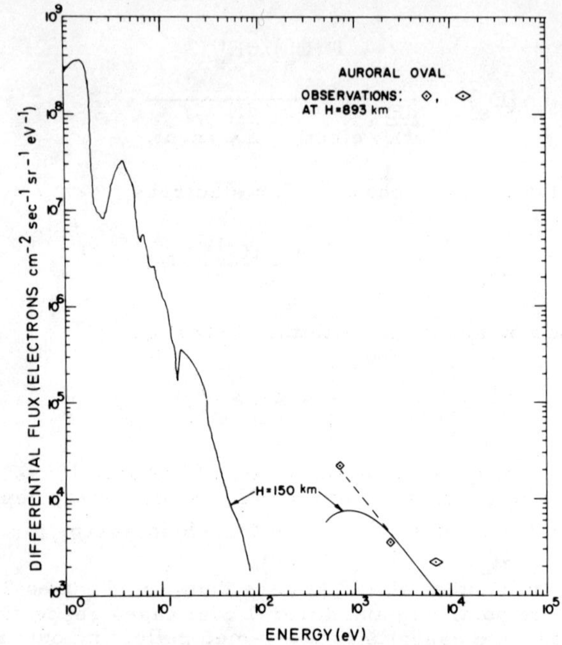

Figure 3a. Differential flux of secondary electrons at 150 km for auroral oval auroras. The primary electron spectra as measured at 893 km by Hoffman, and calculated at 150 km (Rees, 1969) are also shown.

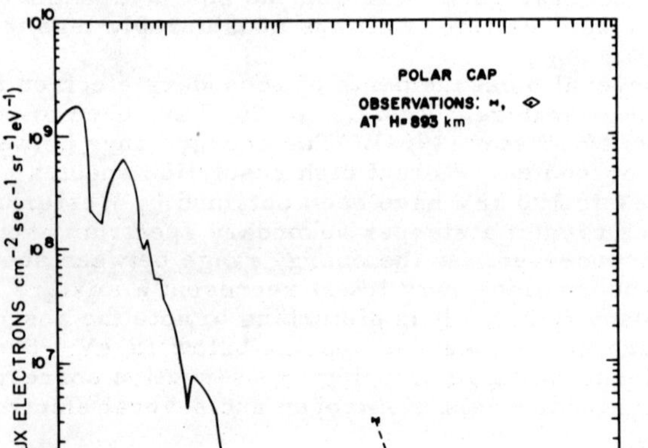

Figure 3b. Differential flux of secondary electrons at 150 km for polar cap auroras. The primary electron spectra as measured at 893 km by Hoffman, and calculated at 150 km (Rees, 1969) are also shown.

AURORAL OVAL				
Altitude (km)	$\eta\,(O'S)^{x}_{el}$ ($cm^{-3}\,sec^{-1}$)	$\eta\,(O'S)^{xx}_{D.R.}$ ($cm^{-3}\,sec^{-1}$)	$\eta\,(O'D)_{el}$ ($cm^{-3}\,sec^{-1}$)	$\eta\,(O'D)_{D.R.}$ ($cm^{-3}\,sec^{-1}$)
95	3.24 (+0)	1.24 (+1)	3.10 (+2)	1.24 (+2)
100	2.83 (+1)	7.8 (+1)	2.10 (+3)	7.8 (+2)
110	4.71 (+1)	9.35 (+1)	2.85 (+3)	9.35 (+2)
120	2.87 (+1)	3.66 (+1)	1.44 (+3)	3.66 (+2)
140	1.46 (+1)	4.98 (+0)	5.97 (+2)	4.98 (+1)
160	1.02 (+1)	2.10 (+0)	3.73 (+2)	2.10 (+1)
200	4.94 (+0)		1.52 (+2)	
250				

POLAR CAP				
	$\eta\,(O'S)_{el}$ ($cm^{-3}\,sec^{-1}$)	$\eta\,(O'S)_{D.R.}$ ($cm^{-3}\,sec^{-1}$)	$\eta\,(O'D)_{el}$ ($cm^{-3}\,sec^{-1}$)	$\eta\,(O'D)_{D.R.}$ ($cm^{-3}\,sec^{-1}$)
95	8.63 (-1)	2.87 (-1)	8.28 (+1)	2.87 (+0)
100	1.24 (+1)	3.20 (+1)	9.20 (+2)	3.20 (+2)
110	6.38 (+1)	1.29 (+2)	3.86 (+3)	1.29 (+3)
120	1.13 (+2)	1.51 (+2)	5.66 (+3)	1.51 (+3)
140	1.95 (+2)	1.17 (+2)	7.76 (+3)	1.17 (+3)
160	2.86 (+2)	1.24 (+2)	9.50 (+3)	1.24 (+3)
200	1.80 (+2)	6.10 (+1)	4.36 (+3)	6.10 (+2)
250	5.89 (+1)	2.46 (+1)	9.95 (+2)	2.46 (+2)

x $\eta(O'S)_{el}$ ≡ Excitation rate of O('S) atoms by electron impact

xx $\eta(O'S)_{D.R.}$ ≡ Excitation rate of O('S) atoms by dissociative recombination

Table 2

There is still a dearth of available data on many of the cross-sections for excitation of states leading to observable emission features in the auroral spectrum (and the dayglow). For illustrative purpose the production rates of O('S) and O('D) have been computed using the cross-sections given by Henry et al. (1969). These are summarized in Table 2. The rate of population of the $O(2\,p^3 3s^5S)$ level which leads to the 1356 Å line observed with rockets and satellite borne instruments is computed using cross-sections given by Dalgarno et al (1969).

Some excited species are produced by dissociative recombination of molecular ions. The concentrations of the various ions must therefore be known if this source of excitation is to be evaluated. For the present

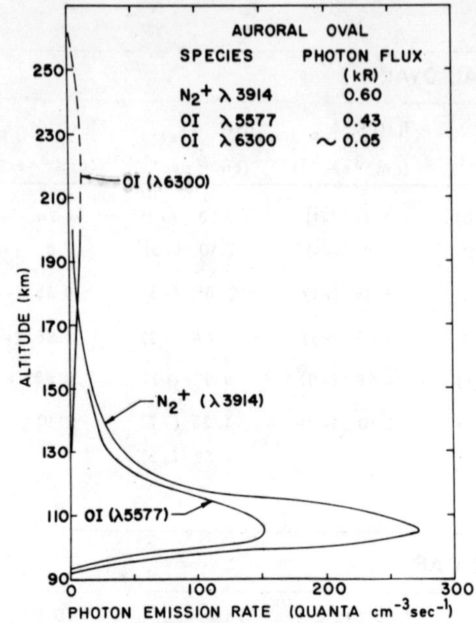

Figure 4. Altitude profiles of photon emission rates of some radiations resulting from the auroral oval precipitation. Integrated fluxes are also shown.

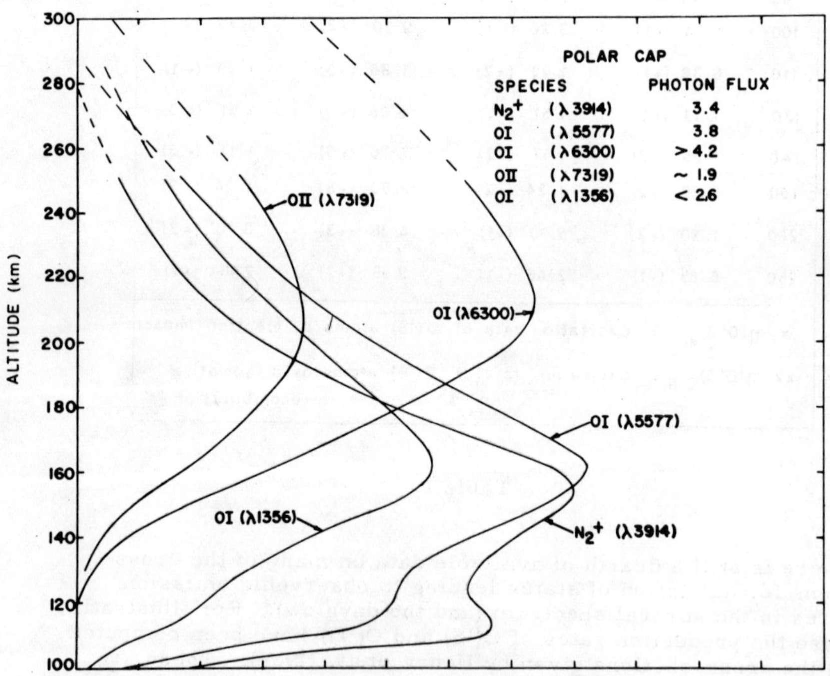

Figure 5. Altitude profiles of photon emission rates of some radiations resulting from polar cap precipitation. Integral fluxes are also shown.

purpose only dissociative recombination of O_2^+ need be considered because a fraction of these recombinations leads to the excited atomic states 'D and 'S. The fraction adopted is 0.5 for 'D and 0.05 for 'S, based on a recent multi-experiment rocket flight by P. Hays (private communication). Ionospheric and laboratory data now appear to be in reasonable accord (Zipf, private communication). The production rates of O'S and O'D atoms for the oval and polar cap precipitation are given in Table 2. If the O_2^+ concentration is always as low as recent rocket borne ion mass spectrometer measurements indicate (Donahue, private communication) the contribution to the production of O'S from dissociative recombination may differ from the present computations. It may be less or more since the rate coefficient for O'S production may be larger than has been assumed here. Theoretical computations of the O'S production rate from observed ionization rates fall short of predicting the observed photon emission rate (Rees et al., 1969). It has recently been suggested by Donahue (private communication) that dissociation of O_2 by electron impact may be a major source of O'S atoms. The required cross-section is large and there are no relevant laboratory measurements available. It can only be stated at this time that the problem of the excitation of the auroral green line, the most prominent emission features in the visible part of the spectrum, has not been completely solved.

The O('D) level is populated by two processes in addition to the ones described above. Cascading from the 'S level provides a small contribution, and another source, important at high altitude, is excitation by electrons in the high energy tail of the hot electron plasma. The rate at which ambient electrons are heated by the secondary electrons at any altitude is

$$H(E_s) = \int_0^{\infty} \left|\frac{dE}{dX}\right|_{electron} \phi(E_s)\, dE_s \qquad (10)$$

The electron temperature is obtained by solving the heat conduction equation for the electron gas, including local sources and sinks (Rees et al., 1967; Walker and Rees, 1968). Neither cascading nor thermal excitation have been included here.

The O'D level is quenched at low altitude, principally by N_2. A rate coefficient of $5 \cdot 10^{-11}$ cm^3 sec^{-1} is adopted (Noxon, 1969). The OI (1356 Å) radiation is absorbed by O_2, the opacity reaching unity at about 120 km; the correction to the zenith intensity will be small above 130 km.

Altitude profiles of photon emission rates attributable to electron precipitation are summarized in Figures 4 and 5. The auroral oval electron flux was small, representing an energy deposition rate of 0.9 ergs cm^{-2} sec^{-1} with correspondingly weak radiation. The 6300 Å radiation is only a crude estimate. The integrated column emission rates are also listed. The polar cap precipitation amounted to some 7.0 ergs cm^{-2} sec^{-1} which would produce well observable radiations. There is a small additional source of 7319/7330 due to electron impact excitation of ground state O^+ ions. The polar cap aurora peaks at an altitude above the average aurora, a direct result of the soft electron spectrum.

7. DISCUSSION

Polar cap optical emissions are described in detail by Sandford in these proceedings. It suffices to compare the results of the calculations presented here with recent observations obtained by Eather (1969) from

Ratio	AURORAL OVAL (predicted)	Λ 60 - 75° (Eather, 1969)	POLAR CAP (predicted)	Λ 75-85° (Eather, 1969)
5577Å / 3914Å	0.7	0.85 - 1.1	1.1	0.65-1.0
6300Å / 3914Å	~0.08	0.16 - 0.32	>1.2	1.3 - 1.9

Table 3

an aircraft which flew many sorties in the latitude range 60° to 90° geomagnetic during January and February 1968. Eather gives intensity ratios of N_2^+ (4278 Å), OI (6300 Å) and OI (5577 Å). These results are converted to the 0-0 band emission of N_2^+ and grouped into latitude intervals corresponding roughly to the auroral oval and the polar cap. They are summarized in Table 3 together with the ratios which result from the present calculations. Of course, theory and observations do not correspond to the same aurora. Calculations were made on "typical" electron precipitation events while observations refer to the average of many latitude surveys. The computed auroral oval 5577 Å flux does not include a contribution from the airglow, which would increase the ratio 5577/3914 from 0.7 to at least 0.9. The polar cap 6300 Å intensity represents a lower limit because thermal excitation and cascading are not included in the computations. The ratio 6300/3914 should therefore be about 25% larger, or about 1.5 instead of 1.2. As indicated earlier, the auroral oval 6300 Å radiation is only a best guess.

Particle precipitation and optical observations in the polar cap ionosphere can be rationalized in terms of the physical processes which occur when the particles interact with the atmosphere. It does not appear necessary to invoke additional forces due to electric fields or winds to explain the gross and slowly varying features of the ionization and the optical radiation which are observed.

Acknowledgements. The research has been supported in part by the Atmospheric Sciences Section, National Science Foundation (GA 1238), and by the National Aeronautics and Space Administration (NGR 06-003-052).

REFERENCES

Aarts, J.F.M., F.J. deHeer, and D.A. Vroom, Emission cross-sections of the first negative band system of N_2 by electron impact, Physica, 40, 197, 1968.

Dalgarno, A., M.B. McElroy, and A.I. Stewart, Electron impact excitation of the dayglow, J. Atmospheric Sci., 1969.

Eather, R.H., Latitudinal distribution of auroral and airglow emissions: the soft auroral zone, J. Geophys. Res., 74, 153, 1969.

Henry, R. J. W., P. G. Burke, and A. L. Sinsailan, Scattering of electrons by C, N, O, N^+, O^+, and O^{++}, Phys. Rev., 178, 218, 1969.

Hoffman, R. A., Low energy electron precipitation at high latitudes, J. Geophys. Res., 74, 2425, 1969.

Kieffer, L. J. and G. M. Duun, Electron impact ionization cross-section data for atoms, atomic ions and diatomic molecules: I Experimental data, Rev. Modern Phys., 38, 1, 1966.

Noxon, J. F., to be published, 1969.

Pfister, W., Auroral investigations by means of rockets, Space Sci. Rev., 7, 642, 1967.

Rees, M. H., Auroral ionization and excitation by incident energetic electrons, Planetary Space Sci., 11, 1209, 1963.

Rees, M. H., J. C. G. Walker, and A. Dalgarno, Auroral excitation of the forbidden lines of atomic oxygen, Planetary Space Sci., 15, 1097, 1967.

Rees, M. H., Auroral electrons, Space Sci. Rev., In press, 1969.

Rees, M. H., A. I. Stewart, and J. C. G. Walker, Secondary electrons in aurora, Planetary Space Sci., 1969.

Seaton, M. H., Electron impact ionization of Ne, O, and N, Phys. Rev., 113, 814, 1959.

Silverman, S. M. and E. N. Lassettre, Collision cross-sections for oxygen in the excitation energy range 10 to 80 eV, J. Chem. Phys., 40, 2922, 1964.

Silverman, S. M. and E. N. Lassettre, Generalized oscillator strengths and electronic collision cross-sections for nitrogen at energies above 10 eV, J. Chem. Phys., 42, 3420, 1965.

Stewart, A. I., In preparation, 1969.

Takayanagi, K. and T. Takahashi, Behaviour of slow electrons in atmospheric gases, Part III: Collision cross-sections of molecular nitrogen, Rep. Ionos. Space Res. Japan, 20, 357, 1966.

Thrane, E. V. and W. R. Piggott, The collision frequency in the E and D regions of the ionosphere, J. Atmospheric Terrest. Phys., 28, 721, 1966.

Walker, J. C. G. and M. H. Rees, Electron densities and temperatures in aurora, Planetary Space Sci., 16, 459, 1968.

Westerlund, L. H., The auroral electron energy spectrum extended to 45 eV, J. Geophys. Res., 74, 341, 1969.

LOW FREQUENCY NOISE EMISSIONS AND THEIR SIGNIFICANCE FOR ENERGETIC PARTICLE PROCESSES IN THE POLAR IONOSPHERE

T. R. Hartz

National Radio Propagation Laboratory
Communications Research Centre, Ottawa, Canada

Abstract. Studies have been made at 200 kHz of broad-band whistler-mode radio noise emissions that are observed at high latitudes with the Alouette II satellite. Average intensities are portrayed in the form of a polar contour diagram that denotes the principal emission regions. These are found to occur at about noon near 78° invariant, with extensions toward the morning and evening hours at somewhat reduced intensity and latitude. The seeming correspondence with the daytime portion of the auroral oval would indicate that the source of the emissions logically can be linked only with the influx of energetic charged particles, and that the noise is generated by the Cerenkov process from impinging electrons with energies of the order of 0.1 to 1.0 keV. This interpretation is discussed with relation to ionization production in the high latitude ionosphere.

1. INTRODUCTION

The Alouette II satellite includes two experiments to measure the background radio noise level within and above the ionosphere (Nelms et al., 1966). This spacecraft has been operating for some 5 or 6 hours each day since its launch in November, 1965, and a detailed description of its equipment and operations can be obtained elsewhere (Franklin and Maclean, 1969). One of the most fruitful experiments on board is the topside ionosonde, which is described in these proceedings by Nelms and Chapman. The automatic-gain-control (AGC) voltage from the ionosonde receiver provides a continuous measure of the radio noise and interference environment for the satellite in the frequency range from about 0.1 to 15 MHz, and these data have been studied at some length (Hartz, 1969). Of particular interest are the various characteristic noise emissions that originate within the ionosphere or magnetosphere and from which information can be obtained on ionospheric processes associated with the influx of energetic particles. This paper will summarize results on one of these noise bands: namely, on the emissions that are observed below the local electron cyclotron frequency, f_H, and the local plasma frequency, f_N, at the satellite.

Also on board this same satellite is a broad-band VLF receiver covering the frequency range from about 50 Hz to 30 kHz (Barrington, 1969). Data are available from this experiment, both on the spectral characteristics of the observed emissions and on their intensity levels integrated over the receiver bandwidth. Some of the relevant VLF statistical results have been included here to help characterize the noise observations in the frequency range below f_H.

The Alouette II orbit has an inclination of about 80°, with approximate perigee and apogee heights of 500 and 3000 km. The orbit precession is such that a 24-hour diurnal sampling of the data can be obtained from a 3-month operational period, but the height variation cannot be obtained independent of the diurnal variation. It was found, however, that no obvious height dependence appeared in the first analysis of the data, and the possibility of a height variation has been ignored in the subsequent analysis.

2. OBSERVATIONS OF NOISE EMISSIONS BELOW f_H

Anomalously high noise levels in this frequency range were first reported by Walsh et al. (1964) on the basis of a rocket-borne experiment. Similar results have been obtained consistently with the Alouette II satellite, though with systematic variations in the intensity and spectrum of the emissions. Figure 1 shows sample recordings of this type of noise, each made during the course of a 30-second sweep of the receiver frequency from about 0.1 to 15 MHz.

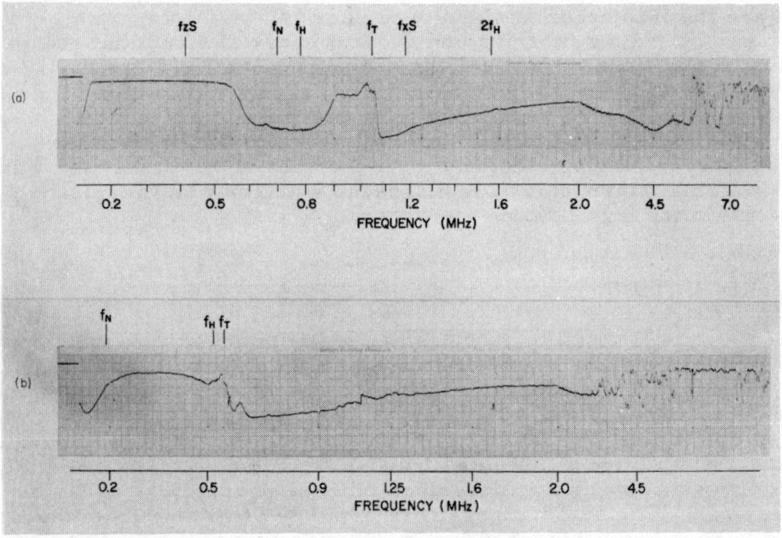

Figure 1. Sample recordings of the AGC voltage from the sweep-frequency receiver on the Alouette II satellite. The noise and interference level appears as a function of frequency, but the intensity scale, which covers some 80 dB, is frequency dependent and cannot be represented conveniently here. Certain characteristic frequencies of the local ionosphere have been identified and marked on each record. The whistler-mode noise band of interest appears below f_H and f_N, and Frame (a) shows this noise type at 0.2 MHz. Frame (b) shows a case where the noise at 0.2 MHz is in a different frequency regime, namely between f_N and f_T, though the portion between 0.1 and 0.2 MHz shows whistler-mode emissions.

Here only part of each record is shown as the higher frequencies are contaminated by interference from ground transmissions. In each case the characteristic frequencies of the local ionosphere are indicated, and the pertinent features to note are the noise bands that appear at the lowest frequencies. In Figure 1(a) this noise is continuous from the receiver limit at approximately 0.1 MHz up to about 0.6 MHz, which is somewhat less than f_H. In Figure 1(b) the lowest frequency components, that is, frequencies less than both f_N and f_H, are much reduced in amplitude; in fact, in this case f_N is below f_H and the emissions at and above 200 kHz must be considered to belong to a different noise type (Hartz, 1969; Walsh et al., 1964) which occurs between f_N and f_T. In this paper only those emissions that appear below f_H and f_N will be considered: these are typified in Figure 1(a) and, as indicated by the general smoothness of the trace, are quite broadband in nature. This noise type showed no obvious dependence on the dipole antenna orientation owing to the spin of the satellite.

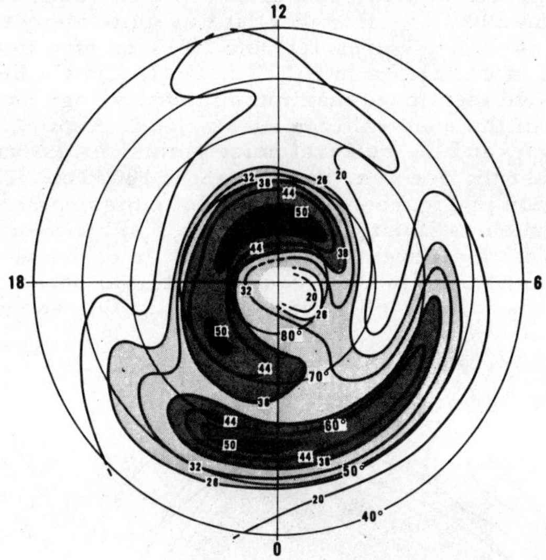

Figure 2. Contour diagram of the average whistler-mode noise intensity at 200 kHz as a function of invariant latitude and local mean time. The contours are labelled in decibel values relative to the receiver threshold.

A preliminary statistical study of the noise at about 200 kHz in recordings obtained at the Ottawa telemetry station showed that maximum intensities occurred in two latitude belts, with peaks at about 55° and 75° invariant (Hartz, 1969). The apparent lack of correlation with the classical auroral zone maximum near 67° geomagnetic prompted a suggested explanation of Doppler-shifted cyclotron radiation from an upward electron flux, rather than a process involving an influx of energetic auroral particles. A more complete analysis, in which the diurnal as well as the latitudinal variation was studied, has now been carried out and the results, as portrayed in Figure 2, cast doubt on the initial explanation. The data included are from two 6-month periods - January to July of both 1966 and 1967 - and represent the relative noise intensity in decibels at a frequency of 200 kHz. These results were obtained at three telemetry stations in the Northern Hemisphere - Resolute Bay, Ottawa, and Ft. Myers - in such a way that an approximately uniform sampling in time and latitude resulted. An average intensity for each 5-degree latitude and 1-hour time sector was then obtained and contour lines were drawn as a function of invariant latitude (= arc cos $L^{-\frac{1}{2}}$) and local mean time. (Note that for the stations involved local mean time does not differ very much from geomagnetic time, except at the highest latitudes.)

Two maximum intensity regions can be seen in Figure 2. The first has a dominant peak at mid-day in the latitude range 75 - 80° and, at a slightly reduced level, the contours extend around into the evening hours with a gradual shift to lower latitudes. The second has peak values between 55° and 60° and extends through most of the night hours, though a pre-midnight maximum is indicated. A significant minimum exists over the pole, though the data sampling was not quite adequate to define it completely at the lowest intensities. Minimum values, though usually non-zero, are also observed at the lower latitudes.

A single frequency of 200 kHz was chosen for convenient data analysis, regardless of the particular f_N values, since the range of f_H values in different parts of the space-craft orbit was such as to be always greater than 200 kHz. However, it was quite apparent that spectral differences - cf. Figures 1(a) and 1(b) - existed in the data, primarily because of variations in f_N. The first, and the higher latitude, of the above mentioned maximum intensity regions in Figure 2 consists of the spectral type illustrated in Figure 1(a). Here f_N is greater than f_H and the relevant noise emissions extend down in frequency to at least the receiver limit at about 100 kHz. The second peak intensity region in Figure 2 is made up of emissions characteristic of low f_N values, such as illustrated in Figure 1(b) between f_N and f_T, and which represents a different type of noise. In cases where the noise characteristics can be seen in the frequency interval between the receiver limit and f_N, the level is typically low in this second region.

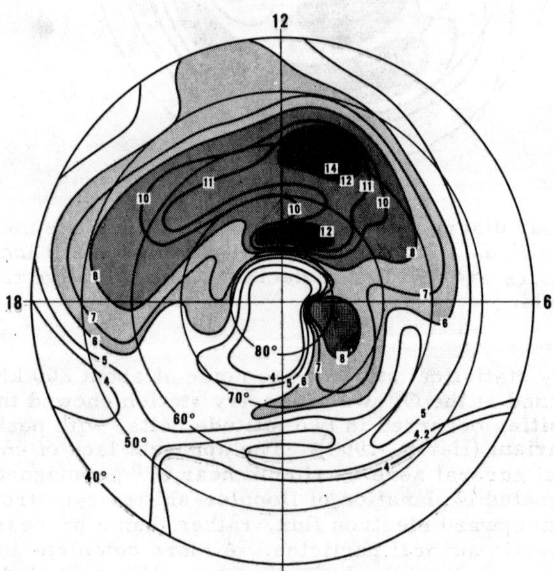

Figure 3. Contour diagram of the average VLF noise intensity as a function of invariant latitude and local mean time. The intensity levels are given in decibels relative to the receiver threshold.

This spectral differentiation between the two peak-intensity regions of Figure 2 is supported by the results obtained by Barrington and Harvey (1969) from an analysis of the emissions recorded by the Alouette II VLF receiver during 1967. Figure 3 shows contours of the average relative intensities over the VLF receiver bandwidth, based on the AGC levels, plotted as a function of invariant latitude and local mean time. While these intensities are somewhat dependent on the spectral characteristics of the noise in the VLF receiver range, they should show general agreement with Figure 2 for the very broad-band emissions. Such is clearly the case near the noon meridian in the 75 - 80° latitude interval. The lack of agreement with the night-time peak between 55° and 60° is also consistent with the absence of the lower frequency components, as noted above.

The VLF results, as also those of Taylor and Gurnett (1968), show high intensity values throughout the daylight hours at latitudes between about 55° and 70°. Such values do not correlate with the 200 kHz results, and spectral studies by Barrington and Harvey (1969), and also by Taylor and Gurnett (1968), indicate that they are restricted to the ELF range.

Such spectral studies have also shown that the emission that are included in the noon peak of about 75 - 80° are at frequencies greater than the lower-hybrid-resonance (LHR) frequency at the satellite, and that they extend up to the high frequency limit of the VLF receiver (Barrington and Harvey, 1969). These results would then indicate that this particular noise band is probably continuous from the LHR up to just below f_H, where this high frequency limit is somewhat variable but in no case has it been found to exceed about 0.9 f_H. The extension of this type of noise toward the morning hours is more noticeable in the VLF data than in the LF results, while the converse appears true for the afternoon extension. However, this apparent inconsistency is only quantitative and is probably very much dependent on the fact that the LF intensities are for a single fixed frequency and the VLF intensities are an average over the band between the LHR and the receiver limit. This conclusion is supported by the work of Gurnett (1966), whose fixed frequency results agreed very well with the afternoon and evening contours shown in Figure 2.

3. INTERPRETATION OF THE HIGH-LATITUDE EMISSIONS

A comparison of the statistical pattern of intense LF noise emissions portrayed in Figure 2 and that of aurorally associated phenomena as summarized by Hartz and Brice (1967) shows immediately some obvious discrepancies. Here there is no evidence of the zone or ring of "drizzle" precipitation of very energetic electrons at about 65°. The eccentric ring or "splash" zone, often called the auroral oval, seems to agree fairly well with the higher latitude noise region, save that the noise intensity peak occurs at about noon and there is a minimum at night in the time and latitude interval where the more intense and energetic particle splashes occur. The auroral phenomena do not show any prominent features at night in the latitude range of 55° - 60°, other than the low-latitude red arcs; this region does correspond, however, to the plasmapause and to the F-layer ionization "trough", but, as the noise is of a different type than that at the higher latitudes and has not been thoroughly studied yet, it will not be discussed further.

In spite of these discrepancies, there is sufficient correspondence with the daylight auroral oval, or the zone in which discrete auroras occur on the day side of the geomagnetic pole, that the most logical explanation for the intense noise emissions at the high latitudes is one involving particle precipitation. However, the more energetic auroral particles, such as those associated with the nighttime phenomena or break-up features, must be ruled out. The process suggested is the Cerenkov radiation process (Jørgensen, 1968) where emissions are generated from incoming electrons with velocities that exceed or are comparable to the phase velocity of the radiation in the medium. Because the broad-band noise emissions apparently are isotropic and continuous in frequency from the LHR to about 0.9 f_H, simple considerations (Brice, 1967) of whistler-mode propagation in the ionosphere suggests that some limitations can be placed on the region of origin of the noise components. The fact that the emissions are not continuous up to the local f_H implies that they arise at some distance from the satellite and are related to the cut-off for the whistler-mode propagation at that remote location. From the known decrease of the geomagnetic field with height, it follows that the electron gyro-frequency drops to about 100 kHz approximately an earth radius above the satellite at these high latitudes. Assuming propagation more or less parallel to the field lines, the observed LF emissions between 100 kHz and the upper band limit could not have originated at more than an earth radius above the space-craft. Consequently a relatively local origin is indicated, as opposed to one, say, in the magnetospheric tail or magnetopause. In the general vicinity of the satellite the parameters of the medium can be determined with the space-craft instrumentation, and the refractive index can be obtained, at least approximately, at that height and immediately above. This is found to be of the order of 10 or more for the frequencies of interest here, which is consistent with the suggested Cerenkov process and the influx of electrons with energies of the order of 1 keV and less.

There is now a growing literature dealing with direct measurements of electrons in this energy range from rockets and satellites at high latitudes, as well as with such indirect measurements as high altitude auroras and F-region ionization production. Johnson et al. (1966) and Sharp and Johnson (1969) reported low-energy ($<$ 5 keV) electron precipitation near noon between about 75° and 80° invariant. Maehlum (1968) and Burch (1968) have reported soft electron (50 eV $< E <$ 10 keV) fluxes between about 75° and 83° invariant near magnetic noon and between about 70° and 78° near midnight. In this "soft" zone the particle spectrum was found to be very steep, with the flux largely dominated by the electrons with energies between 100 eV and 1.3 keV. Hoffman (1969) has obtained results on soft electron "bursts", also with a very steep energy spectrum, at all times of day and at latitudes that generally correspond to the high latitude side of the auroral oval. Some preliminary results of a similar nature have been obtained by Riedler and Borg on the ESRO-I satellite and by Heikkila on the ISIS-I satellite (Private communications). The auroral observations, particularly those on the day side of the geomagnetic pole, reported by Feldstein (1969), Sandford (1968), Akasofu (1968), Lassen (1969), Starkov (1968), and Starkov and Feldstein (1967) among others, indicate that auroras are prevalent throughout the day, that they may even occur a greater percentage of the time on the day side than at night, that they are not as dependent on magnetic disturbance as the nighttime auroras, and they probably occur at significantly greater heights - above about 140 km - than the nighttime auroras.

These various results indicate that there is an influx of electrons at energies of the order of 0.1 to 1 keV into a diurnal ring which, on the day side, is the auroral oval, and on the night side, approximates the high-latitude edge of the oval. The fluxes have been estimated to be greater than 10^9 cm^{-2} sr^{-1} sec^{-1} keV^{-1} at 100 eV (Burch, 1968; Hoffman, 1969; Westerlund, 1969). The precipitation seems to occur during undisturbed times, and the most obvious change during magnetic disturbances seems to be a shift to somewhat lower latitudes.

Accordingly, it appears that the LF noise intensity contours above about 65° latitude agree with the soft electron precipitation on a statistical basis, except probably for the interval between about midnight and 06 hours. In this interval the VLF data of Barrington and Harvey seem to agree much better with the particle observations. However, this portion of the nighttime ionosphere is characterized by low f_N values at the satellite, which tend to fall below 200 kHz for most of the recordings studied. That is, different wave propagation and antenna impedance conditions apply to the relative intensities for the LF noise observed in this region, and it is then not too surprising to find a lack of agreement with both the VLF and the particle observations. A much more reliable measure of the Cerenkov noise intensities is to be expected at a frequency that is always less than both f_N and f_H, though yet above the LHR frequency.

The statistical whistler-mode noise data can be taken as representative of the average influx of low-energy electrons into the ionosphere. Accordingly, one should expect to find evidence of this influx in the ionization produced in the F-region, particularly when the solar ultra-violet illumination is reduced or absent. The findings of Nishida (1967), of Sato and Colin (1969), and of Andrew and Thomas (1969) tend to bear this out. At heights between about 500 and 1000 km they identified a ring of enhanced ionization, similar to the auroral oval, with the higher electron densities on the day side of the pole. This is quite similar to the pattern of broad-band LF and VLF noise emissions.

The foregoing results on noise emissions represent the average situation, which is weighted in favor of undisturbed, or low Kp, conditions. Some LF data, obtained near the noon meridian when Kp was ≥ 4, have been examined and the results indicate that the polar peak - i. e., the intensity peak between 75° and 80° - is reduced at such times. Barrington and Harvey (1969) have found similar results in the VLF data for Kp ≥ 2, suggesting that during disturbances minimum values occur near noon in place of the polar peak. This intensity reduction is not apparent at other times of day in either the LF or VLF data. Further studies of this phenomenon are required, as there have not yet been enough data obtained during occasions when Kp was high. It should be noted, however, that the auroral data of Lassen suggest a minimum occurrence near noon, geomagnetic time (Lassen, 1969), though his results were also not completely definitive.

4. CONCLUDING REMARKS

The broad-band noise emissions observed by Alouette II in the frequency range between the lower-hybrid resonance and the lesser of the electron cyclotron or plasma frequencies have been studied. Average intensities have been obtained from a twelve-month data sample as a function of invariant latitude and local time. The maximum intensity region occurs on the day side of the geomagnetic pole in the invariant latitude interval from 75° to 80°, and the intensity contours extend around into the morning and evening hours with a gradual reduction in both intensity and latitude. The overall pattern approximates that of the auroral oval with a much suppressed nighttime portion; in fact, there are indications of a minimum near the midnight meridian at latitudes of about 68° - 70°. A deep minimum exists over the whole of the polar region within the ring of high intensity, and, except for a nighttime zone of a different type of noise that peaks between 55° and 60°, the noise levels are low at all lower latitudes.

It is concluded that these broad-band emissions arise as Cerenkov radiation and can be related to the influx into the upper atmosphere of electrons having energies in the range 0. 1 to 1 keV. From the general

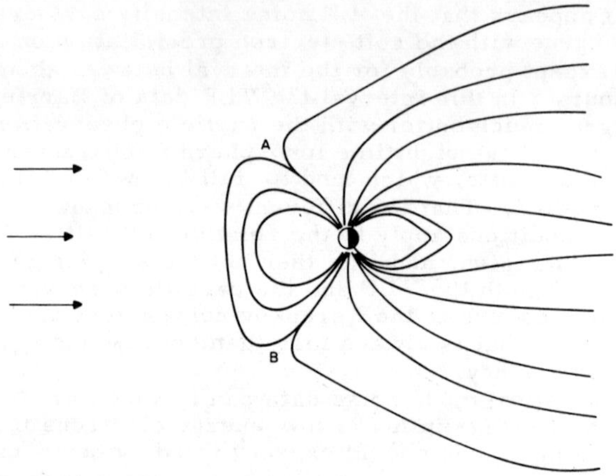

Figure 4. Diagramatic representation of the magnetic field configuration in the noon-midnight meridian of the magnetosphere. The arrows represent the solar wind flow, and the cusps or "neutral" points, marked A and B, indicate regions where material from the solar wind or from the magnetospheric surface may penetrate relatively easily into the magnetosphere.

characteristics of the noise band the region of generation can be localized to the height interval near and not too far above the satellite. The intensity distribution in latitude and time would then strongly suggest that the particle influx is primarily from the sunward side of the magnetosphere in the latitude interval that connects to the magnetospheric cusps or the so-called neutral field points (or more properly neutral lines, since these must have a finite east-west extent) marked in the magnetospheric representation of Figure 4 as A and B. These two indicated regions are thought to be ones where solar wind particles can enter the magnetosphere relatively easily because of the magnetic field configuration.

The results indicate that the inferred electron influx occurs more or less continuously during undisturbed conditions. When Kp is high the preliminary findings suggest that the flux is reduced or cut off near the noon meridian, but at other longitudes well removed from the noon meridian the influx is largely unaffected. As yet no universal time effects or seasonal effects have been observed, but the data are not particularly suited to the detection of such effects.

On the basis of these results, and also those mentioned earlier that refer to direct measurements of the electrons, it appears that some modification can be suggested to the Hartz-Brice pattern of particle precipitation (Hartz and Brice, 1967). The zone they identified with splash-type precipitation, and which is often referred to as the auroral oval, would seem not to be entirely homogeneous and can probably be better represented as made up of two overlapping halves (cf. also Akasofu, 1968). The daytime segment appears to be dominated by a fairly continuous electron influx with energies essentially less than 1 keV, which apparently can be linked with the direct entry of solar wind material at the neutral points on the sunward side of the magnetosphere. The night-time segment appears to relate better to more energetic particles, to electrons with energies of a few keV, and to sporadic processes that can be associated with the plasma sheet in the magnetospheric tail. No sharp division can be indicated that separates these two segments; indeed, this suggested division may prove to be more illusory than real when more information is available, including a better appreciation than now exists for the magnetic field configuration at high latitudes in the magnetosphere.

REFERENCES

Akasofu, S.-I., The magnetosphere and magnetospheric substorm, Ann. Geophys., 24, 507-512, 1968.

Andrews, M.K. and J.O. Thomas, Electron density distribution above the winter pole, Nature, 221, 223-227, 1969.

Barrington, R.E., Ionospheric ion composition deduced from VLF observations, Proc. IEEE, 57, (In press) June 1969.

Barrington, R.E. and R.W. Harvey, Alouette II observations of VLF electric field amplitudes, N.S.N.C.-U.R.S.I. Spring Meeting, Washington, D.C., April, 1969.

Burch, J.L., Low-energy electron fluxes at latitudes above the auroral zone, J. Geophys. Res., 73, 3585-3591, 1968.

Brice, N.M., Ion effects observed in radio wave propagation in the ionosphere, in Electromagnetic Wave Theory, Pergamon Press, New York, pp 197-209, 1967.

Feldstein, Y.I., Polar auroras, polar substorms, and their relationships with the dynamics of the magnetosphere, Rev. Geophys., 7, 179-218, 1969.

Franklin, C.A. and M.A. Maclean, The design of swept-frequency topside ionospheric sounders, Proc. IEEE, 57, (In press) June 1969.

Gurnett, D.A., A satellite study of VLF hiss, J. Geophys. Res., 71, 5599-5615, 1966.

Hartz, T.R., Radio noise levels within and above the ionosphere, Proc. IEEE, 57, In press, June 1969.

Hartz, T.R. and N.M. Brice, The general pattern of auroral particle precipitation, Planetary Space Sci., 15, 301-329, 1967.

Hoffman, R.A., Low-energy electron precipitation at high latitudes, J. Geophys. Res., 74, 2425-2432, 1969.

Johnson, R.G., R.D. Sharp, M.F. Shea, and G.B. Shook, Satellite observations of two distinct day-side zones of auroral electron precipitation, Trans. Am. Geophys. Union, 47, 64, 1966.

Jørgensen, T.S., Interpretation of auroral hiss measured on OGO 2 and at Byrd station in terms of incoherent Cerenkov radiation, J. Geophys. Res., 73, 1055-1069, 1968.

Lassen, K., Polar cap emissions, in Aurora and Airglow, 1968, edited by B.M. McCormac, Reinhold Publ. Corp., New York, 1969.

Maehlum, B.N., Universal-time control of the low-energy electron fluxes in the polar regions, J. Geophys. Res., 73, 3459-3468, 1968.

Nelms, G.L., R.E. Barrington, J.S. Belrose, T.R. Hartz, I.B. McDiarmid, and L.H. Brace, The Alouette II satellite, Can. J. Phys., 44, 1419-1430, 1966.

Nelms, G.L. and J.H. Chapman, The high latitude ionosphere: results from the Alouette/ISIS topside sounders, in The Polar Ionosphere and Magnetospheric Processes, edited by G. Skovli, Gordon and Breach, New York, 1969.

Nishida, A., Average structure and storm-time change of the polar topside ionosphere at sunspot minimum, J. Geophys. Res., 72, 6051-6061, 1967.

Sandford, B. P., Variations of auroral emissions with time, magnetic activity and the solar cycle, J. Atmospheric Terrest. Phys., 30, 1921-1942, 1968.

Sato, T. and L. Colin, Morphology of electron concentration enhancement at a height of 1000 kilometers at polar latitudes, J. Geophys. Res., 74, 2193-2207, 1969.

Sharp, R. D. and R. G. Johnson, Distribution with invariant colatitude of electron and proton precipitation close to noon and midnight, respectively, as found with a low-altitude polar orbiting satellite by the Lockheed group, in Aurora and Airglow 1968, edited by B. M. McCormac, Reinhold Publ. Co., New York, 1969.

Starkov, G. V., Auroral heights in the polar cap, Geomagnetism and Aeronomy, 8, 28-32, 1968.

Starkov, G. V. and Y. I. Feldstein, Change in the boundaries of the oval auroral zone, Geomagnetism and Aeronomy, 7, 48-54, 1967.

Taylor, W. W. L. and D. A. Gurnett, Morphology of VLF emissions observed with the Injun 3 satellite, J. Geophys. Res., 73, 5615-5626, 1968.

Walsh, D., F. T. Haddock, and H. F. Schulte, Cosmic radio intensities at 1.225 and 2.0 Mc measured up to an altitude of 1700 km, in Space Res. IV, North-Holland Publ. Co., Amsterdam, pp 935-959, 1964.

Westerlund, L. H., The auroral electron energy spectrum extended to 45 eV, J. Geophys. Res., 74, 351-354, 1969.

TEMPERATURE AND COMPOSITION STUDIES IN THE POLAR IONOSPHERE

S. J. Bauer

NASA-Goddard Space Flight Center
Greenbelt, Maryland 20771, USA

Abstract. A geomagnetic latitude of ~ 60° appears to represent the demarcation between the midlatitude and polar ionosphere. At this latitude the abundance of light ions (H^+ and He^+) decreases abruptly, leading to a polar ionosphere consisting primarily of O^+. The light ions seem to be escaping via the tail of the magnetosphere; an outward bulk motion of H^+ over the polar cap has ben observed by the ionmass spectrometer on Explorer 31. The plasma density over the polar region decreases much faster with altitude ($N \propto R^{-3}$) than according to a diffusive equilibrium distribution.

The charged particle temperatures, of which the electron temperature T_e has been primarily measured, show a gradient with altitude and latitude. A maximum of T_e is often observed near 60° geomagnetic latitude, with lower temperature over the poles. At an altitude of a few thousand km, the electron temperature in the polar ionosphere is ~ 3000-4000 °K reaching values as high as 6000-8000 °K at an altitude of 6000 km. In addition to the "thermal" plasma, large fluxes of suprathermal electrons have been observed in the polar ionosphere, showing a fine-structure with latitude which seems to be related to the trough and the regions where outward streaming ions have been observed.

1. INTRODUCTION

Since the ionosphere is under solar and geomagnetic control, it is not surprising to find a latitude dependence of ionospheric parameters. Especially, parameters dependent on the magnetic field and energetic particle environment would be expected to increase toward the polar region. It is well known that the ionosphere in the auroral zone is quite different from the mid-latitude ionosphere. Variations in charged particle temperature should therefore be expected, because of the additional heat input by energetic particles at high latitudes. For the composition of the ionosphere, however, no strong latitude variation should be expected, off hand. There is no question, however, that the ionosphere within the polar cap, including perhaps the auroral zone should be different from the low- and mid-latitude ionosphere.

As early as the IGY (Wright, 1962) evidence accumulated from ground-based observations of the ionosphere, that a transition from what one would call a quiet mid-latitude ionosphere occurs at a much lower latitude than expected, i.e., approximately at 45° geographic latitude over the American continent, which is a latitude well below the auroral zone. Later on, this transition was identified in terms of the high-latitude trough (Muldrew, 1965) and also with the L-shell which seems to be related to the plasmapause (Carpenter, 1963). It now appears that a geomagnetic latitude of about 60° represents a boundary between the middle latitude and polar ionosphere. For this reason we will include the latitude range greater than 60° geomagnetic as representative of the polar ionosphere.

In the following, we shall discuss present knowledge regarding the ion composition and charged particle temperatures in the polar ionosphere. The observations pertain primarily to the altitude range above the F2 maximum, i.e., to the topside ionosphere. A number of techniques have led to information about ion composition. However, there is still not a very large amount of data available over the polar cap, because of lack of such observations at very high latitudes. Similarly, this is true for charged particle temperatures. Electron temperatures are available to some extent, but very little information is available about the ion temperatures at higher latitudes. However, it is safe to assume that the electron and ion temperatures are approximately equal in the high-latitude topside ionosphere. It is quite obvious, that the behavior of the polar ionosphere is a direct consequence of the configuration of the magnetosphere, especially the presence of an open tail. This by itself, however, cannot explain some of the observed ionospheric properties. It appears that convective motions within the magnetosphere may actually be the controlling factors in the difference between polar and mid-latitude ionospheric behavior. Generally speaking, one would identify the middle and low-latitude ionosphere with the plasmaphere (Carpenter, 1966), while the polar ionosphere could be identified as closely linked to the plasma outside the plasmaphere and therefore in more or less direct contact with the thermalized solar wind plasma.

2. CHARGED PARTICLES TEMPERATURES

The first indication that the charged particle temperature increases with latitude came from Ariel 1 observations. (Bowen et al., 1964a). Further convincing evidence of this fact came from observations on Explorer 22, Alouette 1, Explorer 31, and the OGO satellites. At higher altitudes, of about 6000 km, this is also indicated by the data from the electron trap on the RAE satellite. Figure 1 shows a comparison of the latitude variation of electron temperature measured on Explorer 22 and on Alouette 2 (Mahajan and Brace, 1969). This figure shows both the latitude gradient of electron temperature, as well as its gradient with altitude. The latter is seen because Explorer 22 measurements were made at an altitude of about 1000 km, while Alouette 2 measurements refer to an altitude of about 2000-3000 km. The average variations of electron temperature for four seasons, as measured with the electrostatic probe on the Alouette 2 satellite covering latitudes of up to about 70° geomagnetic are shown in Figure 2. Other measurements by the GSFC retarding potential analyzer on the OGO 4 satellite show that the high electron temperatures usually encountered in the polar ionosphere often also correspond to times when the I-V curves are very noisy (Figure 3). At such times the background current is greatly elevated indicating the presence of a strong background flux of suprathermal particles. This presence of suprathermal electrons in the polar ionosphere (Maier, 1969), which has a very distinct structure with geomagnetic latitude, seems to be related to the high-latitude trough and to the regions where streaming ions have been observed (Hoffman, 1969). The latter will be discussed

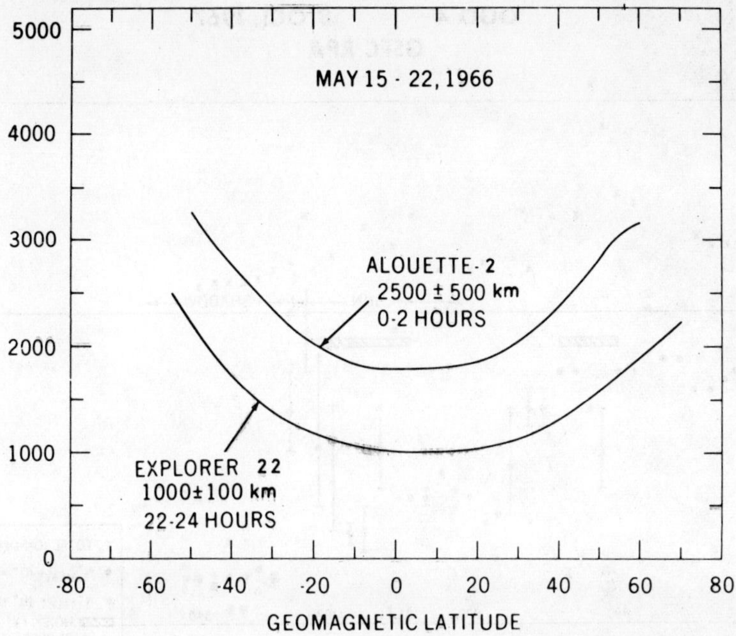

Figure 1. Comparison of electron temperature measurements on Explorer 22 and Alouette 2 (Mahajan and Brace).

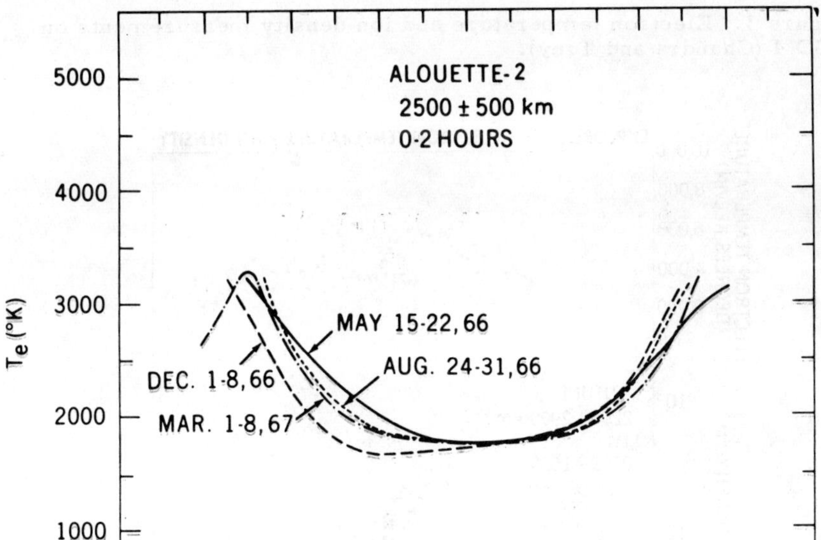

Figure 2. Electron temperature as function of geomagnetic latitude for four seasons (Mahajan and Brace).

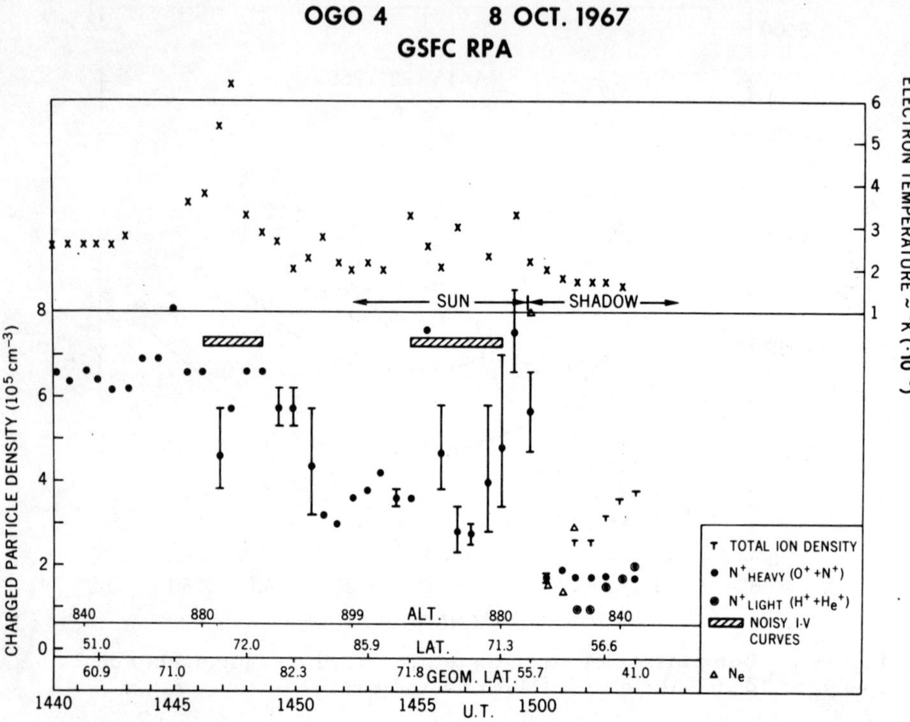

Figure 3. Electron temperature and ion density measurements on OGO 4 (Chandra and Troy).

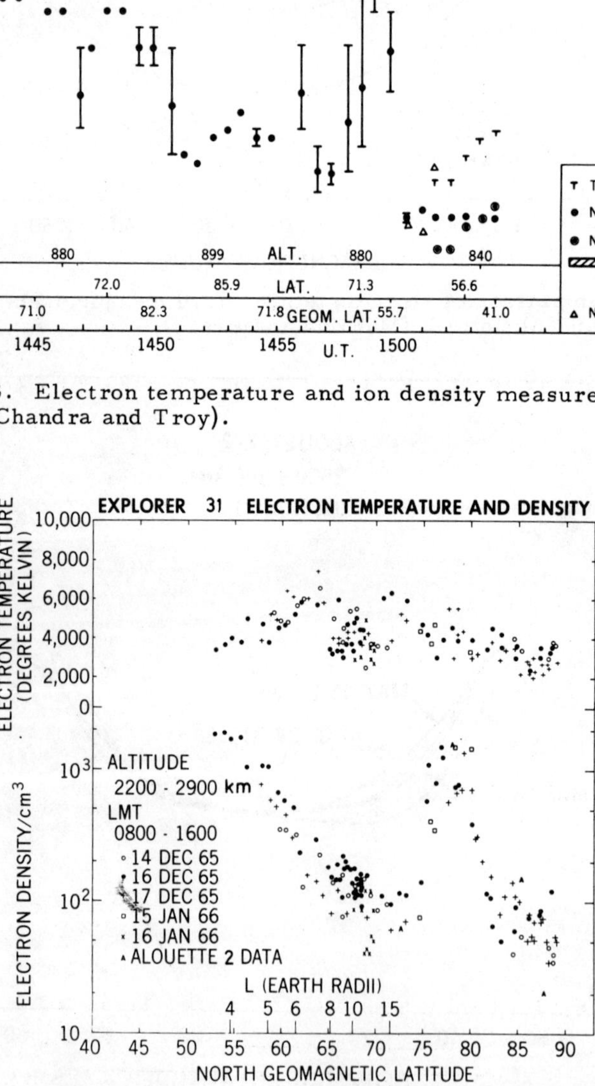

Figure 4. Electron temperature and density measurement over the polar ionosphere (Explorer 31; Donley).

later in conjunction with the ion composition behavior of the polar ionosphere. In the following, we will show some illustrative examples of electron temperature variations in the polar ionosphere obtained by means of the retarding potential analyzer on the OGO and RAE satellites. All satellite measurements of electron temperature indicate an increasing electron temperatures at geomagnetic latitudes greater than about 40°. In some of the data, especially those from Explorer 22, there appears to be an maximum of T_e in the vicinity of about 60° geomagnetic latitude (Figures 1 and 9). Accordingly, the electron temperatures at the poles are lower than at latitudes of ~60° geomagnetic. In polar winter, T_e over the pole is lower than in polar summer, thereas the maximum at ~60° is higher in polar winter than in polar summer (Brace, 1969). The peak near 60° appears to be due to the fact that below this latitude the protonosphere (plasmasphere) acts as a heat reservoir. Electron temperature measurements on Explorer 31, by Donley (1967) during the winter 1965/66 (Figure 4) indicate that at altitudes ~3000 km, T_e ~6000 °K near 65° geomagnetic latitude, while at higher latitudes over the polar cap the electron temperature lies between 2000 and 4000 °K. At altitudes of 5800 km, T_e is greater than 5000-6000 °K for latitudes beyond 40° geomagnetic, reaching values as high as 8000 °K. This is illustrated in Figure 5

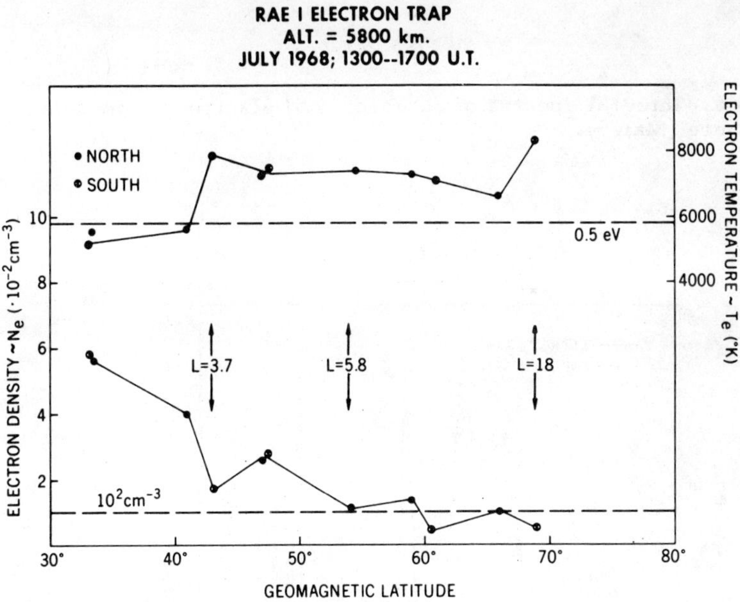

Figure 5. Electron temperature and density measured with the electron trap on RAE-1.

which also shows that the electron temperature is, in general, inversely correlated with the electron density. The same behavior is also apparent from data at somewhat lower altitudes. Electron retarding potential analyzer measurements on Explorer 31 (Maier, 1969) in the polar region indicate the presence of suprathermal electrons, in addition to the thermal plasma whose temperature is of the order of 6000-8000 °K. These suprathermal particles are much greater in number than the high-energy tail of the normal Maxwell-

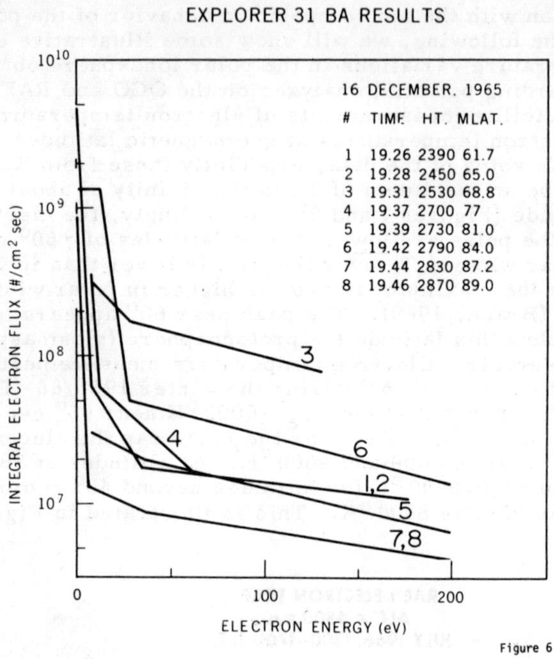

Figure 6. Integral spectra of suprathermal electron in the polar ionosphere (Maier).

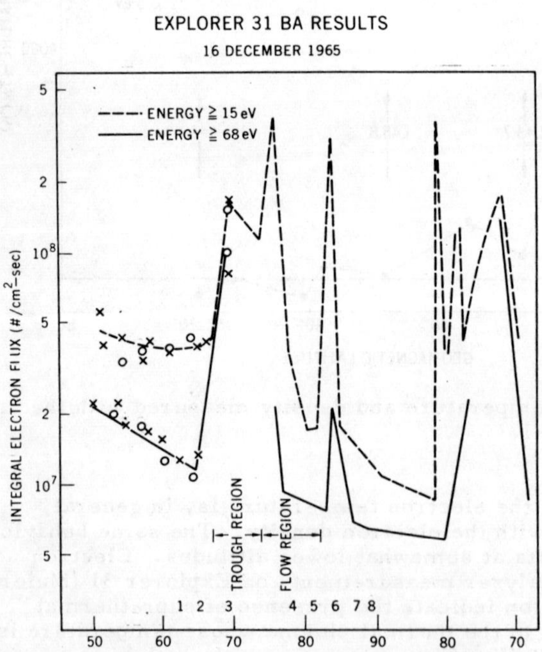

Figure **he polar ionosphere (Maier).**

Boltzmann distribution of the thermal plasma. The integral spectra vary significantly as a function of geomagnetic latitude. This is shown in Figure 6; Curves 1 and 2 correspond to latitudes where the magnetic field lines are still closed; Curve number 3 is just inside the trough. The spectra consist of a thermal component and a high energy tail of the type $dn/dV \propto V^{-3}$ for Curve 3. The power law of the distribution for the closed field lines (Curves 1 and 2) is possibly due to escaping photoelectrons, since during the time of measurement both feet of the magnetic field lines passes through the satellite are sunlit. The contribution of the suprathermal particles in terms of energy density (erg cm^{-3}) or plasma pressure (dyne $cm^{-2} \equiv$ erg cm^{-3}) outside the trough is of the order of $4 \cdot 10^{-12}$ erg cm^{-3}, which is much less than the energy density of the local thermal plasma at middle latitudes. Although suprathermal electrons may be of importance to the heat balance of the topside ionosphere, their contribution to the local pressure or energy density is negligible. In the region of the trough, however, the suprathermal particle pressure is of the order of $3 \cdot 10^{-11}$ erg cm^{-3}, which is comparable to or greater than that of the local plasma; the concentration of the suprathermal electrons is about 1 cm^{-3}. The integral flux of suprathermal electrons of energies greater than 15 eV and greater than 68 eV, as a function of geomagnetic latitude is shown in Figure 7. At about 65° geomagnetic latitude, which corresponds roughly to the position of the high-latitude trough, where the electron density is ~ 30 cm^{-3} compared to more than ~10^{2} cm^{-3} at lower latitudes, a sudden increase occurs in the flux of suprathermal electrons. North of the trough, at magnetic latitudes of 75° to 85° there is a disturbed region of variable electron fluxes. This is also the same region where upward flowing protons were observed by Hoffman (1969). From data where the electron trap position is either within 90° of the magnetic field direction or more than 90° from the field direction, it appears that the flow of suprathermal electrons is not predominantly up or down the magnetic field line. Comparison with other angle intervals leads to the conclusion that the angular distribution of the suprathermal electrons is isotropic. Thus, these suprathermal electrons of energies < 200 eV cannot have a bulk flow velocity which is more than a small fraction of their thermal motion. The numbers indicated in Figure 7 refer to the individual spectra shown in Figure 6. The sharp increase in the flux of suprathermal particles at the high-latitude trough corresponds to a plasma pressure of about $3 \cdot 10^{-11}$ erg cm^{-3} which is comparable to or greater than that of the local thermal plasma. However, north of the trough region, the plasma pressure drops again to about $4 \cdot 10^{-12}$ erg cm^{-3} which is less than that of the ambient thermal plasma. The relatively low plasma pressure lends some support to the suggestion that there could be a supersonic flow of ions, i.e., the polar wind outward from the polar ionosphere as proposed by Banks and Holzer (1968). If the observed sharp decrease in the pressure of suprathermal plasma poleward of the trough region also corresponds to a pressure decrease farther out on the field lines, their low pressure condition required for the polar wind could be satisfied. However, this needs to be verified by further observations.

3. ION COMPOSITION IN THE POLAR IONOSPHERE

A variation of abundance of light ions with geomagnetic latitude was first observed on the Ariel satellite (Bowen et al., 1964b) indicating that the concentration of helium and hydrogen ions decreases with increasing latitude. The same effect was also shown indirectly from topside sounder results which indicated that a latitude gradient in the mean ionic mass exists with a smaller mean ionic mass at lower than

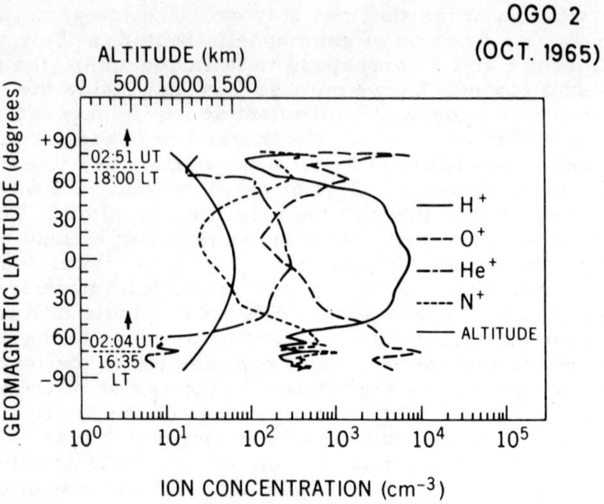

Figure 8. Ion concentration as function of geomagnetic latitude from OGO 2 measurements (Taylor et al.).

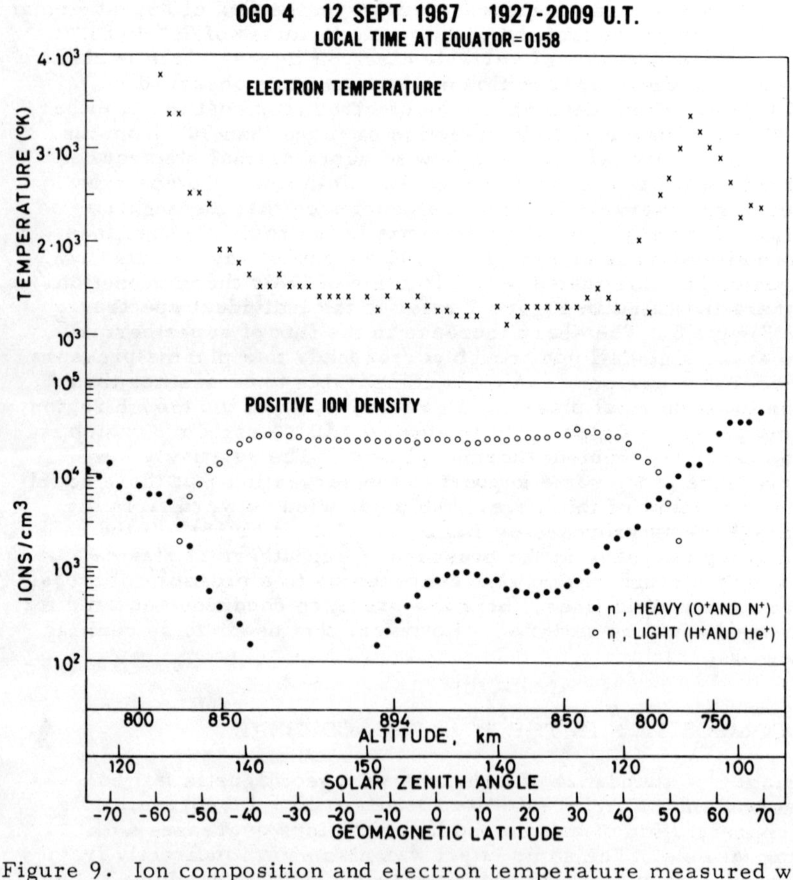

Figure 9. Ion composition and electron temperature measured with the RPA on OGO 4 (Chandra and Troy).

at high latitudes (e.g., Bauer and Blumle, 1964; Watt, 1965). The first direct evidence for a strong latitude dependence of the abundance of light ions especially H^+ (since He^+ could not be detected by this technique) has been observed from the cutoff of VLF noise on the Alouette satellite (Barrington et al., 1966) and from observations of proton whistlers on the INJUN satellite by Gurnett and Shawhan (1965). These results have also been extended more recently. Direct measurements of ion composition by means of ion mass spectrometer and retarding potential analyzers on the OGO satellites have verified this early concept. The ion mass spectrometer measurements of Taylor et al. (1968) on OGO 2 (Figure 8) show distinctly the increase with geomagnetic latitude of the heavy ions O^+ and N^+ and a rather abrupt decrease in the light ions H^+ and He^+ near the latitude of the trough. Similarly, observations of the retarding potential analyzer on OGO 4 (Chandra and Troy, 1969) illustrated in Figure 9 show the concentration of heavy ions ($O^+ + N^+$) increasing with latitude, whereas the concentration of light ions (He^+ and H^+) drops abruptly near 55° to 60° geomagnetic latitude.

Based on the early observations available in 1965 Bauer (1966) speculated that the reason for the absence of H^+ and He^+ results from the fact that these particles are being drained off into the tail of the magnetosphere. Since the field lines issuing from the polar regions are dragged back into the tail, ions which have sufficient kinetic energy to escape the gravitational potential of the earth could travel great distances into the tail and one could therefore expect an outflow of H^+ and He^+. Since only about 0.25 eV is required for H^+ and 1 eV for He^+, these ions can escape from the gravitational field of the earth. Since the electron and ion temperature is easily of that order over the polar regions, as shown here, the concept of an ion-exosphere with escaping ions appears quite plausible. Similar suggestions have been made by Dessler and Michel (1966) and by Nishida (1966). All these concepts, involve thermal escape of particles and essentially argue for a subsonic or sonic flow of ions. For realistic temperatures, the escape fluxes of H^+ and He^+ can be very significant, if we assume reasonable concentrations of H^+ and He^+ at the escape level. For hydrogen ions the escape flux is $\sim 10^8$ cm^{-2} sec^{-1}, i.e., comparable to or greater than the limiting flux of neutral hydrogen (Geisler, 1967). Thus, the thermal escape flux of H^+ is sufficiently large to remove all protons from the polar region.

More recently, however, Banks and Holzer (1968) suggested, that in contrast to the previous concept of thermal escape applicable to neutral particles, there is a supersonic expansion of the ionospheric plasma on the open field lines, resulting in a "polar wind", analogous to the solar wind. As evidence for a supersonic expansion, Banks and Holzer have used the experimental data of Hoffman (1969) shown in Figure 10. Hoffman detected in the polar region upward fluxes of protons having a bulk velocity of the order of 10 to 15 km/sec. Although the idea of a supersonic polar wind is definitely an interesting concept, its verification cannot yet be considered to be firm. (It should be noted that a velocity of the order of 10 to 15 km/sec is very close to the thermal (0.25 eV for H^+) escape velocity of 11 km/sec.)

A very important aspect of the density and composition behavior in the polar ionosphere is the fact that it is not directly tied to the geometry of the magnetosphere, since at about 60° geomagnetic latitude the magnetic field lines are still closed. Only at much higher latitudes ($> 75^\circ$) field lines open into the tail. Therefore, the polar ionosphere seems to be more related to convective motions in the magnetosphere as suggested by Nishida (1966) and by Brice (1967) to explain the whistler knee or plasmapause. This concept invokes convective motions from the tail toward the sunward side in the

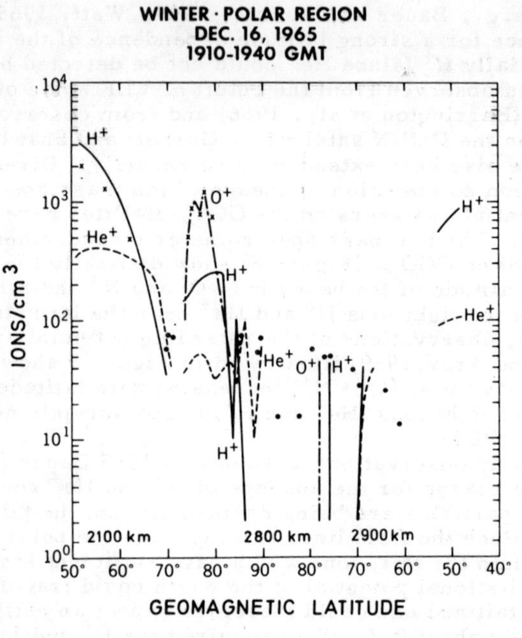

Figure 10

Figure 10. Ion composition over the polar ionosphere measured on Explorer 31 (Hoffman).

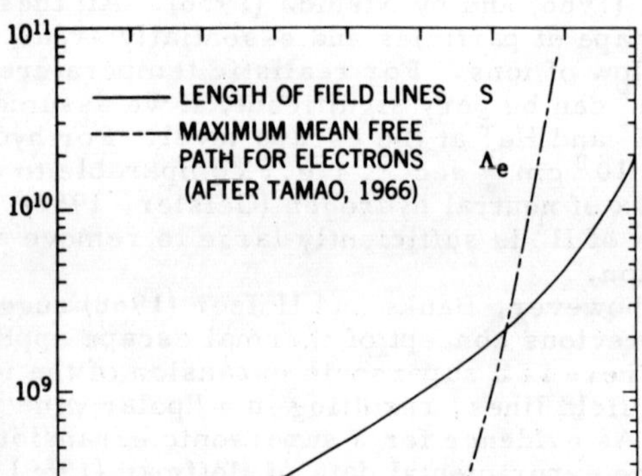

Figure 11

Figure 11. Mean free path for charged particles and length of geomagnetic field lines as function of geomagnetic latitude.

the magnetosphere together with the corotating thermal plasma closein. The plasmapause which generally occurs below auroral latitudes (L = 4) is considered to be the boundary between convective motions and corotation. The ion composition and low density regime of the polar ionosphere obviously has to be looked at from the viewpoint of this model, since at 60° latitude, the magnetic field lines are still closed. In this connection it is also of interest to study the density distribution of the plasma in the low density regime, i.e., an "ion-exosphere". At middle latitudes, diffusive equilibrium is an appropriate representation of the distribution of the thermal plasma in the ionosphere. However, at higher latitudes the simple concept again breaks down for physical reasons. Diffusive equilibrium is a direct consequence of a Maxwellian velocity distribution, which is dependent on collisions. At latitudes beyond about 60° geomagnetic, corresponding to the latitude of the trough and the plasmapause, the electron density in the topside ionosphere can be very low, as determined from Alouette data and also shown here in some of the data from the OGO 4. At this latitude, the mean free path of charged particles becomes about comparable to the length of the field line, indicating that collisions play only a negligible role. In Figure 11 the mean free path for Coulomb collisions is shown together with the length of field lines as function of geomagnetic latitude. Because of Coulomb collisions, the mean free path ($\Lambda_e \sim 10^4\, T_e^2/N$), increases rather rapidly with latitude due to the positive gradient in electron temperature and the rather low plasma density. Thus, at geomagnetic latitudes >60°, a "collisionless" regime or an "ion exosphere" prevails. Such a concept was introduced by Eviatar, Lenchek and Singer (1964). The density distribution for an "ion exosphere" can be derived by applying the laws of conservation of particle energy, together with the first magnetic invariant for the motion of charged particles. This distribution is given by

$$N(R, L) = N_M \exp(-R/H') \cdot \{1 - (1 - \frac{B}{B_M})^{\frac{1}{2}} \exp(-\frac{RB}{H'(B_M - B)})\} \quad (1)$$

where $H = k(T_e + T_i)/m_+ g$ is the scale height, R is a geopotential length parameter and B is the intensity of the magnetic field. The subscript M refers to the reference level below which a Maxwellian distribution applies. The density distribution given by this formula corresponds essentially to a diffusive equilibrium distribution (first term) modified by weighting function (second term) which depends on the intensity of the magnetic field. Figure 12 shows relative plasma density distribution along a field line (in this case L = 5) for diffusive equilibrium, for the "collisionless" or "ion exosphere" model and for two power-law distributions. The "collisionless" model should be representative of the low plasma density region of the magnetosphere, i.e., beyond the plasmapause. At lower altitudes this model can be approximated by the R^{-4} power law, while at higher altitudes at R^{-3} power law appears to represent a somewhat closer fit. It is interesting to note that the plasma density distribution for the polar wind model of Banks and Holzer leads, in the asymptotic case, to a dependence of $N \propto R^{-3} (\ln R)^{-1/2}$. Thus, a R^{-3} power law is also reasonably representative of the asymptotic polar wind solution.

In any event, it is quite obvious that the charged particle distribution over the polar region should decrease much more rapidly with altitude than according to a diffusive equilibrium or hydrostatic distribution. It appears, at least at the present time, that a R^{-3} power law is probably a reasonable approximation to reality, regardless if one accepts an ion exosphere or a supersonic polar wind model for the distribution of the charged particles in the topside polar ionosphere.

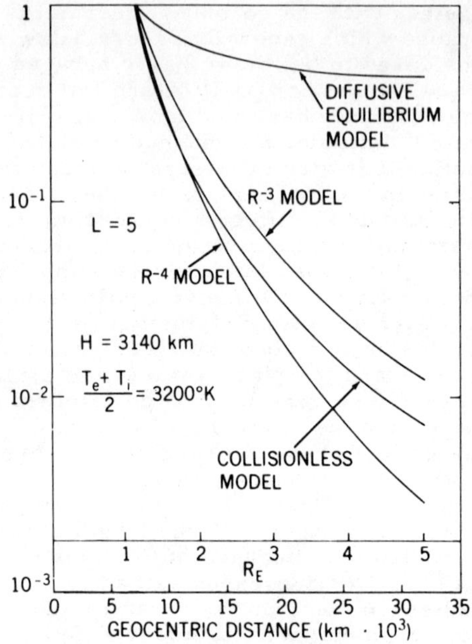

Figure 12. Plasma density distribution along field lines according to diffusive equilibrium, "collisionless" model, R^{-3} and R^{-4} power law.

REFERENCES

Banks, P.M. and T.E. Holzer, The polar wind, J. Geophys. Res., 73, 6846-6859, 1968.

Barrington, R.E., J.S. Belrose, and G.L. Nelms, Ion composition and temperature at 1000 km as reduced from VLF resonances and topside ionograms, in Electron Density Profiles in the Ionosphere and Exosphere, edited by J. Frihagen, North-Holland Publishing Company, Amsterdam, pp. 387-396, 1966.

Bauer, S.J., The structure of the topside ionosphere, in Electron Density Profiles in the Ionosphere and Exosphere, edited by J. Frihagen, North-Holland Publishing Company, Amsterdam, pp. 270-279, 1966.

Bauer, S.J., Hydrogen and helium ions, Ann. Geophys., 22, 247-254, 1966.

Bauer, S.J. and L.J. Blumle, Mean diurnal variation of the topside ionosphere at mid-latitudes, J. Geophys. Res., 69, 3613-3618, 1964.

Bowen, P.J., R.L.F. Boyd, C.L. Henderson, and A.P. Willmore, Measurement of electron temperature and concentration from a spacecraft, Proc. Roy. Soc. London, A281, 514, 1964a.

Bowen, P.J., R.L.F. Boyd, C.L. Henderson, and A.P. Willmore, Ion composition in the upper F-region, Proc. Roy. Soc. London, A281, 504, 1964b.

Brace, L.H., presented at COSPAR, 1969.

Brice, N.M., Bulk motion of the magnetosphere, J. Geophys. Res., 72, 5582, 1967.

Carpenter, D.L., Whistler studies of the plasmapause in the magnetosphere 1, J. Geophys. Res., 71, 693-710, 1966.

Chandra, S. and B. Troy, presented at the AGU-Spring Meeting, Washington, D.C., 1969.

Dessler, A.J. and F.C. Michel, Plasma in the geomagnetic tail, J. Geophys. Res., 71, 1421-1426, 1966.

Donley, J.L., Observations of the polar in the altitude range 2000 to 3000 km by means of satellite borne electrons traps, in Space Res. VIII, North-Holland Publishing Company, Amsterdam, pp. 381-389, 1967.

Eviatar, A.H., A.M. Lenchek, and S.F. Singer, Distribution of density in an ion-exosphere of a non-rotating planet, Phys. of Fluids, 7, 1775-1779, 1964.

Geisler, J.E., On the limiting flux of ionization into the protonosphere, J. Geophys. Res., 72, 81, 1967.

Gurnett, D.A. and S.D. Shawhan, Determination of ion concentration electron density and proton gyrofrequency from dispersion of proton whistlers, J. Geophys. Res., 71, 741-754, 1966.

Hoffman, J.H., The ion-mass spectrometer on Explorer 31 satellite, Proc. IEEE, June 1969.

Maier, E.J.R., Suprathermal electrons in the polar ionosphere, J. Geophys. Res., 74, 2421-2424, 1969.

Muldrew, D.B., E layer ionization troughs deduced from Alouette data, J. Geophys. Res., 70, 2653-26, 1965.

Nishida, A., Formation of the plasmapause, or magnetospheric knee by the combined action of magnetospheric correction and plasma escape from the tail, J. Geophys. Res., 71, 5669-5680, 1966.

Taylor, H.A., H.C. Brinton, M.W. Pharo, and N.K. Rahman, Thermal ions in the exosphere: evidence of solar and geomagnetic control, J. Geophys. Res., 73, 5521-5534, 1968.

Wright, J.W., Diurnal and seasonal changes in the structure of the quiet mid-latitude ionosphere, J. Res. NBS, 66D, 297-312, 1962.

THE POLAR F-REGION

H. Rishbeth

Radio and Space Research Station, Slough, Bucks., U.K.

Abstract. This paper reviews the physical processes operating in the polar F-region, and discusses how they may be related to some leading features of polar F-region phenomena.

Section 1 gives a brief resume of ordinary F-region theory (Section 1.1) and some comments on the time systems appropriate for use in polar regions (1.2).

Section 2 discusses five different physical processes which may be important in the polar F-region: solar photoionization (2.1), particle ionization (2.2), thermal expansion (2.3), electric fields (2.4), and neutral air winds (2.5).

Section 3 deals with some aspects of polar F-region behaviour and their possible causes: the polar F1-layer and the effect on it of movements (3.1 and 3.2), seasonal changes in the F1- and F2-layers (3.3), and the diurnal variations of the F2-layer and their possible explanation in terms of vertical movements (3.4 and 3.5).

Section 4 deals with three specialized theoretical topics that may be of interest in connection with polar phenomena: ion-drag (4.1), the "polar wind" (4.2), and field-aligned electric currents (4.3).

Finally, Section 5 summarizes present ideas on the major topics of interest: the maintenance of the polar winter F2-layer (5.1), and the diurnal and seasonal variations (5.2). It seems likely that the winter maintenance is largely due to ionization by low energy particles; that winds are largely responsible for the diurnal variations; and that atmospheric composition changes may be involved in the seasonal effects. Some parameters about which further experimental data are particularly needed are noted in Section 5.3.

1. INTRODUCTION

This paper is intended as a basic review of the processes likely to be important in the polar F-region. It does not describe F-region morphology in detail, since this would entail great complexity. It concentrates on observations that are judged to be particularly relevant to the physical processes, and the paper should be regarded as a basis for discussion rather than as an authoritative survey.

The questions with which this paper is mainly concerned are the maintenance of the polar ionization (especially in winter), and the variations of the polar F1- and F2-layers. Storm effects are not considered, since too much complexity would be entailed in treating them, and the "troughs" which divide the polar ionosphere from the mid-latitude ionosphere are dealt with in other contributions.

1.1 A Summary of F-Region Theory

It is appropriate to start by summarizing the processes that are thought to control the normal F-region in mid-latitudes, since most of these processes should have some relevance even at polar latitudes. Present F-region theory has been built up from the contributions of many individual workers; the complete theory is of course very complicated, but the following may be regarded as an outline (Rishbeth, 1967; 1968).

The F-layer is controlled by production, loss and transport of ionization. The production is mostly due to solar radiation in the range of roughly 150-911 Å; the peak of the production rate q is found at the level of the daytime F1-layer, around 170 km. Loss of ionization occurs by a two-stage process, which gives rise to a "square-law" loss formula (αN^2) in the lower F-region and a "linear" formula (βN) in the F2-layer. The transition between the "square-law" and "linear" formulas takes place at F1-layer heights. Above the F1-layer the electron density N increases upwards, because β decreases faster than q.

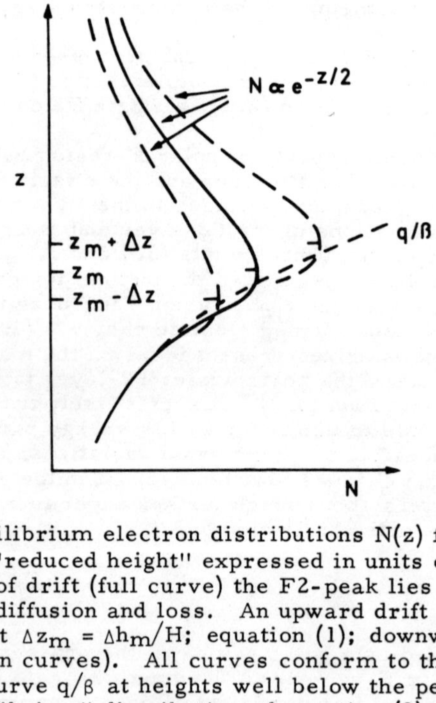

Figure 1. Equilibrium electron distributions N(z) for the F2-layer, height z being "reduced height" expressed in units of scale height H. In the absence of drift (full curve) the F2-peak lies at a level z_m determined by diffusion and loss. An upward drift W raises the peak through a height $\Delta z_m = \Delta h_m/H$; equation (1); downward drift lowers the peak (broken curves). All curves conform to the "photochemical equilibrium" curve q/β at heights well below the peak, and to the "diffusive equilibrium" distribution of equation (2) well above the peak (Rishbeth and Garriott, 1969).

Transport processes, which cause only perturbations in the lower F-region, become dominant at the F2-peak. The level of the peak is normally determined by plasma diffusion, being at the level where $D/H^2 \sim \beta$, where D is the plasma diffusion coefficient and H the neutral gas scale height. The peak can be displaced by movements of ionization, particularly vertical movements: these may be due to neutral air winds (which cause plasma motion along geomagnetic field lines) or electric fields (which cause plasma motion across field lines), although the effects of electric fields may be modified by ion drag (Sec. 4.1). A vertical drift velocity W displaces the peak through a height (Figure 1)

$$\Delta h_m \sim WH^2/D(h_m) \sim W/\beta(h_m) \qquad (1)$$

Wherever the peak is situated, the peak electron density $N_m \sim q(h_m)/\beta(h_m)$, and at night the entire layer decays with a coefficient of order $\beta(h_m)$.

Above the peak the plasma assumes a diffusively controlled distribution which consists of a "diffusive equilibrium" component and a "flux" component. For the simplest case (one ion species, isothermal atmosphere, thermal equilibrium) these two components (respectively) combine to give a distribution

$$N(z) = N_1\, e^{-z/2} + N_2\, e^{-z} \qquad (2)$$

where z is the "reduced height" of the neutral ionizable gas. At great heights the distribution is modified by the presence of light ions.

Calculations based on the above ideas have been moderately successful in explaining the behaviour of the F-region at middle and low latitudes. Numerous problems remain unsolved, but this may well be due to lack of knowledge of how the many parameters (especially those of the neutral atmosphere) vary with position, time and season, rather than to any failure of the basic ideas. Thus one may reasonably expect the same principles to apply to the polar ionosphere, as discussed in Section 2.

1.2 Time in Polar Regions

In describing F-region phenomena, it is necessary to specify what time is used. Universal Time (UT) is the standard time of the Greenwich meridian. Other kinds of time are related to the difference between the sun's longitude and the longitude of the observer, in given sets of axes; it will be sufficient to quote times to within the nearest hour. Thus Local Time (LT) refers to geographical coordinates and is conventionally defined in terms of standard meridian time - the difference between this and apparent solar time being usually trivial - while Geomagnetic Time (GT) refers to geomagnetic dipole coordinates. Some physical phenomena are controlled by the direction of the magnetic field lines; in polar regions the field lines converge on the dip poles and thus it may be useful to define "magnetic dip time" (DT) with reference to these poles. Thus 00 DT and 12 DT occur when the sun lies in the observer's magnetic meridian, respectively on the <u>far</u> side and on the <u>near</u> side of the dip pole. Difficulties can arise with "dip time" at lower latitudes, but for polar regions the idea may be useful. Clearly DT is undefined at the dip poles, LT at the geographic poles and GT at the poles of the dipole field. Moreover, GT and DT do not usually advance uniformly with respect to LT at any given station.

Much of the recent theoretical ionospheric work has been concerned with the Antarctic. Figure 2 shows the disposition of Antarctic stations. The dip pole is near Terre Adelie (67 °S geog.), near the Antarctic circle; at this location solar illumination occurs daily even in midwinter. The polar cap stations, where solar illumination ceases in midwinter, all lie within a limited range of direction from Terre Adelie (about 90°) and it is difficult to discriminate properly between (say) UT and DT controlled phenomena. The geometry of the Arctic is

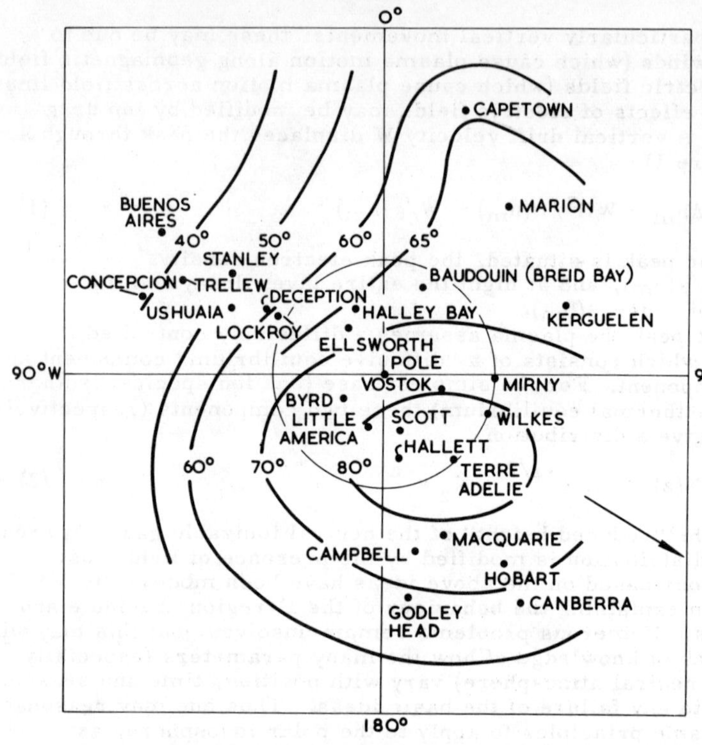

Figure 2. Polar projection map of Antarctic stations, showing the Antarctic Circle and lines of constant magnetic dip. The dip pole is near Terre Adelie (Piggott and Shapley, 1962). The arrow shows the approximate direction at 06 UT of the centre of the "diurnal bulge", situated in low latitudes. As discussed in Sections 2.5 and 3.5, a wind blowing away from the bulge (antiparallel to the arrow), will be directed away from Terre Adelie at this hour at most polar stations.

rather different, but fortunately there is a station near the dip pole (Resolute Bay, 75 °N geog.). Even if data from both polar regions are considered, it may be hard to unravel the UT, LT, GT and DT effects.

2. PHYSICAL PROCESSES IN THE POLAR F-REGION

2.1 Photoionization

Kamiyama (1966) and L. Thomas (1966a) have shown that photoionization can supply the F-region plasma so long as the solar zenith angle $\chi < 103^{\circ}$. Hence at midwinter the problem of maintenance is only serious in a region of about 10° radius surrounding the geographic pole. The peak of production may be above the peak of electron density, in which case the shape of the layer is insensitive to the form of the production function. Naturally maximum production occurs around 12 LT, though the maximum resulting electron density may occur slightly later.

Night sky radiations (He II, Ly β, etc.) can be discounted as a source of F-region ionization (Ogowa and Tohmatsu, 1966).

2.2 Particle Precipitation

Particle precipitation is expected to play a larger part in the polar F-region than at lower latitudes. For production of ionization, protons and electrons of order 10^2 - 10^3 eV are most important at F-region heights (Rees, 1963). Photoelectrons are unlikely to produce much new ionization but may raise the electron temperature, as may particles of higher energies.

The particle input needed to maintain the polar F2-layer may be very roughly estimated as follows. At equilibrium, and neglecting gain and loss due to any horizontal transport, the column production rate is

$$Q = \int_0^\infty q \, dh = \int_0^\infty \beta N \, dh \approx \beta_E N_m Y \quad (3)$$

where the slab thickness $Y \sim 200$ km, and $\beta_E \sim 10^{-4}$ s^{-1} is the effective loss coefficient for the layer. Thus, using c.g.s. units, $Q \sim 2 \cdot 10^3 N_m$. In the vicinity of the dip poles in winter there is a "base level" of ionization, $N_m \sim 5 \cdot 10^4 \, cm^{-3}$ so that $Q \sim 10^8 \, cm^{-2} \, s^{-1}$.

Assuming that 35 eV is required per ionization and that 5% of the energy appears as 6300 Å oxygen red line airglow (as deduced from data given by Dalgarno, 1964a), the required energy input is $6 \cdot 10^{-3}$ erg cm^{-2} s^{-1} and the red line emission is about 100 rayleighs These values seem quite reasonable, so that although quantitative confirmation is needed, low-energy particles seem the most likely source of polar winter ionization. Outside the "cavity" that surrounds the dip pole (Thomas and Andrews, 1969), the winter electron density may be several times greater (i.e., around $3 \cdot 10^5$ cm^{-3}) but even so the required energy input ($3 \cdot 10^{-2}$ erg cm^{-2} s^{-1}) and airglow emission (500 rayleighs) seem quite plausible. Particle precipitation varies widely with space and time, and may well cause the UT variations in the polar winter ionosphere (Maehlum, 1968; Burch, 1968).

It is not possible to make a simple estimate of the N_2^+ emissions (3914 Å, etc.) associated with polar F-region ionization, because the calculation depends on the $O{:}N_2$ concentration ratio which is not at all well known.

Because the rate coefficient of the (O^+, N_2) loss reaction increases rapidly with electron temperature (Schmeltekopf et al., 1967), particle precipitation tends to cause faster loss of plasma as well as enhanced production. It is not obvious which of these competing effects will be more important.

2.3 Temperature Variations causing Thermal Expansion or Contraction

Although little is known of temperature variations in the polar atmosphere, they are of possible interest in that (unlike winds and electric fields) they might cause the vertical motions observed near the dip poles. Actually hmF2 is modified by thermal expansion of the neutral gas or plasma; NmF2 is not much affected (Garriott and Rishbeth, 1963; L. Thomas, 1966b), except that some dependence of NmF2 on electron temperature may result from the electron-temperature-sensitive rate of the loss reactions, just mentioned in Section 2.2.

2.4 Transport of Plasma by Electric Fields

Electric fields perpendicular to magnetic field lines produce plasma drifts that can be described in terms of "magnetic flux tube circulation" (Hines, 1964). F-region ions and electrons drift together, their motion being similar to that of E-region electrons. Consequently the F-region drift is nearly antiparallel to the E-region current, which is mainly electron (Hall) current (but not quite antiparallel because of the contribution of ion (Pedersen) current, which does not correspond to any F-region drift motion).

Various polar current systems, deduced from ground magnetic data, have been suggested by different workers. They represent "equivalent ionospheric systems", which only resemble the actual ionospheric current flow if there is no major contribution from magnetospheric currents. Assuming for the moment that the currents are mainly ionospheric, the quiet-day pattern consists of current flowing straight across the polar cap towards the longitude of 10 hrs local time, though its direction varies with magnetic disturbance (Obayashi, 1967). The F-region drift is therefore directed approximately away from the longitude of 10 hrs so that plasma from lower latitudes drifts across a polar station around this time of day. (This description is too crude to distinguish between GT, DT or even LT). A more detailed description of the drift patterns has been given by Sato and Rourke (1964) for the Antarctic region.

Two serious limitations of electromagnetic drift can be noted. First, the drift speed is thought to be no more than about 0.1 km s^{-1}, so that plasma can only be transported a distance of about 1000 km during its lifetime of order $\beta^{-1} \sim 10^4$ s. Second, it now seems most improbable that electromagnetic drift can move plasma from the mid-latitude ionosphere into the polar ionosphere, because of the existence of the mid-latitude "troughs" associated with the plasmapause. These features probably mark the boundary between the mid-latitude "flux tube convection" which occurs within a closed volume, and the high-latitude convection pattern (Nishida, 1966). These two convection patterns are distinct, and the electric fields must be such as to drive the plasma along the boundary, never across it. It may also be supposed that electric fields cannot be responsible for creating features like the "ridges" and "cavities" observed within polar regions, since the "flux circulation" description implies that electric fields produce motions that are almost divergence-free. The transport due to electric fields is primarily horizontal; some discussion of vertical transport and ion-drag is given in Section 4.1.

2.5 Transport of Plasma by Neutral Air Winds in the Thermosphere

The winds blow across the polar cap, roughly from the centre of the "diurnal bulge" (situated near 14 LT in low latitudes) towards the "antibulge" (near 03 LT), though their direction is modified by Coriolis force (Kohl and King, 1967). They drive plasma along magnetic field lines, and hence alter the height of the F2-layer and the effective rate of loss (Figure 1). The horizontal wind speeds (U) are typically 150 m s^{-1}, but the field-aligned drift speed (U sin I cos I) does not exceed 50 m s^{-1} at most polar locations and is zero at the dip pole. The winds cannot play any other part in maintaining the polar F-region, since they cannot transport plasma across field lines.

The effect produced by the winds is essentially diurnal. Since it depends on the direction of the field lines it should be related to "magnetic dip time". Maximum upward drift occurs around 03 DT and maximum downward drift around 14 DT. Because of the lack of knowledge of the air distribution in high latitudes, present-day calculations depend on the "global" atmospheric models, like that of Jacchia (1965). The F-region variations deduced from such models can only reflect the gross features of the atmosphere and geomagnetic field, even though the polar atmosphere <u>might</u> have quite a complex structure.

Some such complexity is suggested by the observation that the thermospheric density is more sensitive to magnetic disturbance in high latitudes than in lower latitudes (Jacchia et al., 1967), which might be due to auroral heating (Cole, 1962). Present-day satellite drag data does not seem capable of establishing details of high-latitude structure. In addition, Hines (1960) has suggested that the polar atmosphere may rotate independently of the mid-latitude atmosphere, because of the "drag" imposed by the geomagnetic field. If real, this effect might influence the wind system at high latitudes.

3. ASPECTS OF POLAR F-REGION BEHAVIOUR

3.1 The Polar F1-Layer

The F1-layer is observed in summer in polar regions, and is basically solar controlled, as at mid-latitudes. However, other influences must be at work since, at given stations, foF1 does not depend only on solar zenith angle (Piggott and Shapley, 1962). Knecht (1959) has shown that foF1 varies diurnally by ± 5% at the South Pole where, of course, χ remains sensibly constant during a day.

According to simple theory the shape of the N(h) profile depends on the "shape parameter" $G = \beta^2/\alpha q$, evaluated at the height of peak production (e.g., King, 1961). Because this height varies with χ, G is a function of χ and this fact accounts for important features of mid-latitude F1-layer behaviour. However, Piggott and Shapley (1962) show that the appearance of the F1-layer depends also on hmF2 and on other factors, such as magnetic activity.

3.2 Movements of the F1-Layer

It is necessary to consider how movements might affect the F1-layer. To have a significant effect on the N(h) distribution, a vertical drift W must move the plasma vertically through (say) 0.25 H ~ 10 km during its lifetime τ. For the F1-layer τ is roughly the sum of the lifetimes for the "αN^2" and "βN" loss laws, i.e.,

$$\tau = (2\alpha N)^{-1} + \beta^{-1} \simeq 50 \text{ s} + 100 \text{ s} = 150 \text{ s} \qquad (4)$$

for the reasonable values $N = 2 \cdot 10^5 \text{ cm}^{-3}$, $\alpha = 5 \cdot 10^{-8} \text{ cm}^3 \text{ s}^{-1}$, $\beta = 10^{-2} \text{ s}^{-1}$. Hence a significant vertical velocity would be $10 \text{ km}/150 \text{ s} \simeq 70 \text{ m s}^{-1}$, rather larger than electric fields or winds are likely to provide in polar regions.

The electron density can also be influenced by a vertical gradient of vertical drift velocity, which "compresses" the layer (increasing N) if $dW/dh < 0$ and "stretches" it (decreasing N) if $dW/dh > 0$. To a first order the electron density is altered by a factor of $(1 + \tau\, dW/dh)^{-1}$, τ being the lifetime. In the F2-layer W varies little with height (whether it is due to winds or electric fields); but in the F1-layer the horizontal winds have a vertical gradient of order 1 m s^{-1} per km, or 10^{-3} s^{-1}. At the geographic South Pole, with a dip angle $I = 75^\circ$, the corresponding value of $\tau\, dW/dh \simeq 0.04$, which can account for about half of the foF1 variation. This mechanism would give maximum foF1 around 14 DT = 17 UT, whereas the maximum is observed around 21 UT (Knecht, 1959). It should be noted that the critical frequency foF1 corresponds only to an inflexion on the N(h) profile, and the position of this inflexion may be more sensitive to vertical drift and drift gradients than the above simple argument would indicate.

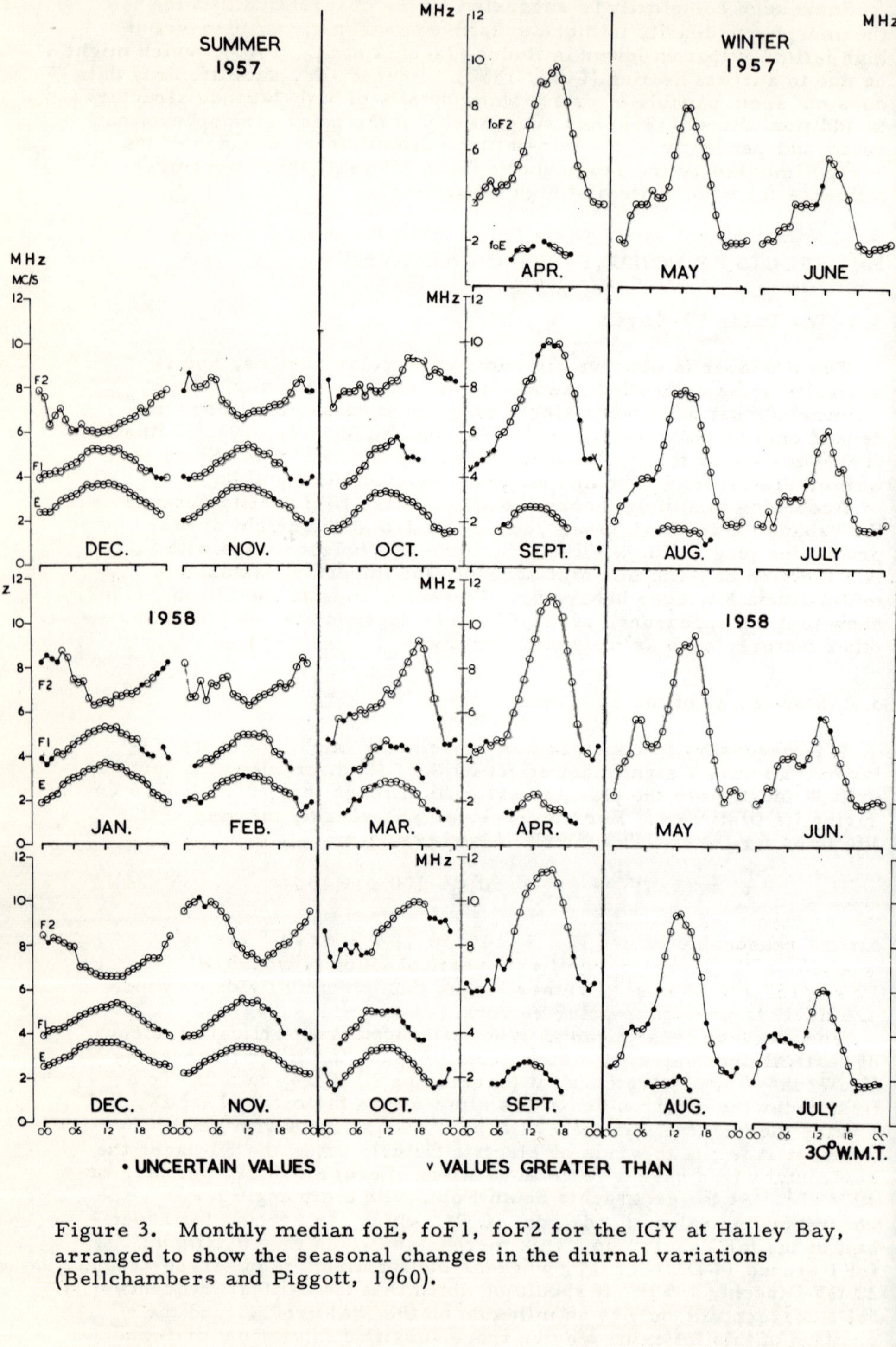

Figure 3. Monthly median foE, foF1, foF2 for the IGY at Halley Bay, arranged to show the seasonal changes in the diurnal variations (Bellchambers and Piggott, 1960).

3.3 Summer-to-Winter Changes in the F1- and F2-layers

The polar F1-layer is a summer phenomenon and its appearance in spring, and disappearance in autumn, seem to accompany the rapid transitions of F2-layer behaviour at these times (Figure 3). Both F1- and F2-layer changes are probably more rapid than can be attributed to the slow variation of χ. The F1-layer changes can be interpreted as changes of the "shape parameter" $G = \beta^2/\alpha q$. The only component of G that seems at all likely to undergo a rapid change is β, which depends on the concentrations of the minor atmospheric constituents N_2 and O_2.

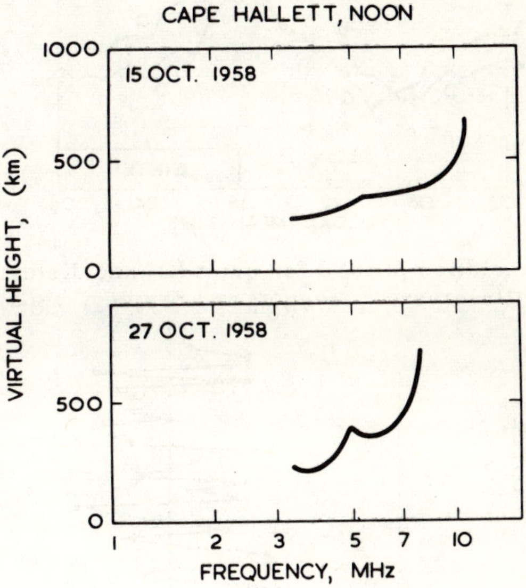

Figure 4. Ordinary ray traces for the F1- and F2-layers, taken from noon ionograms at Cape Hallett, before (15 October 1958) and after (27 October 1958) the onset of a polar stratospheric warming (Bullen, 1964).

Bullen (1964) has suggested that the spring changes of F1-layer shape (increases of G) are triggered by polar stratospheric warmings. The Cape Hallett ionograms shown in Figure 4 show the change in F1-layer structure before and after a stratospheric warming (15 and 27 October 1958). In another example shown by Bullen (9 and 12 October 1957), the change was complete within three days. Not much is known about variations of neutral atmospheric structure in the polar F-region, or about whether stratospheric warmings can affect the structure; but the evidence gives grounds for believing that the polar F1-layer may be as much influenced by atmospheric processes as by movements.

There also exist distinctive "summer" and "winter" patterns in the F2-layer, as shown in Figure 3 for Halley Bay and in Figure 5 for another station, Little America. The electron density is generally lower on disturbed days than on quiet days, at any rate in summer, but the shape of the diurnal variation is much the same. The suddenness of the change from winter to summer conditions is illustrated for one station in Figure 6; Piggott and Shapley (1962) have studied the change for a number of Antarctic stations during the spring of 1958, and showed that it is completed within a few days. The corresponding

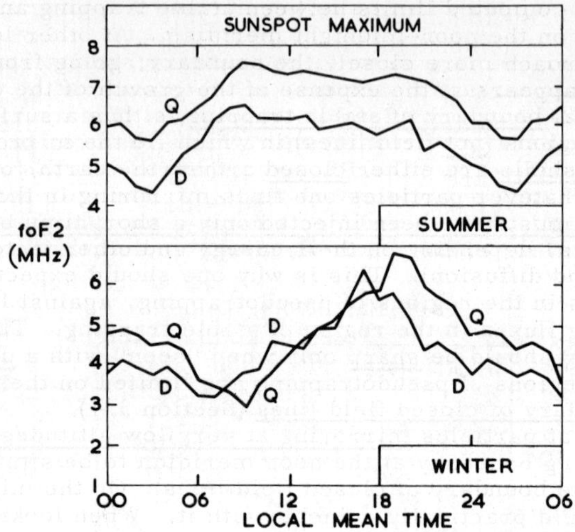

Figure 5. Diurnal variation of foF2 for quiet (Q) and disturbed (D) days during the IGY at Little America (Piggott and Shapley, 1962).

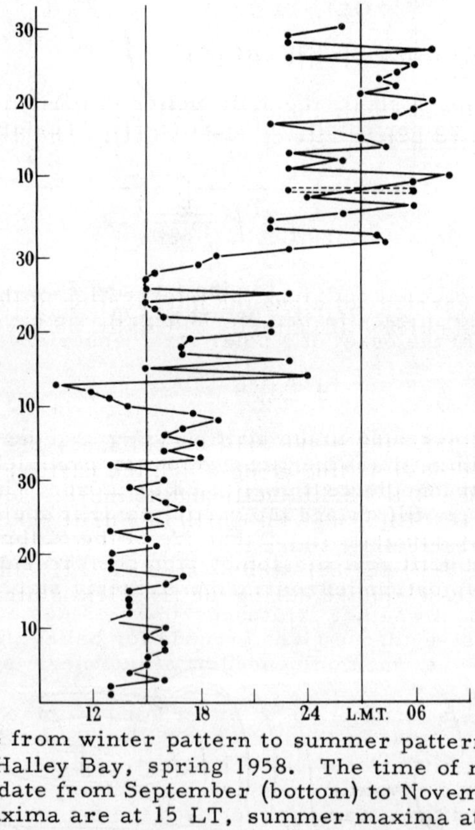

Figure 6. Change from winter pattern to summer pattern of diurnal foF2 variation at Halley Bay, spring 1958. The time of maximum foF2 is plotted against date from September (bottom) to November (top). Typical winter maxima are at 15 LT, summer maxima at 03 LT (Royal Society IGY Antarctic Expedition, Halley Bay, Vol. II, 1962).

change in autumn has not been studied in such detail, but it is thought to be fairly sharp.

There is no feature of the electric field or wind systems, as known at present, that could give rise to such sudden changes. It can only be surmised that major changes in atmospheric structure occur during the spring and autumn, perhaps affecting the ratio of O, O_2 and N_2 concentrations. Such changes have been postulated as a cause of the mid-latitude F2 seasonal anomaly (e.g., Rishbeth and Setty, 1961).

3.4 Latitudinal and Diurnal Variations in the Polar F2-Layer

Most of the extensive data on the polar F2-layer comprise values of foF2 (or NmF2). These are often used in the form of monthly medians, in which case the various dangers of averaging must be guarded against. Some data on hmF2 are available for certain stations (Figures 7 and 8).

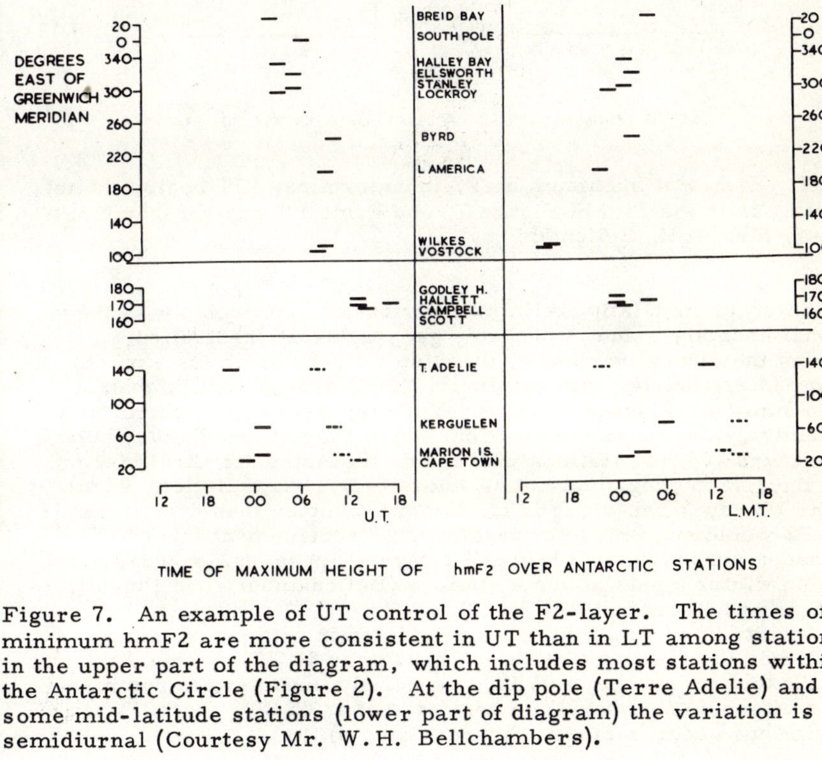

Figure 7. An example of UT control of the F2-layer. The times of minimum hmF2 are more consistent in UT than in LT among stations in the upper part of the diagram, which includes most stations within the Antarctic Circle (Figure 2). At the dip pole (Terre Adelie) and some mid-latitude stations (lower part of diagram) the variation is semidiurnal (Courtesy Mr. W.H. Bellchambers).

There has been much discussion as to whether the polar F2-layer is basically geographically or magnetically controlled, and whether the diurnal variations are LT or UT (e.g., Duncan, 1962; Piggott and Shapley, 1962; Hill, 1963; J.O. Thomas, 1963; Sato and Rourke, 1964). The behaviour is different in different sectors of the Arctic and Antarctic and depends on what parameter is being studied. No detailed survey is made here, but it seems clear that the sharply defined features of the "troughs" (Muldrew, 1965), and other structures at higher latitudes (Thomas and Andrews, 1969), are linked with the geomagnetic field and magnetospheric structure. On the other hand, the daytime distribution of NmF2 in winter may partly be a poleward

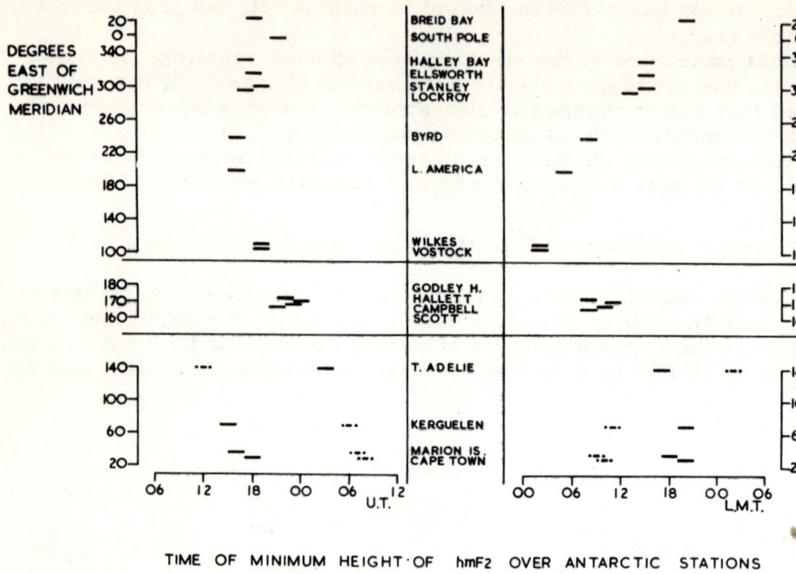

Figure 8. Times of maximum hmF2 in Antarctica. UT control is not so marked as in the data of Figure 7, and some LT control may exist (Courtesy Mr. W.H. Bellchambers).

extension of the mid-latitude NmF2 distribution, connected with the "seasonal anomaly", and is probably geographically controlled. Regarding the diurnal variation, the winter F2-layer shows a much stronger solar control, with maximum NmF2 around 13 LT, than is found in summer (Figures 3 and 5). This apparent solar control may not be entirely due to photoionization, since Piggott and Shapley note that it occurs at some stations well inside the Antarctic Circle (e.g., Halley Bay, 76°S geog.) but not at others (e.g., Cape Hallett, 72°S).

Polar F2-layer behaviour is often more complex than data on NmF2 and hmF2 would suggest. For instance, a "replacement layer" phenomenon, in which two distinct F2-layers appear on ionograms, occurs on winter nights at some stations (Bellchambers and Piggott, 1960; Sato and Rourke, 1964).

UT variations of NmF2 have been detected in both polar regions. In the Arctic the maximum tends to occur near 20 UT. In the Antarctic the maximum tends to be at 06 UT, especially in winter; in summer there is often another maximum besides that at 06 UT. Both UT and LT variations occur in hmF2 (Figures 7 and 8).

3.5 Explaining the Diurnal Variations

The UT variations of foF2, just mentioned, have been variously ascribed to particle precipitation (Duncan, 1962), electromagnetic drifts (Sato and Rourke, 1964) and drift due to winds (King et al., 1967; 1968). Evidence is accumulating that low-energy particles do play a part in the polar F2-layer (Maehlum, 1968) but the wind theory of King et al. has explained some variations of hmF2 and NmF2, including the 20 UT maximum of NmF2 in the Arctic and the 06 UT maximum in the Antarctic. As described in Section 2.4, the pattern of vertical drift depends on the location of the "diurnal bulge" in relation to the dip pole, so that wind effects may be quite well described in terms of "dip time", though in limited regions they may equally be interpreted as UT effects.

The complexity of polar F2-layer behaviour suggests that many processes are at work. The reported preponderance of UT effects in some parameters and LT effects in others (Piggott and Shapley, 1962) implies that the complete physical description cannot be simple. For instance, vertical drifts (whether due to winds or electric fields) should lead to quite a close connection between the NmF2 and hmF2 variations (Figure 1), which is found in some of the data but by no means all. Moreover, as Challinor (1969) has pointed out, the vertical drift due to winds or electric fields should decrease to zero as the dip pole is approached. The extent to which the F2-layer variations actually decrease towards the dip pole is not properly established, but hmF2 variations do occur at Terre Adelie: they seem to be semidiurnal, instead of the diurnal variation usually observed elsewhere (Figure 7). The cause of these hmF2 variations at the dip pole is not obvious, since horizontal drifts or particle precipitation would not be expected to affect hmF2 much.

4. NOTES ON FURTHER THEORETICAL TOPICS RELEVANT TO THE POLAR F-REGION

4.1 The Ion-Drag Time Constant for Suppressing Vertical Electromagnetic Drift

Following earlier suggestions by Cowling and Martyn, Dougherty (1961) showed that if the F-region plasma is moved by electric fields, it tends to accelerate the neutral air because of collisions between ions and neutral particles. The collisions tend to equalize the velocities of the plasma and the air with a certain time constant τ_c, approximately the mean interval between collisions with ions experienced by any one neutral particle. Thus $\tau_c \approx \nu_{ni}^{-1}$, the reciprocal of the neutral-ion collision frequency for momentum transfer. From the data of Dalgarno (1964b) for an assumed temperature of 1000 °K

$$\tau_c = (7.3 \cdot 10^{-10}\ N)^{-1}\ \text{sec} = 4 \cdot 10^5\ N^{-1}\ \text{hrs} \qquad (5)$$

where N(in c. g. s. units) is the ion (or electron) concentration. As a working rule one can say that the plasma and air velocities are effectively the same, for electromagnetic drifts varying diurnally and semidiurnally, if $\tau_c < 2$ hours which implies $N > 2 \cdot 10^5\ cm^{-3}$. This condition is generally met at the F2-peak, except in the polar night.

The significance of this calculation is that the electric fields driving the electromagnetic drifts are not nearly powerful enough to move the neutral air vertically, which would disturb the hydrostatic equilibrium of the air. Consequently, when the plasma and air velocities have been equalized, both must be horizontal. Hence electromagnetic drift will not generally cause much vertical motion (unless τ_c is large, N being small) and its importance in producing changes of height or electron density is therefore limited.

4.2 The Influence of the Polar Wind

According to Banks and Holzer (1968), plasma flows out of the polar ionosphere along geomagnetic field lines. The existence of this "polar wind" requires there to be a substantial "flux" component in the topside electron distribution, with $N_2 > 0$ in equation (2). This implies that the "plasma scale height" ($-N\ dh/dN$) in the topside polar ionosphere is smaller than the value appropriate to diffusive equilibrium. It also implies that a stronger source is needed to maintain the polar F-region than would otherwise be the case. A rough estimate of the latter effect can be made from the work of Gliddon and Kendall (1962), who compare the theoretical behaviour of model F2-layers conforming to the "diffusive equilibrium" distribution ($N_1 \neq 0$, $N_2 = 0$) and the "flux" distribution ($N_1 = 0$, $N_2 \neq 0$) (respectively case II and case I in Gliddon and Kendall's Figure 4). The daytime peak electron densities are the same for the two distributions, within a few per cent, but the "flux" distribution decays twice as fast as the other at night. Probably the existence of the "polar wind" does not make any substantial difference to the problem of maintaining the polar F2-layer.

4.3 The Effect of Field-Aligned Currents on F-Region Structure

Block and Fälthammar (1968) have shown that large field-aligned electric currents can distort the shape of the F2-layer. Since it is possible that such currents flow into or out of the polar ionosphere, it is worth making an estimate of the orders of magnitude involved.

The ions and electrons carrying an electric current experience collisions with neutral particles. If there is to be any substantial effect on the plasma distribution, the collisional forces must be comparable with the other forces acting on the plasma (gravity, pressure gradients, electric polarization field). All these forces are of the order of magnitude of the weight of a positive ion, $m_i g$. Writing m = mass, v = drift velocity, ν = collision frequency, g = gravitational acceleration, and using subscripts e, i, n for electrons, ions, neutrals, the requirement is

either (for an electron current) $m_e v_e \nu_{en} \sim m_i g$

or (for an ion current) $m_i v_i \nu_{in} \sim m_i g$ (6)

At 300 km, using reasonable values $\nu_{in} = 0.5\ s^{-1}$, $\nu_{en} = 20\ s^{-1}$ and assuming the ions are O^+, these conditions require $v_i \sim 20\ m\ s^{-1}$ or $v_e \sim 1.3 \cdot 10^4\ m\ s^{-1}$. For an electron density $N = 3 \cdot 10^5\ cm^{-3}$ the corresponding current density is about 10^{-6} amp m^{-2} for an ion current or $6 \cdot 10^{-4}$ amp m^{-2} for an electron current. Any field-aligned current is likely to be carried mainly by electrons, which are very much more mobile than ions. The field-aligned currents actually flowing in the mid-latitude F-region are not expected to exceed 10^{-9} amp m^{-2} (Reid, 1965; Maeda and Murata, 1965) and there is no obvious reason why much bigger field-aligned currents should flow above the polar ionosphere. Only above the auroral electrojets does it seem likely that field-aligned currents could greatly exceed the estimated mid-latitude values.

The large gradients of electron density existing in the plasma "troughs" must produce Hall currents, but these do not produce any effects resembling those described above.

5. CONCLUSION

5.1 Summary of Present Ideas on the Maintenance of the Polar F-Region

Solar photoionization can maintain the observed level of electron density, within about 10° of the midwinter geographic pole (Section 2.1).

Corpuscular ionization by 10^2 - 10^3 eV electrons may well be sufficient to maintain the winter polar F-region: the required input of about $6 \cdot 10^{-3}$ erg cm^{-2} s^{-1} (which would cause about 100 rayleighs of 6300 Å airglow) seems compatible with observations (Section 2.2). Further investigation is needed regarding the level of ionization in the "cavity" around the dip poles.

Other sources of ionization are very problematical. Electric fields do not seem capable of transferring plasma from the mid-latitude F-region to the polar F-region, largely because they can only move plasma a few hundred kilometres during its lifetime (Section 2.4). Winds do not move plasma across field lines and therefore cannot provide the horizontal transport needed to maintain the winter F-region (Section 2.5).

Influx of plasma along field lines is discounted because of the lack of a suitable source of supply; solar photoionization is quite inadequate because of the very low gas density at heights where polar field lines are illuminated in winter by solar ultraviolet light. There may indeed be an outflux rather than an influx, owing to the "polar wind" (Section 4.2).

5.2 Diurnal and Seasonal Variations

Winds can account for some of the LT and UT variations of NmF2 and hmF2, though other factors probably contribute (Sections 3.4 and 3.5). The seasonal changes are found, not only in the F2-layer, but also in the F1-layer which is controlled by production and loss processes (Section 3.3): this indicates that some change of atmospheric structure occurs, for which corroborative evidence is needed. It is notable that the seasonal changes occur rather abruptly.

5.3 Parameters of which Further Study is Required

The neutral atmosphere. The mid-latitude F-region only began to be well understood when knowledge of the neutral atmosphere was increased, largely by the use of satellites. The same may well be true of the polar F-region. It is necessary to measure the composition and temperature of the atmosphere at F-region heights, with particular regard to possible seasonal variations. Rocket experiments are probably necessary for this purpose. The temperature distribution in high latitudes, which may be quite complex, must be determined if the wind pattern is to be accurately computed. Any direct information on winds, e.g., from the motion of artificial neutral and ionized clouds, is valuable.

The plasma. Again, temperature and composition need to be further studied. The incoherent (Thomson) scatter technique would be particularly useful for this purpose, especially since it also offers the possibility of measuring the plasma drift velocity.

Ionospheric currents. If the movements due to electric fields are to be established, it is necessary to determine the ionospheric current flow, and to find out what currents flow along magnetospheric field lines (Sections 2.4 and 4.3). Measurements with rocket-borne magnetometers, up to at least the top of the current-bearing E-region (say 150 km), are needed for this purpose.

Low energy particles. Although the evidence is now quite strong that the polar night F-region can be maintained by particles, further data are required, particularly in the low-density "cavity" around the dip poles.

Acknowledgment. Discussions with Mr. W. R. Piggott are gratefully acknowledged.

This paper is published by permission of the Director of the Radio and Space Research Station in the Science Research Council of the United Kingdom.

REFERENCES

Banks, P. M. and T. E. Holzer, The polar wind, J. Geophys. Res., 73, 6846-6854, 1968.

Bellchambers, W. H. and W. R. Piggott, The ionosphere over Halley Bay, Proc. Roy. Soc. London, A256, 200-218, 1960.

Block, L. P. and C. -G. Fälthammar, Effects of field-aligned currents on the structure of the ionosphere, J. Geophys. Res., 73, 4807-4812, 1968.

Bullen, J. M., Ionospheric recombination and the polar stratospheric warming, J. Atmospheric Terrest. Phys., 26, 559-568, 1964.

Burch, J. L., Low-energy electron fluxes at latitudes above the auroral zone, J. Geophys. Res., 73, 3585-3591, 1968.

Challinor, R. A., Universal time control of the polar ionosphere, Nature, 221, 941-943, 1969.

Cole, K. D., Joule heating of the upper atmosphere, Aust. J. Phys., 15, 223-235, 1962.

Dalgarno, A., Corpuscular radiation in the upper atmosphere, Ann. Géophys., 20, 65-74, 1964a.

Dalgarno, A., Ambipolar diffusion in the F-region, J. Atmospheric Terrest. Phys., 26, 939, 1964b.

Dougherty, J. P., On the influence of horizontal motion of the neutral air on the diffusion equation of the F-region, J. Atmospheric Terrest. Phys., 20, 167-176, 1961.

Duncan, R. A., Universal time control of the Arctic and Antarctic F-region, J. Geophys. Res., 67, 1823-1830, 1962.

Garriott, O. K. and H. Rishbeth, Effects of temperature changes on the electron density profile in the F2-layer, Planetary Space Sci., 11, 587-590, 1963.

Gliddon, J. E. C. and P. C. Kendall, A mathematical model of the F2-region, J. Atmospheric Terrest. Phys., 24, 1073-1099, 1962.

Hill, G. E. Sudden enhancements of F-layer ionization in polar regions, J. Atmospheric Sci., 20, 492-497, 1963.

Hines, C. O., On the rotation of the polar ionospheric regions, J. Geophys. Res., 65, 141-143, 1960.

Hines, C. O., Hydromagnetic motions in the magnetosphere, Space Sci. Rev., 3, 342-379, 1964.

Jacchia, L. G., Static diffusion models of the upper atmosphere with empirical temperature profiles, Smithsonian Contrib. Astrophys., 8, 215-257, 1965) (also published as Smithsonian Astrophys. Obs. Special Rep. 170).

Jacchia, L. G., J. Slowey, and F. Verniani, Geomagnetic perturbations and upper-atmosphere heating, J. Geophys. Res., 72, 1423-1434, 1967.

Kamiyama, H., The contribution to the formation of the polar F2-region through photoionization of oxygen by grazing solar radiation at great heights, Sci. Reps. Tohoku Univ., Series 5, 18, 1-15, 1966.

King, G. A. M., Analysis of the F1-F2 transition region, J. Geophys. Res., 66, 2757-2762, 1961.

King, J. W., H. Kohl, and R. Pratt, The effect of atmospheric winds on the height of the F2-layer peak at middle and high latitudes, J. Atmospheric Terrest. Phys., 29, 1529-1539, 1967.

King, J. W., H. Kohl, D. M. Preece, and C. Seabrook, An explanation of phenomena occurring in the high-latitude ionosphere at certain Universal Times, J. Atmospheric Terrest. Phys., 30, 11-23, 1968.

Knecht, R. W. Observations of the ionosphere over the south geographic pole, J. Geophys. Res., 64, 1243-1250, 1959.

Kohl, H. and J. W. King, Atmospheric winds between 100 and 700 km and their effects on the ionosphere, J. Atmospheric Terrest. Phys., 29, 1045-1062, 1967.

Maeda, K. and H. Murata, Ionospheric dynamo theory with consideration for magnetospheric current along the geomagnetic lines of force, Rep. Ionos. Space. Res. Japan, 19, 272-285, 1965.

Maehlum, B. N., Universal-time control of the low-energy electron fluxes in the polar regions, J. Geophys. Res., 73, 3459-3468, 1968.

Muldrew, D. B., F-layer ionization troughs deduced from Alouette data, J. Geophys. Res., 70, 2635-2650, 1965.

Nishida, A., Formation of plasmapause or magnetospheric plasma knee, by the combined action of magnetospheric convection and plasma escape from the tail, J. Geophys. Res., 71, 5669-5679, 1966.

Obayashi, T., The interaction of the solar wind with the geomagnetic field during disturbed conditions, Chapter V in Solar-Terrestrial Physics, edited by J. W. King and W. S. Newman, Academic Press, London, pp. 107-167, 1967.

Ogawa, T. and T. Tohmatsu, Photoelectronic processes in the upper atmosphere, II. The hydrogen and helium ultraviolet glow as an origin of the nighttime ionosphere, Rep. Ionos. Space Res. Japan, 20, 395-417, 1966.

Piggott, W. R. and A. H. Shapley, The ionosphère over Antarctica, in Antarctic Research, Geophysical Monograph No. 7, American Geophysical Union, Washington, pp. 111-126, 1962.

Rees, M. H., Auroral ionization and excitation by incident energetic electrons, Planetary Space Sci., 11, 1209-1218, 1963.

Reid, G. C., Ionospheric effects of electrostatic fields generated in the outer magnetosphere, Radio Sci., 69D, 827-837, 1965.

Rishbeth, H., A review of ionospheric F region theory, Proc. IEEE, 55, 16-35, 1967.

Rishbeth, H., On explaining the behavior of the ionospheric F region, Rev. Geophys., 6, 33-71, 1968.

Rishbeth, H. and C. S. G. K. Setty, The F-layer at sunrise, J. Atmospheric Terrest. Phys., 20, 263-276, 1961.

Rishbeth, H. and O. K. Garriott, Introduction to Ionospheric Physics, Academic Press, New York, 1969.

Sato, T. and G. F. Rourke, F-region enhancements in the Antarctic, J. Geophys. Res., 69, 4591-4607, 1964.

Schmeltekopf, A. L., F. C. Fehsenfeld, G. I. Gilman, and E. E. Ferguson, Reaction of atomic oxygen ions with vibrationally excited nitrogen molecules, Planetary Space Sci., 15, 401-406, 1967.

Thomas, J. O., The electron density distribution in the F2 layer of the ionosphere in winter, J. Geophys. Res., 68, 2707-2718, 1963.

Thomas, J. O. and M. K. Andrews, The trans-polar exospheric plasma, 3: A unified picture, Planetary Space Sci., 17, 433-446, 1969.

Thomas, L., The importance of photoionization in the production of the daytime F-region at high latitudes during winter, Planetary Space Sci., 14, 891-899, 1966a.

Thomas, L., Electron density distributions in the daytime F2 layer and their dependence on neutral gas, ion, and electron temperatures, J. Geophys. Res., 71, 1357-1366, 1966b.

PLASMA TRANSPORT IN THE TOPSIDE POLAR IONOSPHERE

Peter M. Banks

Department of Applied Physics and Information Science
University of California, San Diego
La Jolla, California 92037, USA

Abstract. The escape of ionization from the polar ionospheres occurs with sufficient rapidity to greatly suppress the H^+ and He^+ densities, leaving O^+ as the dominant ion to 3000 to 7000 km. Since the O^+ scale height is small relative to that for H^+, the overall electron densities are greatly reduced below plasmaspheric values. Although steady state plasma escape occurs only along open geomagnetic field lines, upwards plasma transport should continue throughout much of the transplasmaspheric region as a time-dependent phenomenon resulting from field tube convection. From models of ionospheric processes in polar regions, predictions are made of ion density profiles consistent with the polar wind theory. With regard to the general features of plasma transport, a discussion is given of the basic features of plasma motions in situations when ion pressure gradients can be neglected. A comparison between neutral gas and plasma behavior under exospheric conditions is also made.

1. INTRODUCTION

Satellite measurements of ion composition and density in polar regions (Barrington et al., 1966; Hoffman, 1968; Donley, 1968; Thomas and Andrews, 1968; Andrews and Thomas, 1969; Nelms and Chapman, 1969; Timleck and Nelms, 1969) have consistently shown peculiar results. Not only is O^+ the dominant ion to very high altitudes, but the ambient topside electron densities are extraordinarily low with values less than 30 cm^{-3} at 3000 km having been reported over the winter pole (Donley, 1968). On a global scale, regions of enhanced plasma densities in the topside ionosphere are restricted to the plasmasphere, a closed, toruslike volume lying about the geomagnetic equator whose outer edges ($L \approx 4 - 6$) are coincident with lines of force of the geomagnetic field, as shown in Figure 1. Outside the sharp plasmasphere boundary, lower electron densities are found at progressively higher latitudes although local features, such as the auroral zones, are present. Typical high latitude ion composition results of J. Hoffman (personal communication) at 3000 km from Explorer 31 data are shown in Figure 2.

The explanation for the peculiar nature of the polar topside ionosphere appears from recent studies to be linked to the open character of the earth's magnetosphere at high latitudes and to the

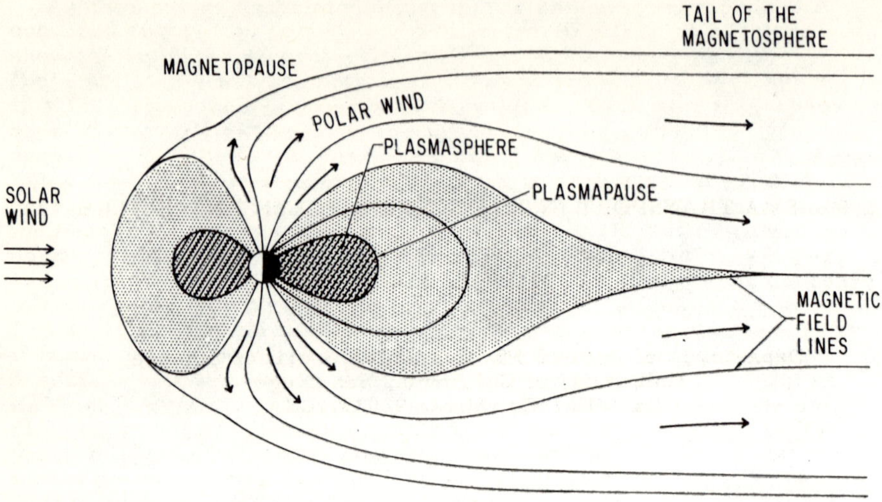

Figure 1. View of the earth's magnetic field and plasma environment. Plasma escape (the polar wind) is present along open field lines. As field tubes convect to lower latitudes in the nighttime sectors, however, upward plasma flow continues until diffusive equilibrium is attained.

process of field tube convection. (Dungey, 1961; Bauer, 1966; Nishida, 1966; Dessler and Michel, 1966; Axford, 1968; Banks and Holzer, 1968). At very high latitudes ($\Lambda \gtrsim 75^{\circ}$) geomagnetic field lines do not form closed loops but rather lead into the tail of the magnetosphere and ultimately connect with the interplanetary field (Dungey, 1961; Axford, 1969). Charged particles, constrained to move along the field lines, must follow along closed dipole-like loops at low latitudes but, at high latitudes for proper conditions, can escape from the earth.

Accepting an open model of the magnetosphere, early arguments were given by Bauer (1966) and Dessler and Michel (1966) that polar ionospheric plasma could gradually seep outwards into the magnetospheric tail via an "evaporative" mechanism. Although the ionospheric dynamics were not considered in detail by Bauer or Dessler and Michel, the implications were clear: Outwards plasma flow would reduce the topside electron densities below the values consistent with diffusive equilibrium.

In more recent work the process of plasma transport in polar regions has been considered in detail (Banks and Holzer, 1968). Using a hydrodynamic approach rather than the principles of evaporation, it has been found that a number of interesting factors (friction, field tube divergence, particle production and loss, partial pressure gradients, and electric fields) conspire to rapidly bring H^+ and He^+ to supersonic speeds and greatly reduce the densities of these light ions below the values appropriate for diffusive equilibrium. Consequently, O^+, which does not escape readily from the upper atmosphere as a result of its larger mass, remains as the dominant ion to altitudes of 3000 to 7000 km with the overall electron density being greatly diminished due to the relatively small scale height of O^+ (as compared with H^+ at mid-geomagnetic latitudes).

In terms of overall dynamics, the polar wind theory of ionospheric composition predicts that in the transplasmaspheric regions:

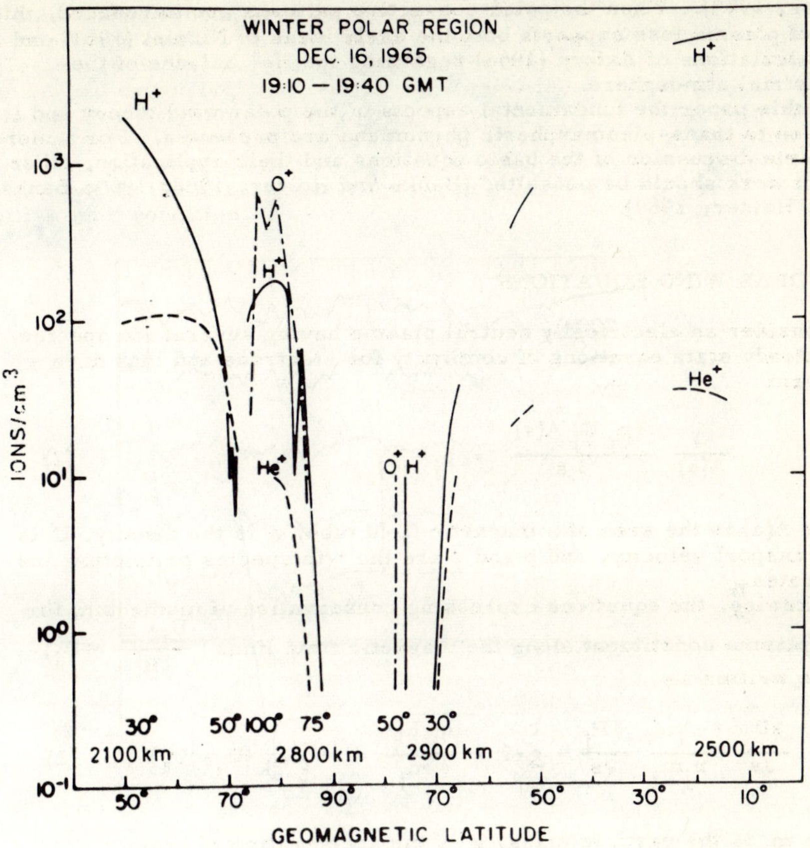

Figure 2. Ion composition measurements from Explorer 31 satellite at about 3000 km altitude in high latitude regions (J. H. Hoffman, personal communication).

a) Upwards fluxes of O^+ on the order of $5 \cdot 10^8$ cm^{-2} sec^{-1} should be present above the F2-peak to support the escape flux of H^+ which begins at greater altitudes.

b) O^+- H charge exchange above about 600 km leads to H^+. Accelerated by its own partial pressure gradient and the ionospheric charge separation electric field (the major contribution arises from O^+- e charge separation since H^+ is almost always a minor constituent at and below its critical point), but retarded by O^+- H^+ friction and particle production, H^+ flows out of the polar ionosphere, rapidly reaching supersonic speeds and continuing on to velocities on the order of 20 to 40 km sec^{-1} or greater, depending upon the magnitude of the currently unknown thermal gradients in the tail of the magnetosphere. H^+ escape fluxes of 5 - 8 $\cdot 10^8$ cm^{-2} sec^{-1} are calculated (normalized to 1000 km) for typical models of summer time (photoionization) polar conditions.

c) He^+, produced through photoionization and lost via ionic reactions with N_2 and O_2, undergoes an acceleration process similar to that for H^+, with the He^+ critical point being found only slightly higher than that for H^+. Escape fluxes on the order of 2 - 4 $\cdot 10^6$ cm^{-2} sec^{-1} have been calculated (Holzer, 1969; Banks and

Holzer, 1969). When the polar convective patterns are introduced, this rate of plasma loss supports both the early ideas of Nicolet (1961) and the calculations of Axford (1968) regarding the He^4 balance of the terrestrial atmosphere.

In this paper the fundamental aspects of the polar wind theory and its relation to trans-plasmaspheric phenomena are presented. For a more complete discussion of the basic equations and their application, other recent work should be consulted (Banks and Holzer, 1968; 1969; Banks, 1968; Holzer, 1969).

2. POLAR WIND EQUATIONS

Consider an electrically neutral plasma having several ion species. The steady state equations of continuity for electrons and ions have the form

$$\frac{1}{A(s)} \frac{\partial n_j U_j A(s)}{\partial s} = p_j - \ell_j \tag{1}$$

where A(s) is the area of a magnetic field tube, n is the density, U is the transport velocity, and p and ℓ are the j-th species production and loss rates.

Likewise, the equations expressing conservation of momentum for each plasma constituent along the magnetic field lines $\left(\frac{\underline{U}_j \cdot \underline{B}}{|\underline{B}|} = U_j \right)$ can be written as

$$U_j \frac{\partial U_j}{\partial s} + \frac{1}{n_j m_j} \frac{\partial P_j}{\partial s} - \underline{g} \cdot \hat{s} - \frac{q_j E_{||}}{m_j} = - \sum_k \nu_{jk} (U_j - U_k) \tag{2}$$

where m_j is the particle mass, $\underline{g}$ is the acceleration of gravity (positive inwards), q_j is the electric charge, $E_{||}$ is the component of $\underline{E}$ parallel to $\underline{B}$, and ν_{jk} is the collision frequency of j-type particles of speed U_j with k type particles of speed U_k. The term $\frac{\partial P_j}{\partial s}$ identifies (2) as the hydrodynamic approximation to the conservation of momentum since, in general, the divergence of the pressure tensor should be introduced. Essentially, the hydrodynamic approximation assumes the j-th species partial pressure is isotropic about the j-th species transport speed, U_j. For a displaced Maxwellian distribution this scalar pressure is related to n_j and temperature through $P_j = n_j k T_j$.

To apply (1) and (2) to polar conditions it is convenient to assume ambipolar motions $(n_e U_e = \sum_j n_j U_j)$ along the magnetic field lines, low speeds with respect to the electron thermal speed $(U_j^2 << k T_e/m_e)$, and to neglect electron collisions.

With these restrictions the electron gas momentum equation reduces to

$$\frac{1}{n_e} \frac{\partial P_e}{\partial s} = - e E_{||} \tag{3}$$

while the j-th ion species momentum equation becomes

$$U_j \frac{\partial U_j}{\partial s} + \frac{1}{n_j m_j} \frac{\partial P_j}{\partial s} - \underline{g} \cdot \hat{s} - \frac{e E_{||}}{m_j} = - \sum_k \nu_{jk} (U_j - U_k) \tag{4}$$

Equations (3) and (4) represent the standard hydrodynamic, ambipolar equations for a plasma where $U_j^2 \ll kT_e/m_e$. In the derivation the actual form of the individual species velocity distribution function enters only in the pressure gradient terms and in the calculation of the collision frequencies, ν_{jk}. Again, in the hydrodynamic approximation individual ion pressures are assumed to be isotropic in a coordinate frame moving with speed U_j, such that the stress tensor reduces to a single scalar pressure gradient parallel to $\underline{B}$ (Dungey, 1958). Depending upon the particular physical model involved, the assumption of a scalar pressure gradient may or may not be valid. As discussed in the final section of this paper, however, for the polar ionosphere conditions are such that for subsonic speeds H^+- O^+ and He^+- O^+ collisions occur frequently enough that the use of the hydrodynamic approximation is well justified. In any case, as shown in Section 3, the results for a plasma model where ionic partial pressure gradients are ignored $\left(\frac{\partial P(H^+)}{\partial s} = 0\right)$ do not differ in any essential detail from the hydrodynamic model.

Continuing the development of hydrodynamic plasma motion, using (3) the parallel electric field can be eliminated from the j-th species momentum conservation equation to give

$$\frac{1}{U_j}\frac{\partial U_j}{\partial s}\left(\frac{U_j^2}{C_j^2} - 1\right) = \frac{g\cdot\hat{s}}{C_j^2} + \frac{1}{A}\frac{\partial A}{\partial s} - \frac{1}{T_i}\frac{\partial T_i}{\partial s} - \frac{1}{C_j^2}\sum_k \nu_{jk}(U_j - U_k) - \frac{1}{Q_j}\frac{\partial Q_j}{\partial s} - \frac{1}{C_j^2 n_e m_j}\frac{\partial P_e}{\partial s} \tag{5}$$

where $C_j = (kT_i/m_j)^{\frac{1}{2}}$ is termed the j-th species sound speed and

$$Q_j = \int_{s_o}^{s} (p_j - \ell_j)\, A\, ds' + n_o U_{jo} A_o \tag{6}$$

is the flux of j-th species ions.

Introducing the j-th ion local Mach number, $M_j = U_j/C_j$, (5) can be rewritten as

$$\frac{1}{M_j}\frac{\partial M_j}{\partial s}\left(M_j^2 - 1\right) = \frac{g\cdot\hat{s}}{C_j^2} + \frac{1}{A}\frac{\partial A}{\partial s} - \frac{1}{C_j^2}\sum_k \nu_{jk}(U_j - U_k) - \frac{1}{Q}\frac{\partial Q}{\partial s} - \frac{1}{2T_i}\frac{\partial T_i}{\partial s}\left(M_j^2 + 1\right) - \frac{1}{T_i}\frac{\partial T_e}{\partial s} - \frac{T_e/T_i}{n_e}\frac{\partial n_e}{\partial s} \tag{7}$$

The different solutions of (7) are shown in Figure 3 for typical summertime polar conditions where O^+ is produced through photoionization, and H^+ is created by O^+- H charge exchange. (A discussion of the parameters used to obtain Figure 3 is given in Holzer, 1969). Both subsonic and supersonic flow corresponding to inwards or outwards plasma motion is available to meet particular sets of boundary conditions. A complete discussion of the choices appropriate to the polar regions is given elsewhere (Banks and Holzer, 1968; 1969) and only a brief description can be given here of the wide variety of possible ambipolar plasma motions.

Since $M(H^+) \geq 0$ as the altitude becomes large in Figure 3, principal interest centers about outwards plasma motion. Neglecting the regions below 700 km where chemical effects compete with transport processes,

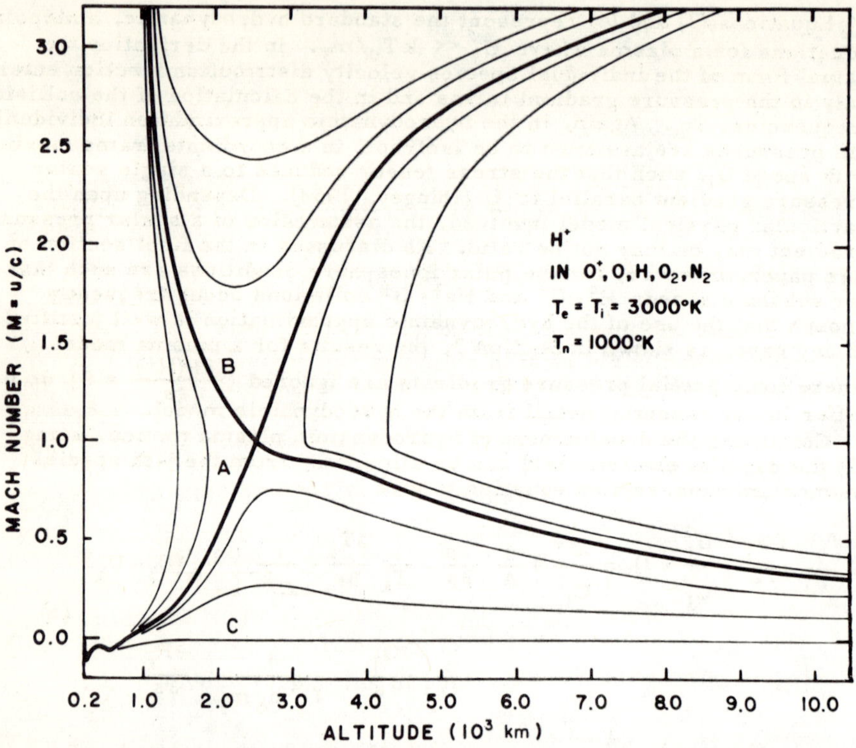

Figure 3. Solutions to the H^+ Mach number equation for $T_n = 1000\ ^oK$, $T_e = T_i = 3000\ ^oK$. For any asymptotic H^+ pressure far from the earth less than that of Curve B, only Curve A is the correct solution for the H^+ Mach number and a supersonic transition point is established at some altitude. For boundary pressures larger than that of Curve B, outwards subsonic flow of H^+ occurs, and at some particular pressure $M(H^+) \rightarrow 0$, corresponding to diffusive equilibrium. For still greater boundary pressures, inwards plasma flow $\left(M(H^+) < 0\right)$ takes place.

various types of solutions are available, corresponding to supersonic flow (Curve A), subsonic flow (the curves lying in the range $0 \lesssim M(H^+) < 1$), and diffusive equilibrium (Curve C). In order to determine which steady state solution applies to a particular physical model, it is necessary to apply appropriate boundary conditions. At low altitudes, for example, it is known that chemical effects rapidly dominate transport and the only appropriate solutions are those which require $M(H^+) \rightarrow 0$ at 200 km altitude or less.

Since Curve A and the subsonic curves described immediately above satisfy this requirement, it is now necessary to apply a second, asymptotic boundary condition to specify exactly a particular solution. Essentially, each of the curves shown in Figure 3 have particular values of plasma pressure at great distances from the earth. If one regards the earth as being surrounded at some large distance by an appropriate boundary sink (or source) of plasma, then by designating a particular value of the boundary plasma pressure, one curve is chosen from the various types of solutions shown in Figure 3. For boundary plasma pressures $P(\infty)$ in the range $0 \leq P(\infty) < P_B(\infty)$, where $P_B(\infty)$ is the asymptotic pressure of the descending critical solution

(Curve B), Curve A is the only correct solution and represents an outwards flux of H^+ with a transition to supersonic flow speeds.

Ideally, Curve A applies over an entire asymptotic range only if $P(\infty) \to 0$ as $s \to \infty$. For $0 < P(\infty) < P_B(\infty)$ the pressure boundary condition is met through a shock transition from Curve A $\left\{M(H^+) > 1\right\}$ to those solutions lying above Curve B but below $M(H^+) = 1$.

When the boundary pressure, $P(\infty)$, is increased beyond $P_B(\infty)$, which typically has the value 50 eV cm^{-3}, one of the infinite set of subsonic escape solutions is obtained, while for some particular value of $P(\infty)$, $M(H^+) \to 0$ giving a distribution of ionization corresponding to diffusive equilibrium. For $P(\infty) > P_c(\infty)$ inwards accretive plasma flow must result and the features for $M(H^+) < 0$, not shown in Figure 3, must be calculated.

To determine the nature of plasma transport solutions appropriate to the polar regions of the earth, one must know the plasma pressures appropriate to the high latitude regions of the tail of the magnetosphere. Although numerous experimental data are not available, Vasyliunas (1969) has reported electron pressure measurements from OGO 3 for $\varepsilon > 200$ eV of less than 2 eV cm^{-3}; that is, far below those pressures required to block a supersonic expansion of H^+ from the ionosphere. Hence, although changing tail plasma pressures can force both inwards and outwards plasma motion, for the most general conditions it appears that the outwards, supersonic expansion mode is most appropriate.

In the ionospheric regions below 1000 km, $n(H^+) \ll n(O^+)$ and $U(H^+) \ll \left(k T_i/m(H^+)\right)^{\frac{1}{2}}$. To indicate the various forces acting upon H^+ (and He^+) the low speed approximation to the H^+ momentum conservation equation can be written from (5) in the form

$$U(H^+) = -\frac{k T_i/m(H^+)}{\nu(H^+, O^+)}\left[\frac{1}{n(H^+)}\frac{\partial n(H^+)}{\partial s} - \frac{m(H^+)\, \mathbf{g}\cdot\hat{\mathbf{s}}}{k T_i} + \frac{T_e/T_i}{n_e}\frac{\partial n_e}{\partial s} + \frac{1}{T_i}\frac{\partial (T_e + T_i)}{\partial s}\right] + v(O^+) \tag{8}$$

ignoring H^+ inertia. Since $n_e \simeq n(O^+)$, (8) indicates that, in addition to acceleration due to the H^+ partial pressure gradient, the O^+ - e charge separation electric field also acts to accelerate H^+ out of the ionosphere. Within the plasmasphere where $U(H^+) \simeq 0$ for diffusive equilibrium this upwards acceleration is generally matched by a positive H^+ pressure gradient $\left(\frac{\partial n(H^+)}{\partial s} > 0\right)$, but in polar regions this is not the case and the resulting outwards flow of H^+ is retarded only through H^+ - O^+ friction (low speeds) and inertia (high speeds).

Plots of H^+ speed as a function of altitude for different temperatures in the hydrodynamic approximation are shown in Figure 4. The details of the model atmospheres, ionization rates, and other parameters used in obtaining the results of Figure 4 are given by Holzer (1969).

3. FEATURES OF THE HYDRODYNAMIC AND ZERO PRESSURE GRADIENT APPROXIMATIONS

In Section 2 emphasis has been placed upon the features of plasma motion under the hydrodynamic approximation; that is, when a scalar pressure term is included in the ion equation of momentum conservation. Although the introduction of an anisotropic pressure can be readily made in the equations of motion (Dungey, 1958), in the present case it is most

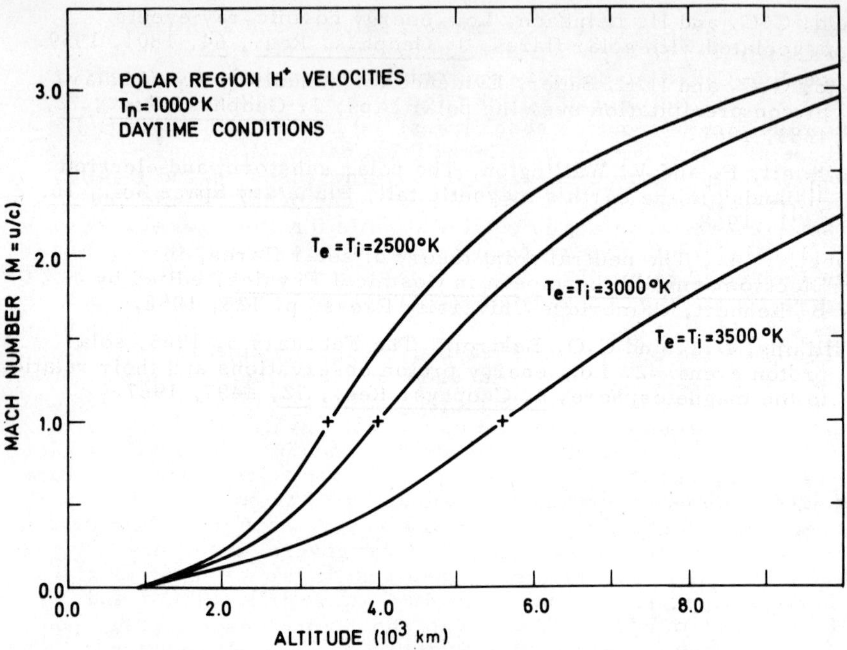

Figure 4. H^+ Mach number as a function of altitude for several plasma temperatures and $T_n = 1000\,^oK$. Higher temperatures increase $n(O^+)$ and move the H^+ critical point outwards.

expedient to discuss immediately the extreme example of plasma motions where, as a result of initial conditions, the j-th species ion pressure is always zero ($P_j = 0$) such that no pressure gradient relating to the j-th ion species enters the momentum conservation equation. It will be shown that such a radical approximation does not lead to results far different from the hydrodynamic approach and that the general form of the solutions for the j-th species velocity along $\underline{B}$ involve both subsonic and supersonic conditions, along with a well defined critical point.

3.1 Neutral Gas Evaporation

Consider first the standard model of evaporation used to describe the altitude profile of light gases in the terrestrial atmosphere. The momentum conservation equation for such a gas in the radial direction is

$$U_k \frac{\partial U_k}{\partial r} + \frac{1}{n_k m_k} \frac{\partial P_k}{\partial s} - \underline{g} \cdot \hat{s} = - \sum_m \nu_{kn} U_k \qquad (9)$$

where the hydrodynamic pressure gradient is included for the moment and k-th species friction with a stationary neutral atmosphere is present.

To apply (9) to H and He in the earth's upper atmosphere certain approximations are made. First, below the exospheric level collisions with the major constituents of the neutral atmosphere occur sufficiently rapidly that the hydrodynamic approximation is assumed to be valid.

In addition, the restraining effect of collisions is sufficiently large to ensure that k-th species inertia effects are small. Thus, the k-th species radial equation of motion becomes

$$U_k = \frac{k\,T_k/m_k}{\sum\limits_m \nu_{km}} \left[\frac{1}{n_k} \frac{\partial n_k}{\partial r} - \frac{g \cdot \hat{s}}{k\,T_k} + \frac{1}{T_k} \frac{\partial T_k}{\partial r} \right] \tag{10}$$

which may be used with the equation of continuity

$$\frac{1}{A} \frac{\partial n_k U_k A}{\partial r} = 0 \tag{11}$$

(H and He production in the thermosphere is very small or nonexistent) to obtain the k-th species density profile below the exobase.

The second order linear differential equation for $n_k(r)$ has two linearly independent solutions corresponding to $U_k = 0$ and $U_k \propto 1/\Sigma\,\nu_{k\ell}$ and may be expressed as

$$n_k = A\,n_k^1(r) + B\,n_k^2(r) \tag{12}$$

where

$$n_k^1(r) \rightarrow U_k = 0 \tag{13}$$

$$n_k^2(r) \rightarrow U_k \propto 1/\Sigma\,\nu_{k\ell} \tag{14}$$

To find $n_k(r)$ in the upper atmosphere requires that boundary conditions be applied. Usually, for H and He, n(H) and n(He) are specified at some low altitude (100 - 120 km) while an mean escape transport velocity $\left(U_E(H),\ U_E(He)\right)$ is given at the exospheric level. Through the calculation of A and B the final expression for $n_k(r)$ can be obtained in terms of the static solution (n_k^1) and the flowing solution (n_k^2).

To determine U_k at the exospheric base certain approximations to the escape of light particles must be used. Thus, at the exospheric level a fraction of the k-th species particles have, by virtue of the particle velocity distribution function, velocity components such that if collisions do not occur they may escape the earth's gravitational field. To describe this effect, termed evaporation, (9) may be used to obtain the radial equation of motion:

$$U_k \frac{\partial U_k}{\partial r} = g \cdot \hat{r} \tag{15}$$

Equation (15) represents adiabatic motion since, after leaving the exospheric level, particles of the k-th species experience only conservative forces with no additional energy input. For $U_k^o \geq \sqrt{2\,g_o\,r_o}$ escape occurs as a result of a true evaporative process.

3.2 Motion of Plasma with $\frac{\partial P_i}{\partial s} = 0$

Consider now the possible motions of a plasma in an atmosphere composed of electrons and a single ion species. Below some ion exospheric level the hydrodynamic equation developed in Section 2 must apply to the ion species (subscript i) whose escape we wish to describe.

First, consider the process when $P_i = 0$ and $\frac{\partial P_i}{\partial s} = 0$ at all altitudes with $n_e = n_i$. In this case the ion equation of motion follows from (3) and (4) directly as

$$U_i \frac{\partial U_i}{\partial s} + \frac{1}{n_e} \frac{\partial P_e}{\partial r} - \underline{g} \cdot \hat{s} = - \sum_k \nu_{\ell k} U_\ell \qquad (16)$$

where, for realistic conditions, a scalar electron pressure and ion-neutral collisions must be included. Comparing (15) and (16) the differences between plasma and neutral gas escape become clear. As a result of the charge separation electric field induced by gravity between the ion gas and electrons, the motion of the ions is not adiabatic. Energy is continually supplied to the ion species from the electric field and substantial kinetic energy can be given to the ions. It is for this reason that the ion evaporative calculations of Bauer (1966) and Dessler and Michael (1966) are incorrect. As shown by (16) the introduction of energy to the ion gases occurs continually over wide altitude ranges independent of collisions and ions which have insufficient kinetic energy at some ion exospheric level to escape the earth's gravitational field may, in fact, participate in a plasma escape flow.

To reduce (16) further, for ambipolar conditions $U_e = U_i$ and, using (1), the ion equation of motion can be written as

$$\frac{1}{U_i} \frac{\partial U_i}{\partial s} \left(\frac{U_i}{C_i^2} - 1\right) = \frac{\underline{g} \cdot \hat{s}}{C_i^2} + \frac{1}{A} \frac{\partial A}{\partial s} - \frac{1}{C_i^2} \sum \nu_{ik}(U_i - U_k) - \frac{1}{Q_i} \frac{\partial Q_i}{\partial s} \qquad (17)$$

where $C_i^2 = k\,T_e/m_i$.

Comparison of (17) and (5) leads to the surprising conclusion that the equation of motion for the zero pressure ion gas has almost the same form as that for hydrodynamic conditions. As a consequence, subsonic and supersonic solutions are available in both approximations as well as a critical point. In the hydrodynamic equation the critical point occurs when $U_i^2 = k(T_e + T_i)/m_i$, while in the zero ion pressure gradient case it occurs at $U_i^2 = k\,T_e/m_i$. In fact, the solutions to (17) are very similar to those illustrated in Figure 3 and, in order to obtain physical results, just the same considerations of boundary conditions must be made to describe the different modes of motion.

Suppose now that the light ion gas (subscript ℓ) is not a major constituent and that $n_e \sim n_i >> n_\ell$. Will the above results still apply for zero pressure gradient conditions? Manipulation of (1), (3), and (4) quickly gives the result

$$\frac{1}{U_\ell} \frac{\partial U_\ell}{\partial s} \left(\frac{U_\ell}{C_\ell^2}\right) = \frac{\underline{g} \cdot \hat{s}}{C_\ell^2} - \frac{1}{n_e} \frac{\partial P_e}{\partial s} - \frac{1}{C_\ell^2} \sum_k \nu_{\ell k} (U_\ell - U_k) \qquad (18)$$

showing that the minor constituent is again accelerated by the charge separation electric field, affected by gravity, and retarded through ion-ion and ion-neutral friction.

While still a minor ion, the zero pressure ℓ-th species has no critical point and the velocity can be determined from (18) with an appropriate boundary velocity at a low altitude. Because the pressure gradient term is missing $\left(\frac{\partial P_\ell}{\partial s} = 0\right)$, explicit forms for the density and velocity can be obtained readily.

At some altitude, however, the assumptions leading to (18) breakdown and the light constituent is no longer a minor species. When this occurs the electron pressure again is coupled to the ℓ-th species speed and the continuity equation and the momentum equation must be solved simultaneously to find U_ℓ and n_ℓ with the result that a critical point dividing subsonic and supersonic flow $\left(\text{with respect to } (kT_e/m_\ell)^{\frac{1}{2}}\right)$ is again found.

The net effect of considering the ℓ-th species as a minor zero pressure ion does not lead to a different physical picture of plasma escape than described before in terms of hydrodynamic equations or a dominant zero pressure species: By neglecting ℓ-th species pressure a slightly different velocity boundary condition is given to Figure 3.

With regard to the applicability of the hydrodynamic approximation to the polar ionosphere and H^+ in particular, it can be shown (Banks and Holzer, 1969) that at the critical point $\left(M(H^+) = 1\right)$

$$\left.\frac{\lambda(H^+ - O^+)}{H(O^+)}\right|_{\substack{\text{critical}\\ \text{point}}} = \frac{1.18\, T_i}{T_e} \tag{19}$$

to a high degree of approximation. Thus, at and below the critical point collisions between H^+ and O^+ are frequent enough to ensure that the H^+ pressure is relatively isotropic about $U(H^+)$.

Above the critical point, $\lambda(H^+ - O^+)/H(O^+) > 1$ and the H^+ exosphere must begin. However, because $M(H^+) \geq 1$ in this region pressure gradients rapidly become unimportant in determining the speed of the flow. As a consequence, the hydrodynamic equation predicts quite accurately the character of the H^+ flow over a very extended range of conditions including the ion exosphere.

4. ION COMPOSITION OF THE POLAR REGIONS

Calculations of ion composition for steady outwards flows of H^+, He^+, and O^+ have been obtained for the coupled ion hydrodynamic equations of motion using implicit numerical techniques. The specific method involves the simultaneous solutions for the equations of continuity and momentum as applied to O^+, H^+, and He^+. In Figure 5 results from Holzer (1969) are shown for the conditions of Figure 3, i.e., $T_n = 1000\,^{\circ}K$, $T_e = T_i = 3000\,^{\circ}K$ with full photoionization production of O^+ $\left(I(O^+) = 5 \cdot 10^{-7}\ \text{sec}^{-1}\right)$ and He^+ $\left(I(He^+) = 10^{-7}\ \text{sec}^{-1}\right)$. Ion losses include O^+ and He^+ reactions with N_2 and O_2.

With regard to H^+, outwards flow begins only above 600 km. Due to $H^+ - O^+$ collisions, $\frac{\partial \ln n(H^+)}{\partial s} = \frac{\partial \ln n(O^+)}{\partial s}$ in situations of limiting ion flow (Hanson and Patterson, 1963; Geisler, 1967) so that H^+ remains a minor constituent to 7000 km. Although the H^+ escape altitude was not given in Banks and Holzer (1968), from Figures 4 and 5 it can be found that below 600 km the H^+ flow is downwards: Only above

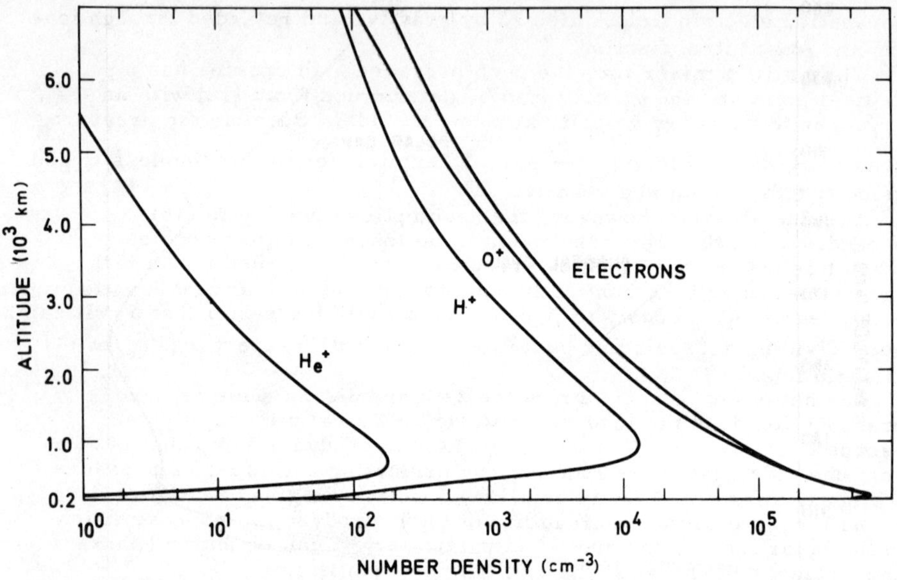

Figure 5. Polar region ion density profiles for $T_n = 1000\,^oK$, $T_e = T_i = 3000\,^oK$. O^+ and He^+ are created through photoionization while H^+ is a product of O^+ - H charge exchange. The H^+ and He^+ critical points are located at 4100 km and 4500 km, respectively, while the individual ion escape fluxes are $\Phi(H^+) = 5.8 \cdot 10^8$ cm^{-2} sec^{-1}, $\Phi(He^+) = 1.8 \cdot 10^6$ cm^{-2} sec^{-1}, $\Phi(O^+) = 5 \cdot 10^5$ cm^{-2} sec^{-1}, all normalized to 1000 km to take account of field tube divergence.

600 km (in the present model) can O^+ - H charge exchange lead to a net loss of H^+.

As shown in Figure 4, increases in electron and ion temperature tend to move the H^+ critical point outwards due to the O^+ density factor in $\nu(H^+, O^+)$ and the O^+ - e charge separation electric field. Nonetheless, as mentioned in Section 3, it can be shown that $\lambda(H^+) \leq H(O^+)$ at and below the critical point so that the hydrodynamic approximation (isotropic partial pressure about the transport speed) remains valid.

The escape flux of H^+ requires an upwards flow of O^+ starting near the F2-peak. For Figure 5, the flux normalized to 350 km is $8 \cdot 10^8$ cm^{-2} sec^{-1}. This value is sufficiently close to the maximum efflux rate of O^+ set by O^+ - O collisions that the O^+ scale height is approximately that of neutral atomic oxygen. Through the charge exchange conversion of O^+ to H^+, the upwards flow of O^+ eventually becomes small and the net ionospheric loss rate of O^+ in the present model is on the order of $5.8 \cdot 10^5$ cm^{-2} sec^{-1}. A profile of the expected vertical ion transport velocities is given in Figure 6.

5. FORMATION OF THE PLASMAPAUSE AND RELATED EFFECTS

The geomagnetic field of Figure 1 represents a time snapshot. At high latitudes field lines, and their associated plasma, are continually swept back into the tail of the magnetosphere where they eventually reconnect, migrate to lower magnetic latitudes, and move towards the sun to begin the convection process anew. (A review of this subject is

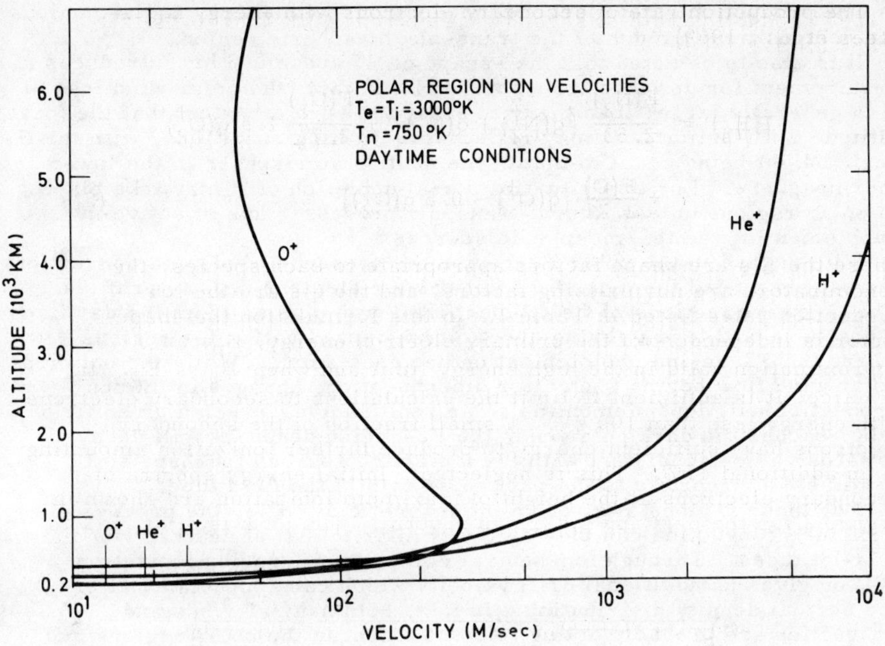

Figure 6. Polar region ion velocity profiles for $T_n = 750\,^{\circ}K$, $T_e = T_i = 3000\,^{\circ}K$ with daytime (photoionization) conditions. The H^+ velocity is negative below 450 km, corresponding to the ion flow needed to maintain the transition between transport and chemical equilibrium conditions.

given by Axford, 1969). Nishida (1966) pointed out that loss of polar plasma, when associated with magnetospheric convection and the low replenishment rate of topside ionization, would lead to the formation of a plasmapause at a geomagnetic latitude corresponding to the lowest boundary of field line convection.

Just before disconnection a given field tube has a considerable plasma content. After disconnection the light ion species (H^+, He^+) are lost to the tail of the magnetosphere and interplanetary space, but only at a finite rate determined by ionospheric conditions such that the present steady state equations apply only after about one hour. By the time the field tube has traversed some portion of the open field regions (perhaps after one hour) at a horizontal speed of about 100 m sec^{-1}, low ionization densities are present along the field tube with full outward flow conditions being present.

Following reconnection plasma continues to leave the ionosphere; a shock front must propagate back along the field line at a speed of 20 - 40 km sec^{-1} to the ionosphere before the upwards flow can be reduced. In the meantime, the field tube itself has moved to lower geomagnetic latitudes in the nighttime regions of the earth and may even have begun its sunwards march, still with plasma flowing outwards attempted to establish diffusive equilibrium. In fact the exact point at which a shock front begins to propagate back along the field lines towards the earth may begin before reconnection takes place when the appreciable plasma pressures of the plasma sheet/neutral sheet are reached.

From this point of view the predictions of the polar wind theory are not restricted to open field lines; time dependent effects also result in upwards plasma transport, reduced electron densities, and anomalous ion composition in much of the trans-plasmaspheric region.

It is also to be noted that the escape of H^+ above 600 km introduces a requirement for upwards transport of H. In fact, the polar wind loss of H is generally larger than thermal escape to such an extent that the local altitude distribution of H may reflect flux limiting conditions, with the H scale height being equal to that of the neutral atmosphere in the lower thermosphere. For this situation a real depletion of H may take place in polar regions unless lateral transport processes are effective in supplementing the thermospheric sources.

With regard to the polar F2-region the presence of plasma convection introduces a need for dynamic solutions for electron density profiles. The poleward motions of field tubes filled with plasma from the daytime ionosphere will tend to bring about a transient situation for regions where the F2-region photoionization becomes small. While the outwards polar wind flow begins soon after disconnection, changes in the ion production will depend upon the solar zenith angle. When a given field tube reaches the dark polar cap, the F2-region densities will decay towards values consistent with the ionization sources (perhaps soft particles) appropriate to the particular path of the field tube.

For those regions away from the magnetic dip pole, the process of field tube convection can play a role similar to that of neutral winds at mid-latitudes. Through ion-neutral collisions F2-region ionization will be given an additional drift velocity which can appreciably alter the normal density distribution. In fact, both neutral winds and convection are probably acting to some extent in the polar regions and should be included in the dynamic calculation of F2-region densities.

Following this same idea, it should also be noted that during polar substorms plasma convection speeds can be considerably enhanced with immediate effect, through ion-neutral relative motions, upon the F2-region densities surrounding the poles, even in the absence of particle precipitation.

Observations of polar region densities (Timleck and Nelms, 1969; Nelms and Chapman, 1969) consistently indicate a deep ionization minimum just outside the nighttime plasmapause boundary, followed by a return to high plasma densities in the region corresponding to auroral precipitation. If convection is acting to move plasma towards the night side of the earth, then it is somewhat puzzling that the deep trough exists unless some extraordinarily effective means of ionization removal is present. Although detailed numerical calculations of this effect have not yet been made, it appears that ambipolar transport may not be adequate to explain the deep troughs; outward ion flows enhanced by electric currents may be present.

Certainly, however, once the trough is formed in the midnight region, convection of field tubes can result in the typical horseshoe pattern of reduced ion densities between 1800 and 0600 just below the nighttime auroral region.

Finally, in the light of Section 3, it is clear that the description of plasma escape from a gravitating body is essentially the same for both the hydrodynamic and the zero ion pressure gradient approximations. Neither case is adiabatic and the resulting velocity and density predictions bear little resemblance to the true evaporation of neutral gases. In both the hydrodynamic and zero pressure cases the final description of the plasma escape process depends upon a choice of the asymptotic plasma pressure. Both subsonic and supersonic escape flows are available, as well as a zero velocity "diffusive equilibrium" solution which can be obtained from (18) with $U_\ell = 0$.

Acknowledgments. Discussions with W. I. Axford and T. E. Holzer have been of great value in the preparation of this paper. This work was supported, in part, through NASA Grant NGR-05-009-075 and by the Advanced Research Projects Agency of the Department of Defense and was monitored by the U. S. Army Research Office-Durham under Contract DA-31-124-ARO-D-257.

REFERENCES

Andrews, M. K. and J. O. Thomas, The trans-polar exospheric plasma, 2: The electron density distribution above the winter pole, Nature, In press, 1969.

Axford, W. I., The polar wind and the terrestrial helium budget, J. Geophys. Res., 73, 6855, 1968.

Axford, W. I., Magnetospheric convection, Rev. Geophys., In press, 1969.

Banks, P. M., The polar wind and its implications, in Aurorae and Airglow, 1968, edited by B. M. McCormac, Reinhold, New York, 1969.

Banks, P. M. and T. E. Holzer, The polar wind, J. Geophys. Res., 73, 6846, 1968.

Banks, P. M. and T. E. Holzer, High latitude plasma transport: The Polar Wind, to be published, 1969.

Barrington, R. E., J. S. Belrose, and G. L. Nelms, Ion composition and temperature at 1000 km as deduced from VLF resonances and topside ionograms, in Electron Density Profiles in the Ionosphere and Exosphere, edited by J. Frihagen, North-Holland Publishing Company, Amsterdam, pp. 387-396, 1966.

Bauer, S. J., The structure of the topside ionosphere, in Electron Density Profiles in the Ionosphere and Exosphere, edited by J. Frihagen, North-Holland Publishing Company, Amsterdam, pp. 270-280, 1966.

Dessler, A. J. and F. C. Michel, Plasma in the geomagnetic tail, J. Geophys. Res., 71, 1421, 1966.

Donley, J. L., Observations of the polar ionosphere in the range 2000 to 3000 km by means of satellite borne electron traps, NASA-GSFC rept. X-615-67-335, 1967.

Dungey, J. W., Interplanetary magnetic field and the auroral zones, Phys. Rev. Letters, 6, 47, 1961.

Dungey, J. W., Cosmic Electrodynamics, Cambridge University Press, Cambridge, 1958.

Geisler, J. E., On the limiting daytime flux of ionization into the protonosphere, J. Geophys. Res., 72, 81, 1967.

Hanson, W. B. and T. N. L. Patterson, Diurnal variation of the hydrogen concentration in the exosphere, Planetary Space Sci., 11, 1035, 1963.

Hoffman, J. H., Ion composition measurements in the polar region from the Explorer 31 satellite, Trans. Am. Geophys. Union, 49, 253, 1968.

Holzer, T. E., Effects of plasma flow on density and velocity profiles in the polar ionosphere, in The Polar Ionosphere and Magnetospheric Processes, edited by G. Skovli, Gordon and Breach, New York, 1969.

Nelms, G. L. and J. H. Chapman, The polar ionosphere: Results from the Alouette/ISIS topside sounders, in The Polar Ionosphere and Magnetospheric Processes, edited by G. Skovli, Gordon and Breach, New York, 1969.

Nicolet, M., Helium, an important constituent in the lower exosphere, J. Geophys. Res., 66, 2263, 1961.

Nishida, A., Formation of a plasmapause, or magnetospheric plasma knee by combined action of magnetospheric convection and plasma escape from the tail, J. Geophys. Res., 71, 5669, 1966.

Thomas, J. O. and M. K. Andrews, Transpolar exospheric plasma, 1, Plasmasphere termination, J. Geophys. Res., 73, 7307, 1968.

Timleck, P. and G. L. Nelms, Electron densities less than 100 electron cm^{-3} in the topside ionosphere, Proc. IEEE, June 1969.

Vasyliunas, V. M., Low energy particle fluxes in the geomagnetic tail, in The Polar Ionosphere and Magnetospheric Processes, edited by G. Skovli, Gordon and Breach, New York, 1969.

EFFECTS OF PLASMA FLOW ON DENSITY AND VELOCITY PROFILES IN THE POLAR IONOSPHERE

Thomas E. Holzer

Department of Applied Physics and Information Science
and
Institute for Pure and Applied Physical Science
University of California, San Diego
La Jolla, California 92037, USA

Abstract. The basic theory of plasma escape from the polar ionosphere through the polar wind (Axford, 1968) has been discussed by Banks and Holzer (1968) and Banks (1969). These authors have shown that ionospheric density and velocity profiles associated with outward plasma flow may be distinctly different from profiles associated with diffusive equilibrium conditions. In this paper, density and velocity profiles are presented for several model polar ionospheres in which the H^+ and He^+ plasmas flow away from the earth with supersonic speeds. The parameters of these ionospheric models are the electron temperature, the ion temperature, the model neutral atmosphere, and the solar flux. There are certain characteristics common to all of the polar wind profiles which distinguish them from typical profiles associated with diffusive equilibrium conditions. Some possible techniques for observing such characteristics in the topside polar ionosphere are discussed. Finally, the effects of the polar wind on the terrestrial helium budget (Axford, 1968) and on the distribution of neutral hydrogen in the polar atmosphere are briefly considered.

1. INTRODUCTION

As the solar wind flows past the earth, it sweeps high-latitude geomagnetic field lines into the anti-solar direction to form the magnetospheric tail. Plasma pressure in this tail region can be significantly lower than pressures measured inside the plasmapause. Such a pressure difference leads to the expectation that the dynamic properties of the high-latitude ionosphere may differ considerably from those of the mid-latitude and low-latitude ionospheres. Indeed, high-latitude ionospheric density measurements have shown several peculiar features not observed within the plasmapause (e.g., low electron densities and anomalous ion composition). It now appears possible that some of these features might be explained by outward transport of high-latitude ionospheric plasma.

The theory of the polar wind (Axford, 1968) predicts that light ionic constituents in the high-latitude ionosphere will escape from the earth in a supersonic hydrodynamic flow. Preliminary calculations (Banks and Holzer, 1968; Holzer and Banks, 1968) have shown that such a supersonic flow produces effects in the polar ionospheric density profiles which are qualitatively similar to some of the peculiar features observed by satellites. In the preceding paper, Banks (1969) has reviewed this theory of plasma flow in the topside ionosphere. He points out that depending on the plasma pressure in the tail of the magnetosphere, the theory will predict supersonic or subsonic outward plasma flow, diffusive equilibrium, or subsonic or supersonic inward flow. The supersonic outflow solution corresponds to low tail plasma pressures, and it is this solution with which the present paper will be concerned.

Section 2 contains a series of density and velocity profiles corresponding to various models of the polar ionosphere, all of which are associated with the supersonic escape of the H^+ and He^+ plasmas. The parameters characterizing these models are the electron temperature, the ion temperature, the model neutral atmosphere, and the solar flux. Of course, the plasma pressure in the geomagnetic tail is a fifth parameter of these models, but since only supersonic flow (low plasma pressure) solutions are considered here, this parameter is not varied.

None of the profiles presented can be expected to be an accurate representation of all the high-latitude ionosphere. In the first place, magnetospheric convection processes make the polar ionosphere a highly non-steady system, while the theory used in creating the models discussed here is based on a steady state assumption. In addition, the high-latitude ionosphere is certainly not a homogeneous region, and one can expect a wide variation over the polar cap of all the parameters mentioned above. Nevertheless, the density and velocity profiles for all the models have in common certain characteristics which are not found in typical diffusive equilibrium profiles. These characteristics should be observable in the polar ionosphere (Section 3) if it is a region of supersonic plasma outflow, and in fact, some of them have already been observed. Yet the range of applicability of the polar wind theory to the high-latitude ionosphere cannot be determined until many more observations have been made.

The purpose of this paper is threefold. First, it is to provide the observationalist with an idea of some characteristics of polar wind density and velocity profiles which are distinctly different from the corresponding characteristics of diffusive equilibrium profiles. Second, it is to present some typical polar wind escape fluxes of H^+ and He^+ and to examine briefly some possible effects of these fluxes on the neutral hydrogen distribution in the polar atmosphere and on the terrestrial helium budget (Axford, 1968). Third, it is to give the reader a physical feeling for the complex interaction of forces which can bring about a supersonic plasma expansion in the polar ionosphere.

2. POLAR WIND MODELS OF THE HIGH LATITUDE IONOSPHERE

The model atmospheres discussed in this section are composed of electrons and the following ions and neutrals: O^+, H^+, He^+, O, O_2, N_2, He, and H. O^+ ions are produced through photoionization and H^+-O charge exchange, and they are lost through O^+-H charge exchange and dissociative recombination with N_2. H^+ ions are produced through O^+-H charge exchange. He^+ ions are produced through photoionization, and they are lost through dissociative charge transfer with N_2 and O_2. Table 1 lists the production and loss rates and collision frequencies which were used in our models. Three model neutral atmospheres (composed of O_2, N_2, O, He, and H), corresponding to T_n = 750 °K,

1000 °K, and 1500 °K, were used. These neutral atmospheres were calculated for mid-latitude conditions, so that they may not be good representations of polar atmospheres. However, consideration of the effects on our ionospheric models of varying the composition of the neutral atmosphere (varying T_n) is useful.

The ion velocity and density profiles for a given model are obtained through the self-consistent solution of the six equations (two each for O^+, H^+, and He^+) given by (1) and (7) in the preceding paper (Banks, 1969). The solution is obtained by using an iterative numerical technique. As was seen in equations (1), (2), and (7) in Banks' paper, the shape of the profiles is governed by the polarization electric field (the electron pressure gradient), frictional interactions, production and loss effects, gravitational acceleration, and the divergence of magnetic flux tubes (In the models presented here, it is assumed that $A \propto r^{-3}$, where r is the distance from the center of the earth.). The relative importance of these effects will be discussed with regard to the individual model ionospheres.

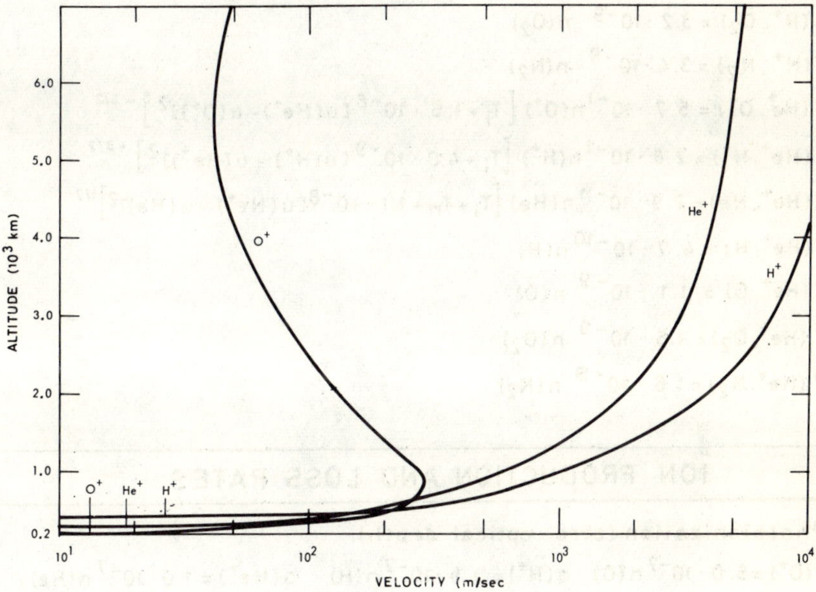

Figure 1. Daytime polar region ion velocity profiles for $T_e = T_i$ = 3000 °K and T_n = 750 °K. Note the rapid expansion of the H^+ and He^+ plasmas and the peak in the O^+ velocity profile near 900 km. The peak arises from the upwards flux of O^+ above the F2 peak which is necessary to supply the large H^+ escape flux.

Figure 1 shows velocity profiles for the three ionic constituents of a model ionosphere in which the ion and electron temperatures are equal and constant ($T_e = T_i$ = 3000 °K), a 750 °K model neutral atmosphere is used, and a typical daytime solar flux is assumed. The "daytime condition" for the solar flux arbitrarily refers to an altitude independent atomic oxygen photoionization rate of $5 \cdot 10^{-7}$ sec^{-1}. In Figure 1, it is seen that the H^+ and He^+ plasmas both expand rapidly, reaching their critical points at 2300 km and 2500 km, respectively. The O^+ plasma reaches a peak upward flow velocity of about 300 m sec^{-1} at an altitude of 900 km. This peak is one of the distinctive characteristics of polar

ION COLLISION FREQUENCIES

$$V(O^+,O) = 1.6 \cdot 10^{-11}\, n(O)\, \left[T_i + T_n + 4.3 \cdot 10^{-8}\, [u(O^+) - u(O)]^2\right]^{1/2}$$

$$V(O^+,N_2) = 6.9 \cdot 10^{-10}\, n(N_2)$$

$$V(O^+,O_2) = 6.6 \cdot 10^{-10}\, n(O_2)$$

$$V(O^+,H^+) = 7.5 \cdot 10^{-2}\, n(H^+)\, \left[T_i + 4.7 \cdot 10^{-9}\, [u(O^+) - u(H^+)]^2\right]^{-3/2}$$

$$V(O^+,He^+) = 1.4 \cdot 10^{-1}\, n(He^+)\left[T_i + 1.6 \cdot 10^{-8}\, [u(O^+) - u(He^+)]^2\right]^{-3/2}$$

$$V(H^+,O^+) = 1.2\, n(O^+)\, \left[T_i + 4.7 \cdot 10^{-9} [u(H^+) - u(O^+)]^2\right]^{-3/2}$$

$$V(H^+,He^+) = 1.1\, n(He^+)\, \left[T_i + 4.0 \cdot 10^{-9}\, [u(H^+) - u(He^+)]^2\right]^{-3/2}$$

$$V(H^+,H) = 1.0 \cdot 10^{-10} n(H) \left[T_i + T_n + 2.7 \cdot 10^{-9} [u(H^+) - u(H)]^2\right]^{1/2}$$

$$V(H^+,He) = 1.05 \cdot 10^{-9} n(He)$$

$$V(H^+,O_2) = 3.2 \cdot 10^{-9}\, n(O_2)$$

$$V(H^+,N_2) = 3.4 \cdot 10^{-9}\, n(N_2)$$

$$V(He^+,O^+) = 5.7 \cdot 10^{-1} n(O^+) \left[T_i + 1.6 \cdot 10^{-8} [u(He^+) - u(O^+)]^2\right]^{-3/2}$$

$$V(He^+,H^+) = 2.8 \cdot 10^{-1} n(H^+) \left[T_i + 4.0 \cdot 10^{-9} [u(H^+) - u(He^+)]^2\right]^{-3/2}$$

$$V(He^+,He) = 2.9 \cdot 10^{-11} n(He) \left[T_i + T_n + 1.1 \cdot 10^{-8} [u(He^+) - u(He)]^2\right]^{1/2}$$

$$V(He^+,H) = 4.7 \cdot 10^{-10} n(H)$$

$$V(He^+,O) = 1.1 \cdot 10^{-9}\, n(O)$$

$$V(He^+,O_2) = 1.5 \cdot 10^{-9}\, n(O_2)$$

$$V(He^+,N_2) = 1.6 \cdot 10^{-9}\, n(N_2)$$

ION PRODUCTION AND LOSS RATES

Photoionization (zero optical depth):

$q(O^+) = 5.0 \cdot 10^{-7} n(O)$ $\quad q(H^+) = 1.5 \cdot 10^{-7} n(H)$ $\quad q(He^+) = 1.0 \cdot 10^{-7} n(He)$

Accidentally resonant charge transfer ($H + O^+ \rightleftarrows H^+ + O$):

$$q(H^+) = \ell(O^+) = 4.3 \cdot 10^{-11} (T_i/16 + T_n)^{1/2}\, n(H)\, n(O^+)$$

$$q(O^+) = \ell(H^+) = 3.8 \cdot 10^{-11}\left[(T_i + T_n/16 + 1.2 \cdot 10^{-8}\, [u(H^+) - u(O)]^2\right]^{1/2}$$

Dissociative recombination ($O^+ + N_2 \rightarrow NO^+ + N$):

$$\ell(O^+) = 1.3 \cdot 10^{-12}\, n(N_2)\, n(O^+)$$

Dissociative charge transfer ($He^+ + N_2 \rightarrow He + N^+ + N$):

$$\ell(He^+) = 1.3 \cdot 10^{-9}\, n(N_2)\, n(He^+)$$

$$\ell(He^+) = 1.1 \cdot 10^{-9}\, n(O_2)\, n(He^+)$$

Table 1. Ion frictional interactions and production and loss rates. $\nu(sec^{-1})$ is a collision frequency, u $(cm\ sec^{-1})$ is a bulk flow velocity, T_i (°K) and T_n (°K) are the ion and neutral atmospheric temperatures, and q $(cm^{-3}\ sec^{-1})$ and ℓ $(cm^{-3}\ sec^{-1})$ are ion production and loss rates. The collision frequencies have been calculated using the cross-sections of Banks (1966) and Holzer and Banks (1968). For frictional interactions which are important in high velocity regions, the collision frequencies have been modified so as to be valid in both the limits of high and low relative bulk flow velocities. Photoionization rates for zero optical depth are taken from Nicolet et al. (1969). The O^+ rate of $5 \cdot 10^{-7}\ sec^{-1}$ is the criterion which determines "daytime" conditions throughout this paper. Accidentally resonant charge transfer rates were taken from Holzer and Banks (1969). The dissociative recombination rate was taken from Rishbeth (1964). Dissociative charge transfer rates were taken from Ferguson et al. (1964).

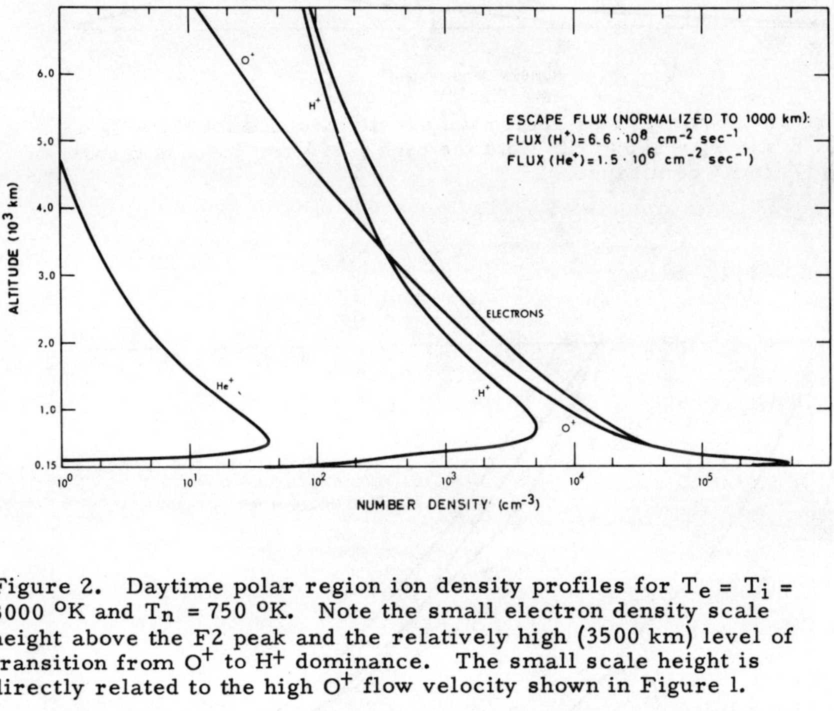

Figure 2. Daytime polar region ion density profiles for $T_e = T_i =$ 3000 °K and T_n = 750 °K. Note the small electron density scale height above the F2 peak and the relatively high (3500 km) level of transition from O^+ to H^+ dominance. The small scale height is directly related to the high O^+ flow velocity shown in Figure 1.

wind ionospheres. It arises from the flux requirement placed on the O^+ plasma by the escape flux of the H^+ plasma, since the primary source of H^+ ions is the O^+-H charge exchange reaction.

The density profiles for the model ionosphere of Figure 1 are shown in Figure 2. These profiles have several interesting aspects. In the first place, the F2 peak occurs at a relatively low altitude, about 200 km. For a few hundred kilometers above the peak, the O^+ density decreases very rapidly-- in fact, the scale height of the O^+ profile in this region is about one-sixth as great as the diffusive equilibrium scale height corresponding to the same ionospheric parameters. This small scale height is directly associated with the rapid upward flow velocity of the O^+ plasma shown in Figure 1. Another interesting feature of these profiles is the relatively high

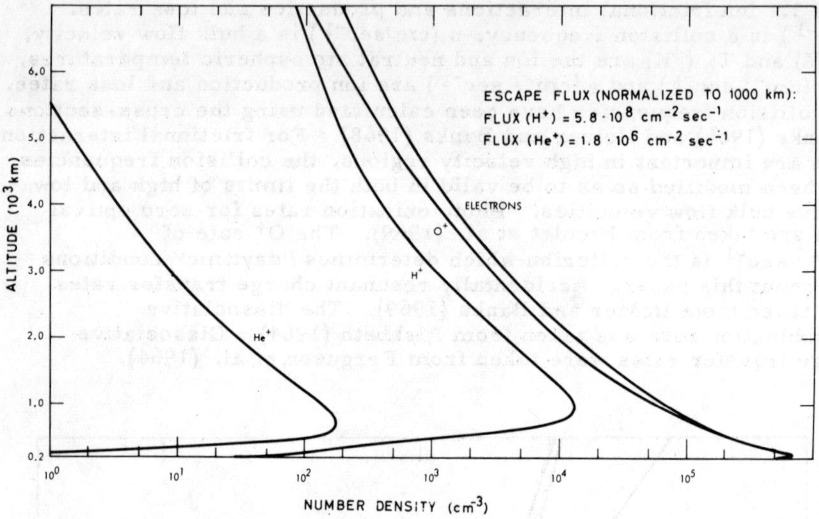

Figure 3. Daytime polar region ion density profiles for $T_e = T_i = 3000\ ^oK$ and $T_n = 1000\ ^oK$. Note the high (7500 km) level of transition from O^+ to H^+ dominance.

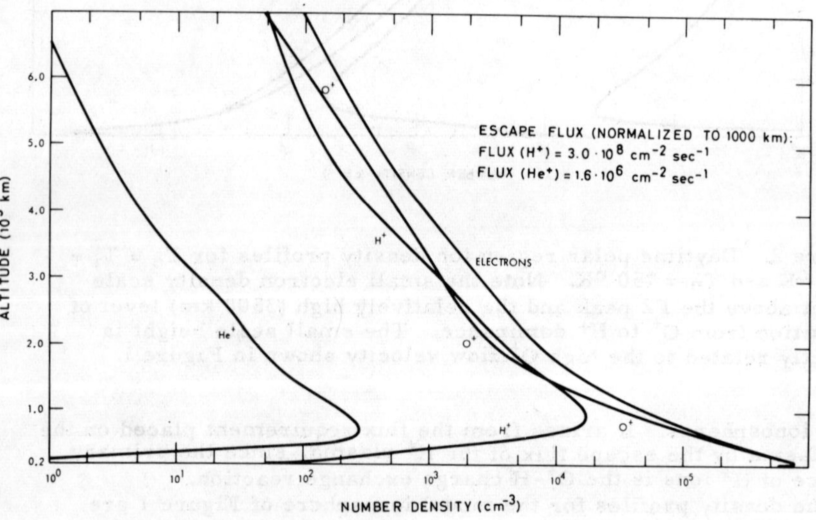

Figure 4. Daytime polar region ion density profiles for $T_e = 3000\ ^oK$, $T_i = 1250\ ^oK$, and $T_n = 1000\ ^oK$. The complicated shape of the curves (three crossings of the H^+ and O^+ curves) arises from a combination of friction, production, and loss effects.

(3500 km) level of transition from O^+ to H^+ dominance. This high level is a consequence of the tendency for the density profile of the flowing plasma (H^+) to parallel the density profile of its scattering medium (O^+) in the region where friction is an important dynamical force. Since above the critical point (2300 km for H^+) the importance of friction relative to the polarization electric field decreases rapidly, the H^+ plasma begins to seek a scale height which is much larger than that of O^+ (cf. Figure 2), and the transition soon occurs. A final point concerning Figure 2 is the listing of H^+ and He^+ fluxes in the figure legend. The fluxes are normalized to 1000 km, which means that they have been calculated at 10,000 km, where production and loss effects are negligible, and then have been multiplied by the ratio of the area of a flux tube at 10,000 km to its area at 1000 km. If the fluxes listed in Figure 2 had been normalized instead to 300 km, they would have the values $FLUX(H^+) = 9 \cdot 10^8\ cm^{-2}\ sec^{-1}$ and $FLUX(He^+) = 2 \cdot 10^6\ cm^{-2}\ sec^{-1}$.

In Figure 3, the ionospheric model has been changed by raising the temperature of the neutral atmosphere from 750 °K to 1000 °K. The principal effect of this temperature increase is to decrease the abundance of atomic hydrogen, while increasing the abundance of the other neutral constituents. These density changes result for O^+ in an increased production rate through photoionization and an increased loss rate through N_2 dissociative recombination, but a decreased loss rate through O^+-H charge transfer. The net result of these effects is to raise the F2 peak by about 75 km, to increase the absolute abundance of O^+ throughout the topside ionosphere, and to decrease the relative abundance of H^+ throughout. The increased friction (increased O^+ density) raises the H^+ critical point to 4100 km. The higher H^+ critical point and the increased O:H ratio raise the transition level from O^+ to H^+ dominance to 7500 km. The fluxes of H^+ and He^+ remain nearly the same as when T_n = 750 °K (cf. Figure 2).

If the ion temperature is lowered from 3000 °K to 1250 °K, the profiles shown in Figure 4 result. Lowering the ion temperature has the effect of driving the accidentally resonant charge exchange process $H^+ + O \rightleftarrows H + O^+$ to the left, and consequently, it has the effect of increasing the H^+ concentration relative to the O^+ concentration. Thus the transition level to H^+ dominance is brought down to 1500 km. However, since the H^+ critical point is still near 4100 km, H^+-O^+ frictional effects bring about a transition back to O^+ dominance at 3000 km, and the final transition to H^+ dominance is effected well above the critical point at about 7000 km (cf. Figure 3).

Figures 1 to 4 have shown profiles for all of the ionic constituents (H^+, He^+, O^+) in the model ionospheres in order to emphasize certain physical phenomena. For the rest of the models, discussed below, only H^+ velocity profiles and electron density profiles will be shown. These models are intended to show the effects of (i) varying the electron and ion temperature together; (ii) varying models of the neutral atmosphere; (iii) varying the atomic oxygen photoionization rate; and (iv) introducing an ion temperature gradient.

Figures 5 and 6 show the H^+ velocity and electron density profiles for a model in which the electron and ion temperatures are held equal and varied from 3500 °K to 2500 °K. Crosses mark the locations of the H^+ critical point in the velocity profiles and the levels of transition between O^+ and H^+ dominance in the density profiles. The higher plasma temperatures lead to a larger O^+ scale height, which, in turn, decreases the charge separation electric field and increases the frictional resistance to the flow of the H^+ plasma. Both of these effects lead to an increase in the altitude of the H^+ critical point, while the former leads to an increase in the O^+ density and hence, in the H^+ density. It is interesting to note that the H^+ escape flux is larger for the high density, slowly expanding model than for the low density, rapidly expanding model.

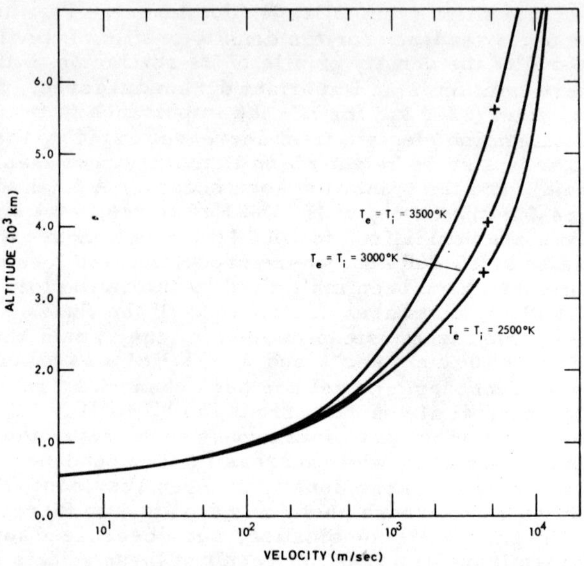

Figure 5. Daytime polar region H^+ velocity profiles for T_n = 1000 °K and several values of T_e and T_i. A decrease in electron and ion temperatures leads to a faster expansion due to complementary effects of the electric field and H^+-O^+ friction. Crosses mark the H^+ critical point.

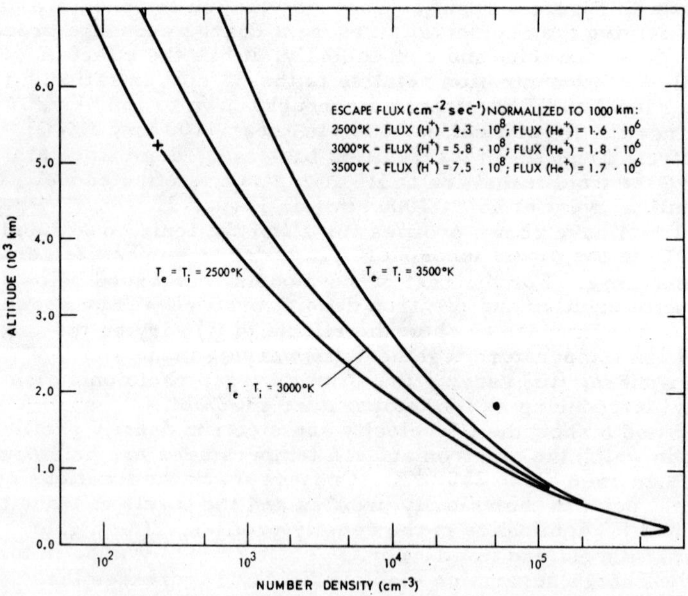

Figure 6

Figure 6. Daytime polar region electron density profiles for T_n = 1000 °K and several values of T_e and T_i. Crosses mark the level of transition from O^+ to H^+ dominance. Note that the H^+ escape flux is larger for a high density, slowly expanding model than for a low density, rapidly expanding model.

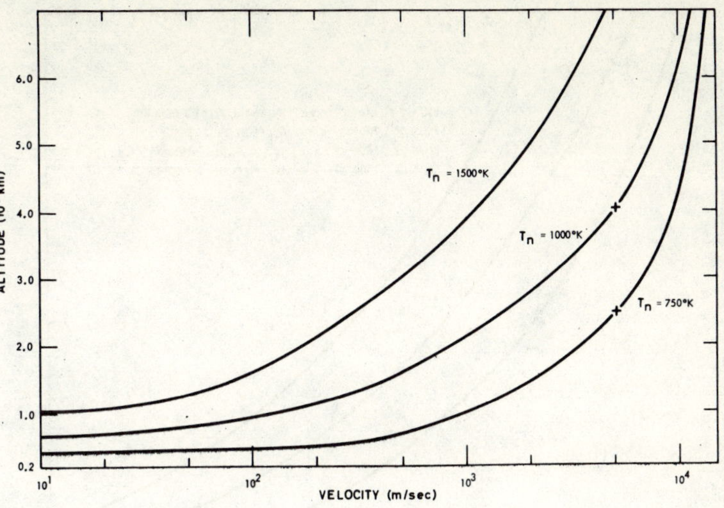

Figure 7. Daytime polar region H^+ velocity profiles for $T_e = T_i = 3000\,^{\circ}K$ and several values of T_n. Crosses mark the H^+ critical point. The expansions are faster for lower values of T_n due to decreased frictional resistance.

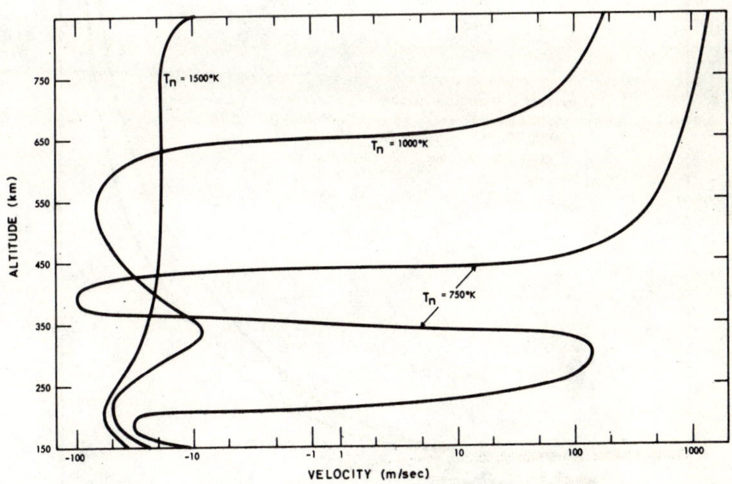

Figure 8. Daytime polar region H^+ velocity profiles for $T_e = T_i = 3000\,^{\circ}K$ and several values of T_n with special emphasis on low altitudes. Note the discontinuity in logarithmic velocity scale between - 1 and 1. Fine structure due to electric field, production, and loss effects disappears at high values of T_n due to very large frictional effects.

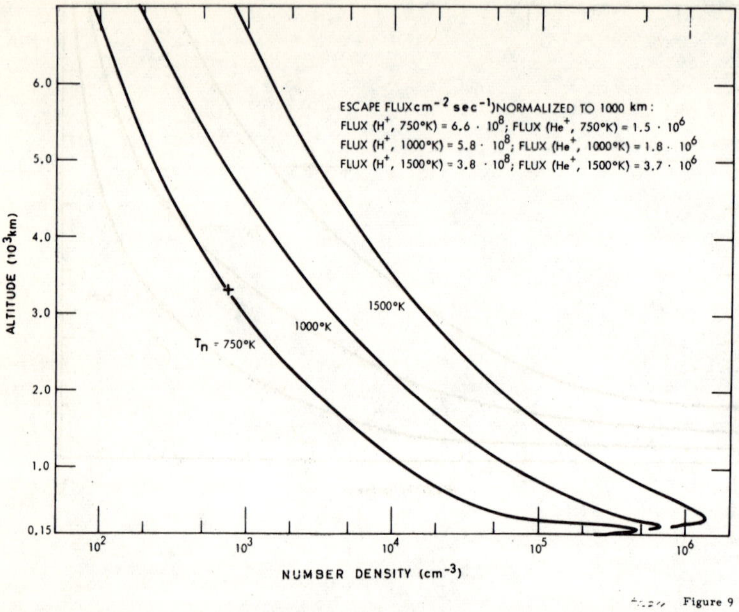

Figure 9. Daytime polar region electron density profiles for $T_e = T_i = 3000\ ^oK$ and several values of T_n. Crosses mark the level of transition from O^+ to H^+ dominance. Note that the H^+ flux is larger for a low density, rapidly expanding model than for a high density, slowly expanding model.

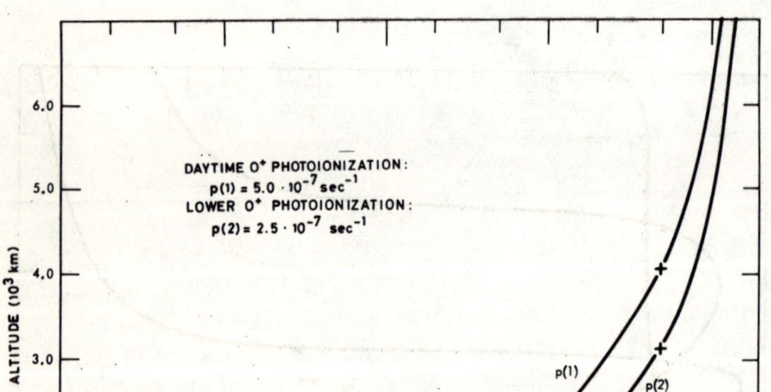

Figure 10. Polar region H^+ velocity profiles for $T_e = T_i = 3000\ ^oK$ and $T_n = 1000\ ^oK$ and two oxygen photoionization rates. Crosses mark the H^+ critical point.

Restoring the ion and electron temperatures to 3000 °K and varying the temperature of the neutral atmosphere between 750 °K and 1500 °K leads to the profiles shown in Figures 7, 8, and 9. As was stated previously, increasing the neutral gas temperature decreases the abundance of atomic hydrogen and increases the abundance of the other neutral constituents. Consequently the effect of this temperature increase is to increase greatly the frictional resistance to H^+ plasma flow (through H^+-O^+ collisions) and to increase slightly the H^+ abundance. Thus a higher T_n leads to a slower expansion (cf. Figure 7). In Figure 8 the region of inward H^+ flow, which has been ignored in the previous velocity profiles, is shown in detail. The structure in the profiles is due primarily to production, loss, and electric field effects. As T_n increases, the large increase in frictional resistance to H^+ flow (through H^+-N_2, H^+-O_2, and H^+-O^+ collisions) tends to diminish and finally almost destroy this fine structure. (Note that since the horizontal scale is logarithmic, there is actually a discontinuity in the graph between -1 and 1.) Once the H^+ velocity makes its final transition to positive values, then all H^+ ions produced above this altitude will escape in the supersonic expansion. Thus it is not surprising that, contrary to the results of the previous models (cf. Figure 6), in these models the rapidly expanding low density ionosphere produces a greater H^+ flux than the slowly expanding high density ionosphere (cf. Figure 9).

Returning now to a 1000 °K model neutral atmosphere and varying the atomic oxygen photoionization rate between $5 \cdot 10^{-7}$ sec^{-1} and 10^{-7} sec^{-1}, the profiles shown in Figures 10 and 11 are obtained. The effects of lowering the oxygen photoionization rate on the H^+ velocity profile are qualitatively similar to the effects obtained either by lowering together T_e and T_i or by lowering T_n. However, reducing the photoionization rate brings about a density decrease but little altitude decrease in the F2 peak, while lowering T_e and T_i leaves the peak relatively unaffected, and lowering T_n decreases both the density and the altitude of the peak.

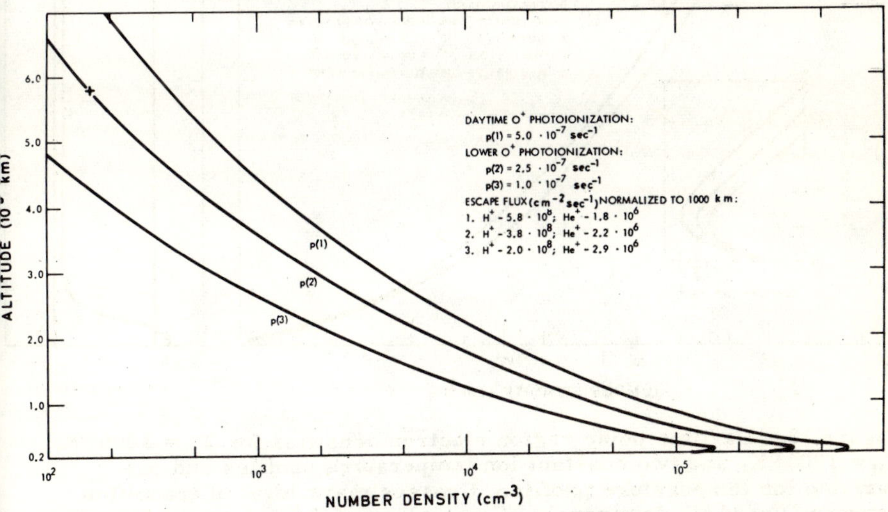

Figure 11. Polar region electron density profiles for $T_e = T_i = 3000$ °K and $T_n = 1000$ °K and several oxygen photoionization rates. Crosses mark the level of transition from O^+ to H^+ dominance. Note that the H^+ escape flux is larger for a high density, slowly expanding model than for a low density, rapidly expanding model.

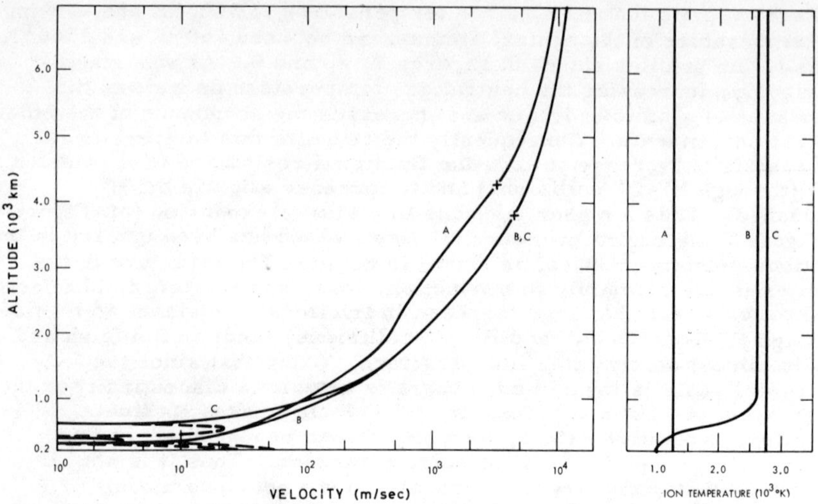

Figure 12. Daytime polar region H^+ velocity profiles for T_e = 3000 °K, T_n = 1000 °K, and two constant ion temperature profiles and one variable ion temperature profile. Crosses mark the H^+ critical point. The profiles indicate that the magnitude of T_i is more important than the gradient in T_i.

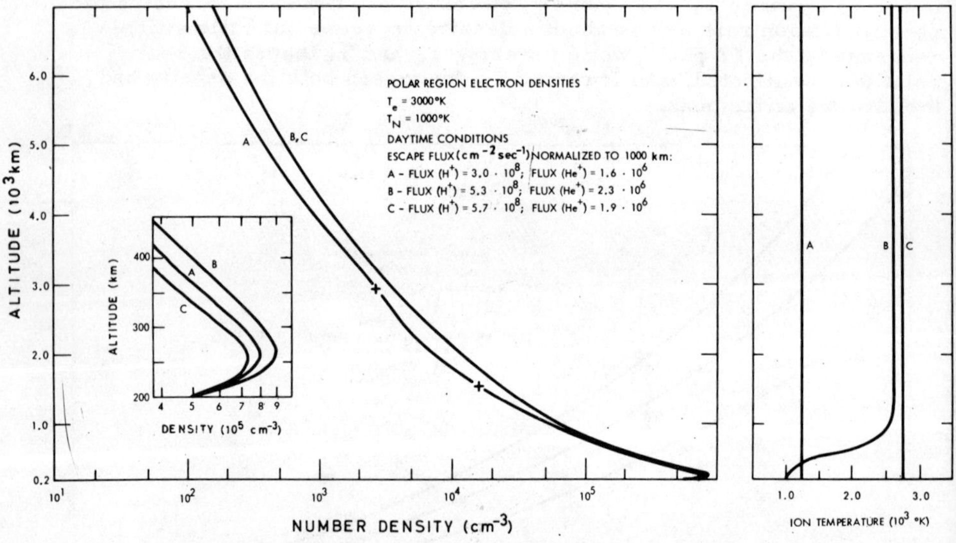

Figure 13. Daytime polar region electron densities for T_e = 3000 °K, T_n = 1000 °K, and two constant ion temperature profiles and one variable ion temperature profile. Crosses mark level of transition between O^+ and H^+ dominance. Comparison of the inset profiles with the full profiles indicates that the magnitude of T_i is more important than the gradient in T_i.

The decrease in the oxygen photoionization rate from $5 \cdot 10^{-7}$ sec^{-1} to 10^{-7} sec^{-1} leads to a factor of three decrease in the H^+ escape flux.

The final model to be considered here involves the introduction of an ion temperature gradient. In Figures 12 and 13 the three profiles shown correspond to (A) a constant 1250 °K ion temperature, (B) an ion temperature which varies between 1000 °K and 2600 °K, and (C) a constant 2750 °K ion temperature. The electron and neutral gas temperatures are held constant at 3000 °K and 1000 °K, respectively In Figure 12 it can be seen that curve B follows closely curve A up to 500 km and follows closely curve C above 1500 km. From the topology of these curves it appears that the magnitude of the ion temperature is far more important in determining H^+ flow characteristics than is the magnitude of the ion temperature gradient. Further support of this statement is obtained by comparing the relative positions of curves A, B, and C in the inset in Figure 13 with their positions at higher altitudes in the profile. However, before more definite conclusions can be drawn about the effect of temperature gradients, several more models involving various types of positive and negative gradients must be considered.

3. DISCUSSION

Three points will be considered in this section. The first involves a determination of the applicability of the polar wind theory. To this end, the distinctive features of a polar wind ionosphere are summarized, and experiments which would permit their observation are proposed. The other two points involve a consideration of the effects of the polar wind on the terrestrial helium budget (Axford, 1968) and on the neutral hydrogen distribution in the polar ionosphere.

In studying the density and velocity profiles shown in Section 2, one can find several features which seem to be characteristic of polar wind model ionospheres, but which are not, in general, characteristic of diffusive equilibrium models. Specifically, these features are: (i) the low altitude (200 km to 275 km) of the F2 peak; (ii) the relatively small electron density scale height above the F2 peak; (iii) a large region of heavy ion dominance (usually up to more than 3000 km); (iv) relatively low electron densities in the topside ionosphere; (v) a large upward flow velocity (greater than 100 m sec^{-1}) of the O^+ plasma for a few hundred kilometers above the F2 peak; and (vi) large upward flow velocities of light ions, increasing monotonically with altitude and becoming supersonic (at a few km/sec) somewhere between 2000 km and 4500 km.

As Banks (1969) has noted, there have already been satellite measurements of some of these features. Further satellite experiments employing topside sounders and detectors measuring local ion and electron densities and temperature and ion flow velocities (cf. Hoffman, 1968) would be useful. Another valuable set of observational data could be obtained from a rocket program based in the polar cap in which altitude profiles of ion and electron densities and temperature could be measured. If the rockets were to have sufficiently small velocities at apogee, then point measurements of ion flow velocities could also be obtained. A more expensive, but very useful, long-term method of polar cap observation would be an incoherent backscatter station. Such a station would permit continuous observation of such things as the F2 peak, the electron density profile just above the peak, O^+ flow velocities just above the peak, and soft electron precipitation in the polar ionosphere.

The observational programs described above could provide possible verification of the polar wind theory, as well as permitting more accurate calculations of polar wind ionospheric models.

However, with the qualitative models presently available, it is still possible to discuss some semi-quantitative consequences of the polar wind. The first of these is the escape of helium ions and the balance of the terrestrial helium budget, a problem which has been considered most recently by Axford (1968). The helium escape flux through the polar wind which is required to satisfy Axford's hypothesis is about $(6 \pm 3) \cdot 10^6$ cm^{-2} sec^{-1}. The escape flux of He^+ in the models discussed in Section 2 is only about $2 \cdot 10^6$ cm^{-2} sec^{-1}. However, the models may be predicting fluxes which are low by as much as a factor of 5 for the following reasons. The model neutral atmospheres used in Section 2 are applicable to typical mid-latitude conditions (Nicolet et al., 1969). It appears, though, that there may exist a winter polar helium bulge (Keating and Prior, 1967; 1968; Jacchia and Slowey, 1967) corresponding to a helium density in the winter polar region which is a factor of 5 larger than predicted (Müller and Hartmann, 1969). This effect would raise the average polar wind model He^+ flux by a factor of 2.5. Another possible factor of 2 error in these fluxes arises from uncertainties in the atomic helium photoionization rate and the He^+ loss rate. It would appear, then, that the average polar wind He^+ flux should lie in the range $3 - 10 \cdot 10^6$ cm^{-2} sec^{-1}. Since the flux required by Axford lies within this range, it does not seem unlikely that his hypothesis will be verified by future polar region observations.

Another consequence of light ion escape through the polar wind relates to the neutral atomic hydrogen distribution in the polar atmosphere. It was seen in the previous models that escape fluxes of H^+ ions can be as high as 10^9 cm^{-2} sec^{-1} (normalized to 300 km). Since neutral hydrogen diffuses vertically through the atmosphere with a flux of no more than about $3 \cdot 10^8$ cm^{-2} sec^{-1}, it is not obvious how such large polar wind fluxes can be supported. Poleward lateral flow of atomic hydrogen probably cannot help to support fluxes of more than a few times 10^8 cm^{-2} sec^{-1}. Thus it appears possible that the neutral hydrogen distribution in the polar atmosphere will show a depletion due to polar wind efflux of H^+ ions.

From the discussion in this section, it is clear that any conclusions drawn regarding the effect of the polar wind on neutral hydrogen, helium, or polar ionospheric density and velocity profiles are accurate only within the limits of our rather poor knowledge of the polar environment. If, however, the future should bring more extensive polar cap observational programs, then more quantitative calculations and more definitive conclusions will become possible.

Acknowledgments. I am grateful to P. M. Banks and W. I. Axford for many helpful discussions during the preparation of this paper. I wish also to thank A. J. Dessler for several stimulating comments which have led to a deeper understanding of this work. This work was supported, in part, through NASA Grant NGR-05-009-075 and by the Advanced Research Projects Agency of the Department of Defense and was monitored by the U.S. Army Research Office - Durham under Contract DA-31-124-ARO-D-257.

REFERENCES

Axford, W. I., The polar wind and the terrestrial helium budget, J. Geophys. Res., 73, 6855, 1968.

Banks, P. M., Collision frequency and energy transfer, ions, Planetary Space Sci., 14, 1105, 1966.

Banks, P. M. and T. E. Holzer, The polar wind, J. Geophys. Res., 73, 6846, 1968.

Banks, P.M., Plasma transport in the topside ionosphere, in The Polar Ionosphere and Magnetospheric Processes, edited by G. Skovli, Gordon and Breach, New York, 1969.

Ferguson, E.E., F.C. Fehsenfeld, D.B. Dunkin, A.L. Schmeltekopf, and H.I. Shiff, Laboratory studies of helium ion loss processes of interest in the ionosphere, Planetary Space Sci., 12, 1169, 1964.

Holzer, T.E. and P.M. Banks, The polar wind, Trans. Am. Geophys. Union, 49, 749, 1968.

Holzer, T.E. and P.M. Banks, Accidentally resonant charge exchange and ion momentum transfer, Planetary Space Sci., 17, 1074, 1969.

Hoffman, J.H., Ion composition measurements in the polar region from the Explorer 31 satellite, Trans. Am. Geophys. Union, 49, 253, 1968.

Jacchia, L.G. and J.W. Slowey, Diurnal and seasonal-latitudinal variations in the upper atmosphere, Smithsonian Astrophys. Obs., Special Rep. 242, 1967.

Keating, G.M. and E.J. Prior, The distribution of helium and atomic oxygen in the lower exosphere, Trans. Am. Geophys. Union, 48, 188, 1967.

Keating, G.M. and E.J. Prior, The winter helium bulge, Space Res., 8, 982, 1968.

Müller, D. and G. Hartmann, A mass spectrometric investigation of the lower thermosphere above Fort Churchill with special emphasis on the helium content, J. Geophys. Res., 74, 1287, 1969.

Nicolet, M., G. Kockarts, and P.M. Banks, Introduction to the study of aeronomy, to be published in Handbuch der Physik (Geophysik), Vol. 49, 1969.

Rishbeth, H., A time-varying model of the ionospheric F2-layer, J. Atmospheric Terrest. Phys., 26, 657, 1964.

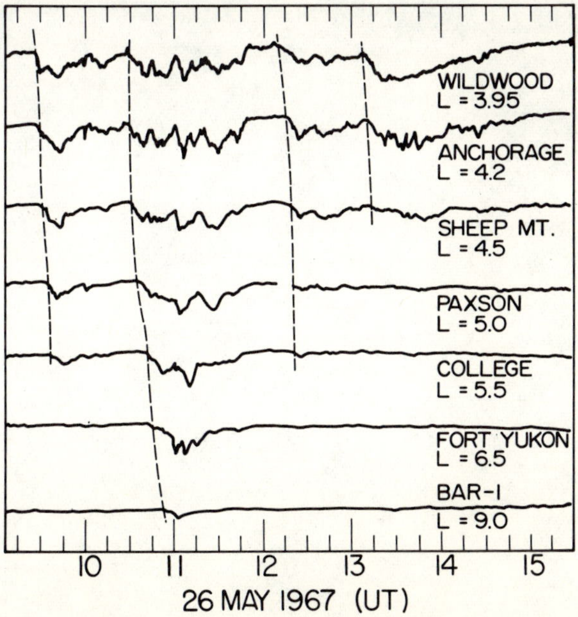

THE POLAR EXOSPHERIC PLASMA

J. O. Thomas and M. K. Andrews

Physics Department
Imperial College, London, UK

Abstract. Some of the main features of the trans-polar plasma distribution in the upper F-region as deduced from Alouette 1 topside sounder data are discussed. The results are related to recent satellite measurements of incoming particles of magnetospheric origin, particularly those in the low energy range (~ 1000 eV), and to recent work on the polar wind.

1. INTRODUCTION

Even a cursory examination of early work on the nature of the polar ionosphere, as revealed by experiments conducted from the ground, convincingly demonstrates the additional complexity of the plasma distribution at high latitudes. Among the factors leading to this complexity one can itemize, with the benefit of hindsight, a number of inter-mingling characteristics:

(i) The presence of a solar-controlled contribution to the plasma by photo-ionisation, basically similar to that responsible for the low-latitude ionosphere.

(ii) An additional source of plasma, namely incoming energetic particles, producing ionisation by collision over an extended volume of the upper atmosphere generally having the shape of a torus or ring.

(iii) The (variable) juxtaposition of these two separate sources of ionisation may lead to a plasma density which, if observed at a given point above the ground, appears to change with time in a complicated way. Nevertheless, the basic trans-polar electron density pattern resulting from this juxtaposition can, and has been, determined recently (Thomas and Andrews, 1968; Andrews and Thomas, 1969; Thomas and Andrews, 1969) by selecting for study those satellite data which were recorded at times when the effects of the two ionisation sources were most likely to be clearly separated in space. The resulting basic distribution shows two relatively plasma-free zones, one in the form of a crescent, and the other located over the polar region itself (the polar depletion (Thomas et al., 1966), or polar cavity). Though the latter may reflect the torus-like shape of the plasma source, this region of depletion at high latitudes is probably enhanced by dynamic loss along the lines described recently in papers on the polar plasma efflux (Banks, 1969; Banks and Holzer, 1968; Holzer, 1969).

Of the two sources of plasma, clearly the second is of more interest as regards ionosphere-magnetosphere coupling, and it is best studied during the polar winter, when the solar U. V. source is largely absent, and during quiet magnetic times. Pronounced magnetic activity has been found to alter the region drastically compared with its basic, quiet-time behaviour. Accordingly, the data in the survey referred to, the conclusions from which are briefly outlined here, were taken from the winter of 1962/63, when solar activity was very low.

2. THERMAL PLASMA

The most consistently observed ionospheric feature at high latitudes is an abrupt decrease in the electron density as measured at a particular altitude, by up to an order of magnitude or more, in a horizontal north-south distance of sometimes less than 100 km. At higher latitudes, at least over the altitude range of 300 - 1000 km seen by Alouette 1, the density rises to higher values, though here the ionisation appears to be rather disturbed. This effect, called the main trough by Muldrew (1965), appears to be the demarcation between the solar produced mid-latitude ionosphere, and that at high latitudes which is aurorally associated. On such an interpretation, the low latitude boundary of the trough is the projection of the plasmapause (Carpenter, 1966) into the ionosphere. During the late morning hours,

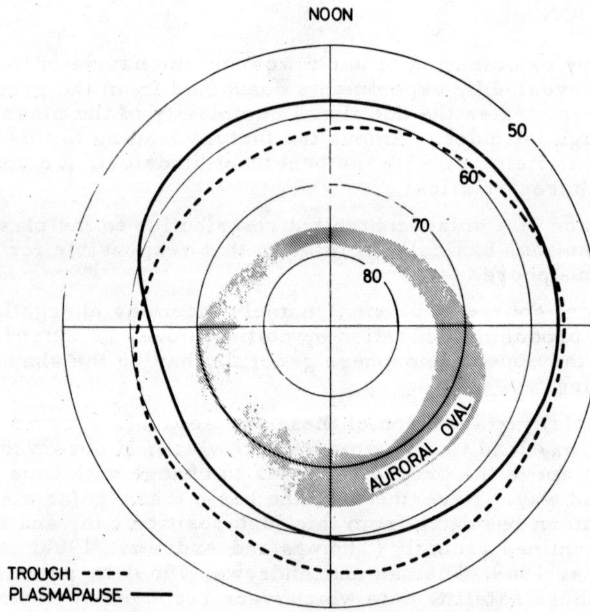

Figure 1. The relative position, in invariant latitude and local time, of the trough boundary, the auroral oval, and the plasmapause projected along a dipole field line to the ionosphere. Positions given are those expected for moderate magnetic activity (Kp = 3). Both trough and plasmapause loci occur below the auroral oval, and while they are approximately the same size, there is a systematic discrepancy between them (Thomas and Andrews, 1968).

the trough as such is not clearly defined, but there remains a feature in the curves of density versus latitude which can be associated with the trough found at other times. Using Alouette 1 data from a very quiet solar period, the trough boundary has been located at all local times (Thomas and Andrews, 1968), and is shown in Figure 1, which also includes the position of the plasmapause (Carpenter, 1966), projected back along magnetic field lines from the equatorial plane to the high latitude ionosphere. There are sufficient similarities in the behaviour of the trough and plasmapause to believe the interpretation placed on the two features here is basically correct, but there are systematic differences in position which have been discussed elsewhere (Thomas and Andrews, 1968).

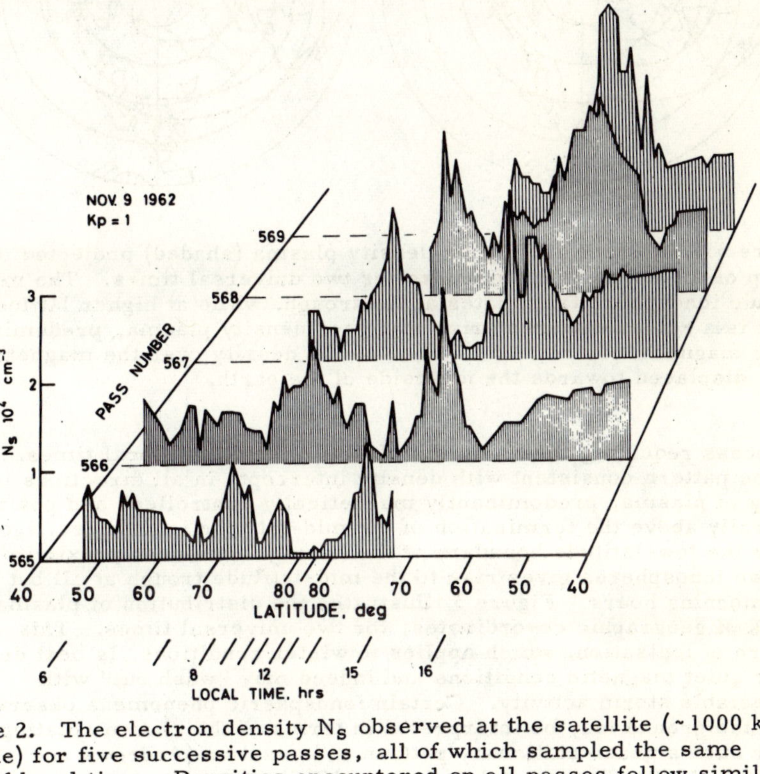

Figure 2. The electron density N_s observed at the satellite (~1000 km altitude) for five successive passes, all of which sampled the same span of local time. Densities encountered on all passes follow similar trends, despite the lag of universal time between them.

At higher latitudes, and particularly at quiet times, the density profiles obtained along satellite tracks at a particular altitude can be surprisingly repeatable from one pass to the next. Such a sequence is illustrated in Figure 2, which shows the density at the vehicle (~1000 km above the earth) during a series of dayside passes. The afternoon trough is plainly visible at about 1500 LT and 70° latitude, while two peaks of density, separated by a region of very low density plasma over the pole, are apparent at higher latitudes. All the passes in this figure are at the same angle to the earth-sun line, though different geographic longitudes are being sampled on different passes, as a result of the earth's rotation. The repeatability of the density profile for long periods of universal time suggests there is a pattern of polar ionisation under which the earth rotates. This density distribution has been investigated using satellite paths at various angles to the earth-sun line,

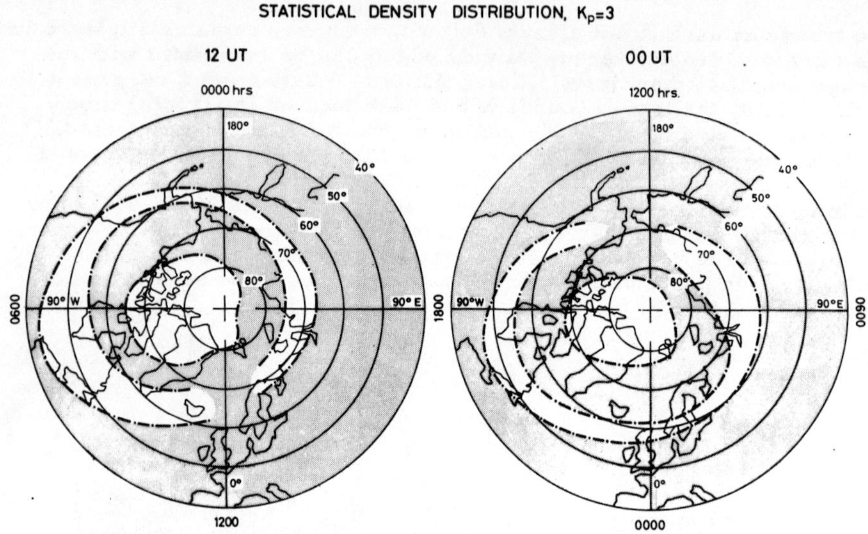

Figure 3. The zones of higher density plasma (shaded) projected over a map of the northern hemisphere for two universal times. The mid-latitude ionosphere terminates at the trough, while at higher latitudes there is a rather disturbed ring of higher density plasma, predominantly under magnetic control, and a zone of low density over the magnetic pole, displaced towards the nightside of the earth.

a process requiring three months of data to cover all local times.

The pattern consistent with density intercepts in all directions is a ring of plasma, predominantly magnetically controlled, and positioned generally above the termination of the mid-latitude ionosphere, i.e., above the low-latitude boundary of the trough. The near proximity of the two ionospheres gives rise to the mid-latitude trough at all but the late morning hours. Figure 3 illustrates the distribution of plasma, in terms of geographic co-ordinates, for two universal times. This pattern of ionisation, which applies to winter conditions, is best defined under quiet magnetic conditions and indeed may "wash out" with appreciable storm activity. Certain ionospheric phenomena observed from the ground may be interpreted in terms of the station rotating under such a quasi-stationary pattern of ionisation (Andrews and Thomas, 1969; Thomas and Andrews, 1969).

3. RELATIONSHIP WITH PARTICLE FLUXES

It has been pointed out that the disturbed nature of the polar ionosphere, including such features as thin, magnetically aligned sheets of ionisation, strongly suggests collision ionisation. To produce significant amounts of F-region ionisation, it is known that fluxes of electrons with initial energies of about 1 keV are required. Recently, such particles have been measured directly (Evans et al., 1967; Burch, 1968; Hoffman, 1969). Significantly, these fluxes also show a cavity over the pole, where their intensity drops by at least two orders of magnitude. The high latitude boundaries, Λ_D and Λ_N in invariant latitudes, of the soft electron fluxes (Maehlum, 1968) in the noon-midnight meridian are compared in Table 1 with the equivalent day and night boundaries of the plasma density ring.

Observation	Λ_D	Λ_N	$\Lambda_D - \Lambda_N$	Cavity size
Electrons, .08-21 keV, winter 1963	78°	72°	6°	30°
Electrons, .05- 1 keV, summer 1967	83°	72°	11°	25°
Plasma density, Kp 3, winter 1962	76°	73°	3°	31°

Table 1

The positional agreement between the precipitation and the enhanced densities strongly suggests cause and effect. Calculations, which must include the effects of plasma diffusion above 200 km altitude, indicate that the fluxes reported by Burch (1969) are sufficient to produce collision ionisation densities equal to those observed. From Alouette data for the period of 1962/63, densities at the F-layer peak were found to be ~ $.5 \cdot 10^5$ cm^{-3} in the trough and the polar depletion. In the nightside peaks, typical densities were in the range $1.5 - 2.5 \cdot 10^5$ cm^{-3}, rising to $3.5 - 4.5 \cdot 10^5$ cm^{-3} on the dayside. Because of a softer spectrum in the dayside fluxes, this day-night asymmetry in electron densities could have been inferred from precipitation measurements, again suggesting that low energy electrons are responsible for much of the observed polar ionisation.

It is emphasised that this density distribution refers to a period of time when ion production due to solar pho-ionisation was negligible, and also to a period of very low solar activity. There is some evidence, from both auroral and ionospheric observations, that the polar cavity may have been absent during the IGY period, which was characterised by extremely high solar activity. During the summer months, there is no evidence of the polar cap depletion in density, though the fact that the flux cavity is still present indicates that photo-ionisation is probably obscuring the pattern of peaks and troughs present in winter.

4. POLAR DENSITY PROFILES

The foregoing results have been based on densities observed at two altitudes, the altitude of the vehicle, and that of the F-layer peak. More information can be obtained on the changes occuring across the pole after performing real height-density analyses of ionograms in the region.

Figure 4 gives graphically the results of such an analysis of a sequence of ionograms, during which the satellite crossed the morning trough from mid-latitudes, and entered a polar peak of ionisation. The scale height H of the topside plasma, defined by

$$H = - N \frac{dh}{dN}$$

may be shown to carry information basically on the ratio $(T_i + T_e)/m_i$, where T_i, T_e refer to ion and electron temperatures, and m_i is the mean ionic mass. The sudden fall in scale height at great heights on entering the trough is consistent with a rise in mean ionic mass, so that the ionosphere at high latitudes is composed mainly of oxygen ions. This abrupt change from mid-latitudes is illustrated in the profiles A and B. The former alone shows the curvature near 600 km caused by a change in composition to ions lighter than O^+ at greater heights. At the same time, the steeper slope of profile B below 500 km, where the dominant ion must be O^+, implies the temperature of the polar

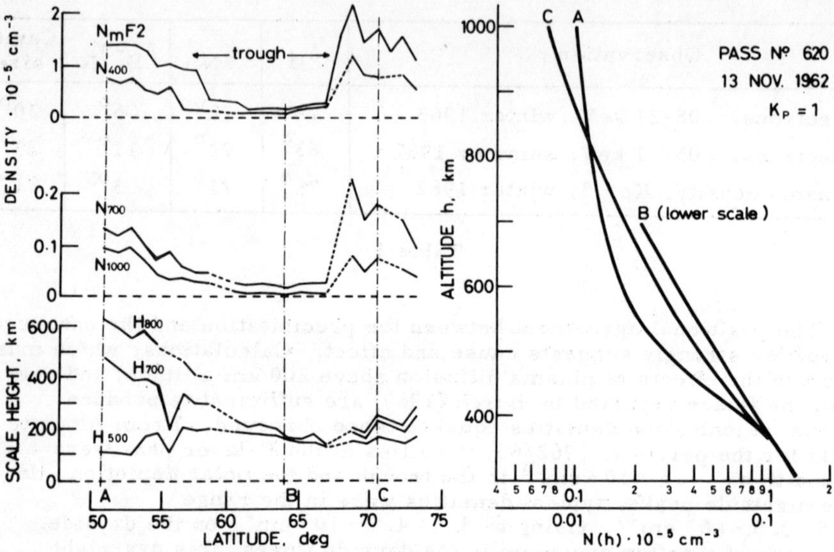

Figure 4. The results of real height-density analyses of a sequence of ionograms during which the satellite crossed the mid-latitude trough and entered a polar peak of ionisation. From the data, it appears that the ionosphere in and above the trough consists of a hotter, predominantly oxygen plasma, at all altitudes up to the vehicle.

ionosphere is greater than at mid-latitudes. Profile C, taken inside a polar peak, is interesting as it indicates the composition of the ionisation peak is remarkably similar to the trough, i.e., predominantly hot O^+, even though the densities differ by an order of magnitude.

At magnetic latitudes above about 80°, profiles corresponding to the night hours frequently show a fall of scale height, below 500 km, compared with values at latitudes a few degrees south, to values difficult to explain on the assumption that diffusive equilibrium applies. Such a depression of H below 500 km, however, is what might be expected as a consequence of the polar wind (Banks, 1969), where O^+ ions flow upwards through the neutral atmosphere to replenish the outflowing protons by charge exchange with neutral hydrogen atoms. In doing so, the scale height of the flowing O^+ plasma becomes depressed below that expected in diffusive equilibrium, and is nearer that of the neutral atmosphere.

5. FUTURE WORK

It appears that the overall questions of the winter source of polar ionisation, and the morphology of that ionisation are becoming known, largely as a result of the data coverage afforded by satellite-borne experiments. There remain questions of detail; for example in the matter of the exact relationship between the trough boundary and the plasmapause, if the coupling between the two is to be along field lines. While magnetospheric circulation (Axford, 1969) is able to predict the existence of a plasmapause, it appears that the density gradients observed, both in the ionosphere and the equatorial plane, lack an adequate explanation.

At highest latitudes, the interesting question of the extent to which plasma takes part in a polar wind type of escape is unknown, as is the extent to which the polar depletion is a result of plasma removal or simply the absence of the ionising source of precipitation. It is likely that satellites such as Alouette 2, reaching altitudes well above those of Alouette 1, will be useful in providing solutions to these as yet unanswered questions.

REFERENCES

Andrews, M.K. and J.O. Thomas, Electron density distribution above the winter pole, Nature, 221, 223, 1969.

Axford, W.I., Magnetospheric convection, Rev. Geophys., 7, 421, 1969.

Banks, P.M., Plasma transport in the topside polar ionosphere, in The Polar Ionosphere and Magnetospheric Processes, edited by G. Skovli, Gordon and Breach, New York, 1969.

Banks, P.M. and T.E. Holzer, The polar wind, J. Geophys. Res., 73, 6846, 1968.

Burch, J.L., Low-energy electron fluxes at latitudes above the auroral zone, J. Geophys. Res., 73, 3585, 1968.

Burch, J.L., Satellite measurements of low energy electrons precipitated at high latitudes, in The Polar Ionosphere and Magnetospheric Processes, edited by G. Skovli, Gordon and Breach, New York, 1969.

Carpenter, D.L., Whistler studies of the plasmapause in the magnetosphere, J. Geophys. Res., 71, 693, 1966.

Evans, J.E., R.G. Johnson, R.D. Sharp and J.B. Reagan, Recent results from satellite measurements of low-energy particles precipitated at high latitudes, Space Sci. Rev., 7, 263, 1967.

Hoffman, R.A., Low-energy electron precipitation at high latitudes, J. Geophys. Res., 74, 2425, 1969.

Holzer, T.E., Effects of plasma flow on density and velocity profiles in the polar ionosphere, in The Polar Ionosphere and Magnetospheric Processes, edited by G. Skovli, Gordon and Breach, New York, 1969.

Maehlum, B.N., Universal-time control of the low-energy electron fluxes in the polar regions, J. Geophys. Res., 73, 3459, 1968.

Muldrew, D.B., F-layer ionization troughs deduced from Alouette data, J. Geophys. Res., 70, 2635, 1965.

Thomas, J.O. and M.K. Andrews, Transpolar exospheric plasma, J. Geophys. Res., 73, 7407, 1968.

Thomas, J.O. and M.K. Andrews, The trans-polar exospheric plasma, 3: A unified picture, Planetary Space Sci., 17, 433, 1969.

Thomas, J.O., M.J. Rycroft, L. Colin and K.L. Chan, The topside ionosphere, II, in Electron Density Profiles in Ionosphere and Exosphere, edited by J. Frihagen, North-Holland Publishing Company, Amsterdam, p. 322, 1966.

THE HIGH LATITUDE IONOSPHERE: RESULTS FROM THE ALOUETTE/ISIS TOPSIDE SOUNDERS

G. L. Nelms

Communications Research Centre (formerly DRTE), Ottawa, Canada

and

J. H. Chapman

Department of Communcations, Ottawa, Canada

Abstract. The major structures of the high latitude topside ionosphere and some of their causal processes are reviewed. Emphasis is given to the relationship between plasma flow in the magnetosphere and the structure of the high latitude ionosphere.

1. INTRODUCTION

Satellite-borne topside sounders have provided a great deal of information on the global structure of the ionosphere at and above the peak density of the F-layer. This paper reviews some of the high latitude results from the topside sounders in Alouettes 1 and 2; some facts about the sounder experiments and the satellite orbits, that are of importance to the measurements of the ionosphere, are given in the remainder of this section.

To the present, four topside sounder satellites have been placed in orbit; Alouette 1 in 1962, Explorer 20 in 1964, Alouette 2 in 1965 and ISIS 1 in 1969. The Alouette and Explorer 20 satellites contain sweep frequency and fixed frequency sounders respectively, ISIS-1 contains both. All four satellites are in near polar orbits, with inclinations of 80° or more. Alouette 1 is in a circular orbit at 1000 km altitude, Explorer 20, Alouette 2 and ISIS-1 are in elliptical orbits with perigees and apogees of 870 and 1010, 500 and 3000, and 570 and 3520 km, respectively.

The sweep frequency sounders in Alouette 1 and 2 cover the frequency range from 0.5 to 11.5 MHz and 0.15 to 14.5 MHz, respectively, with cycle times of 18 and 30 seconds per ionogram (Franklin and Maclean, 1969). The satellites have orbital velocities of about 7 km per second, providing an ionogram sampling distance of about 130 km for Alouette 1 and 200 km for Alouette 2. They are both in 80° inclination orbits. Each of the satellites is therefore capable of making detailed measurements of distribution of ionization in the topside ionosphere, at all heights between the satellite and the peak density of the F-layer, along an approximately north-south line, from $80\,^{o}N$ to $80\,^{o}S$. Such "pole to pole" passes are obtained in about 1 hour of elapsed time, although the local mean time along the sub-satellite track varies by 12 hours: about 5 hours between the inclination latitude and 60° latitude and 1 hour between 60° and the equator. By combining such data collected over successive orbits, the global structure of the ionosphere at a given set of local times is revealed; by combining the data over a period of 3 to 4 months, the diurnal variation (with superimposed seasonal variation) over a wide range of latitudes can be obtained.

The resonance beat phenomena (Hagg, 1967; Hagg et al., 1969) observed on the topside sounder ionograms provides an entirely new experimental technique for making measurements of the density of a tenuous plasma. Because of its importance for studies of the high latitude ionosphere, some details of the mechanism are given below.

The frequencies of the resonance spikes observed on the topside sounder ionograms yield information on f_N, the plasma frequency, and on f_H, the electron gyrofrequency at the satellite or various harmonics and combinations of those frequencies. The cut-off frequencies for the various modes of propagation (i.e., the lowest frequency at which the mode can propagate at the satellite) are also determined by f_N and f_H.

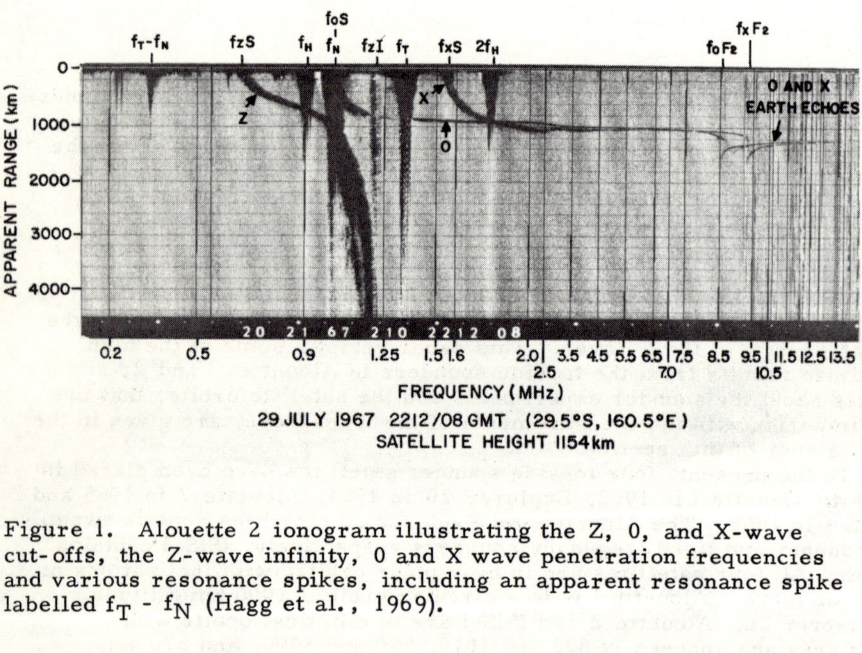

Figure 1. Alouette 2 ionogram illustrating the Z, 0, and X-wave cut-offs, the Z-wave infinity, 0 and X wave penetration frequencies and various resonance spikes, including an apparent resonance spike labelled f_T - f_N (Hagg et al., 1969).

Thus there are a variety of ways (Hagg et al., 1969) of measuring combinations of the electron plasma frequency f_N (and hence, of N) and the electron gyrofrequency (Figure 1 and Table 1). In practice, the electron gyrofrequency is known accurately, for the region below a few thousand kilometers altitude, from spherical harmonic models of the earth's field or from the observed gyrofrequency resonances, and the various plasma phenomena can be used to improve the spatial resolution of the measurements of electron number density.

Table 1 lists the cut-offs and resonance spikes (Lockwood, 1963; Calvert and Goe, 1963) that are commonly observed. Many other resonances are also observed; the reader is referred to Hagg et al. (1969) and Muldrew (1969) for a review of them.

Type	Frequency	Plasma relations	
<u>Cut-offs</u>			
Z wave	fzS	$f_N^2 = fzS^2 + fzSf_H$	(1)
X wave	fxS	$f_N^2 = fxS^2 - fxSf_H$	(2)
0 wave	f_N		
Z infinity	fzI	$f_N^2 = \dfrac{fzI^2\,(fzI^2 - f_H^2)}{fzI^2 - f_H^2\ \sin^2\phi}$	(3)
<u>Resonance Spikes</u>			
plasma frequency	f_N		
electron gyrofrequency	$f_H, 2\,f_H, \cdots n\,f_H$		
upper hybrid frequency	$f_T, 2\,f_T$	$f_N^2 = f_T^2 - f_H^2$	(4)

ϕ is the angle between the wave normal and the direction of the earth's magnetic field.

Table 1

For the Alouette 2 orbit, f_H (and hence fxS and f_T) is always above 0.22 MHz, but at high latitudes and altitudes f_N is often less than 0.15 MHz (N less than 280 electron/cc), the low frequency limit for the Alouette 2 sounder. From equations (2) and (4), as f_N approaches zero, fxS approaches f_T which approaches f_H. Since the frequency resolution under these circumstances is usually no better than the bandwidth of the transmitter pulse (~ 15 kHz), the frequencies at which fxS, f_T, and f_H occur are too closely spaced for separate identification. However, for electron densities less than about 140 electron/cc, $f_T - f_H$ is less than 10 kHz, and the transmitter pulse will energize both resonances simultaneously. The difference frequency, Δf, may then be observed as a beat or modulation pattern (Figure 2) on the closely spaced f_T and f_H resonances (Hagg, 1967). Since f_H can be determined from the $2f_H$ spike (~ 2% accuracy) or from a spherical harmonic field analysis, f_N can be calculated from a knowledge of Δf, according to the equation

$$f_N^2 = (2\,f_H + \Delta f)\,\Delta f \qquad (5)$$

consequently,

$$N = 18.6\ m\,f_H \qquad (6)$$

where m is the number of cycles of modulation observed per 200 km of apparent range, N is the electron number density in electrons/cc and all frequencies are in MHz.

Beats also occur on resonance spikes under other conditions (cf. for example, the f_T resonance of Figure 1) but the resonance beats of interest for measurement of the low electron densities can be isolated relatively easily from the following:

(i) The basic criterion for the observation of the "low density" beats is that the f_H and f_T resonances are separated by less than the bandwidth of the telemetry system (10 kHz).

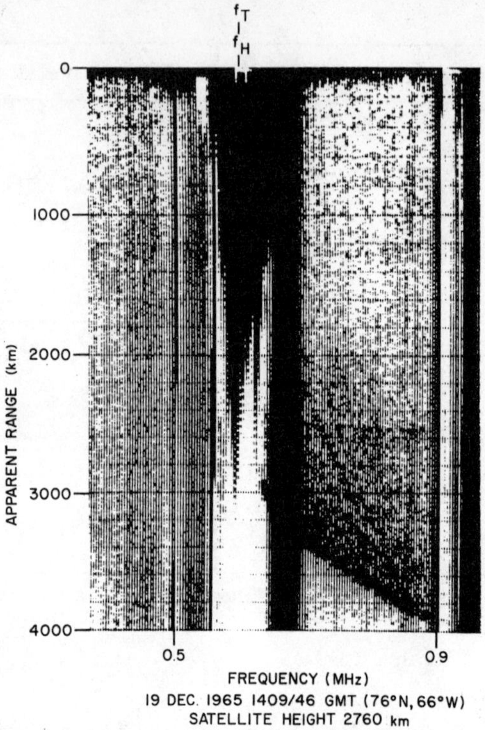

Figure 2. An Alouette 2 ionogram illustrating resonance beats on the combined f_T and f_H resonances (Hagg, 1967).

(ii) Usually, the beats observed on isolated resonances exhibit a series of beat frequencies throughout the spike (cf. for example, the f_T resonances of Figure 1), whereas the low density beats exhibit the same beat frequency throughout the resonance (cf. Figure 2).

Some empirical rules that are used to ensure that the appropriate beats are used are that (i) the f_N resonance must be at a frequency less than 0.2 MHz, (it may be less than the starting frequency of the ionogram, and so not be observed) (ii) the f_H and f_T resonances must be superimposed, and (iii) the beat modulation must appear on the low frequency edge of the superimposed f_H, f_T resonance spike. It is of interest that, as expected, resonance beats of the same form as those used for the low density studies also are observable when the $2f_H$ and f_T resonances are superimposed. Although the electron density is then several thousand electrons/cc, it can, of course, be calculated from this beat frequency; when this is done it corresponds to the electron density obtained from the observed f_N resonance and from the observed fxS (Hagg, Private communication).

Electron densities as low as 8 and as high as 140 electrons/cc have been measured from the Alouette 2 ionograms using this method. The upper limit is set by the telemetry bandwidth and the electron gyrofrequency, and should be about 145 electrons/cc for the Alouette 2 orbit (Timleck and Nelms, 1969); the lower limit is set by f_H and the duration of the f_T and f_H resonance spikes, and should be of the order of 1 electron/cc for the Alouette 2 orbit. However, for beats of less than 1 cycle per 1000 km (N ≈ 5 electrons/cc) identification of the modulation pattern on Alouette 2 ionograms would probably become difficult.

2. RESULTS

The ionized region above the main peak of the F-layer can be studied from many points of view, and because of the complexity of the region, no single simple theory is likely to be adequate for an understanding of the region. Nevertheless, in this region the prominence of field and plasma flow effects increases with height, and it is useful to review current knowledge in terms of the major types of structure that are observed, and to try to relate these observations to the main restricting forces present in the magnetosphere.

Although by no means a surprise, one of the striking results from the topside sounders is the extent of the control that the earth's magnetic field exerts over the distribution and movement of ionization. Since details have been in the literature for some time, only the principal results concerning irregularities will be summarized.

2.1 Occurrence of Irregularities in the Topside Ionosphere

Irregularities are observed in two distinct areas of the topside ionosphere, a polar zone and an equatorial zone, and both occurrence and propagation characteristics are somewhat different in the two regions (Figures 3a-b). Although this paper is mainly concerned with the polar ionosphere, scattering in the equatorial ionosphere is of interest because of the insight it gives on processes in the polar regions.

Spread F has been categorized into two types, "weak" and "strong". Although the terms are subjective, the two types give distinctive patterns on the ionograms, and the boundaries of the regions over which they are observed are reasonably well defined. Weak spread F is usually observed on ionograms recorded at geomagnetic latitudes greater than about 50° to 60° during the day time (Knecht and VanZandt, 1963; Petrie, 1966), and about 45° during the night (King et al., 1967). During the day the transition from ionograms exhibiting spread F to no spread F is often abrupt and occurs over a few degrees of latitude.

During magnetically quiet conditions, the region of strong spread echoes at about noon local time (Petrie, 1966) and 1000 km altitude is confined to geomagnetic latitudes north of 75 ^{o}N. The region moves from north to south before midnight and south to north after midnight. It reaches a minimum latitude of about 75 ^{o}N around midnight. The latitudinal variations of the strong spread echoes are similar to those of sporadic E, visual aurora, radio blackouts and horizontal deviations of the earth's magnetic field.

In the polar zone, it is difficult to compare bottomside and topside data because auroral absorption and polar blackout often obscure spread F observations made from the ground (Calvert and Warnock, 1969). At other latitudes the major features of the two occurrence patterns agree (Lockwood and Petrie, 1963). The major differences are secondary maxima above the F-region peak during the day in the equatorial region and generally greater occurrence above the F-region peak than below.

2.2 Spread F and Ducting in the Topside Ionosphere

Spread F is particularly interesting in the equatorial region, because of the insight it gives us into the structure of irregularities of ionization

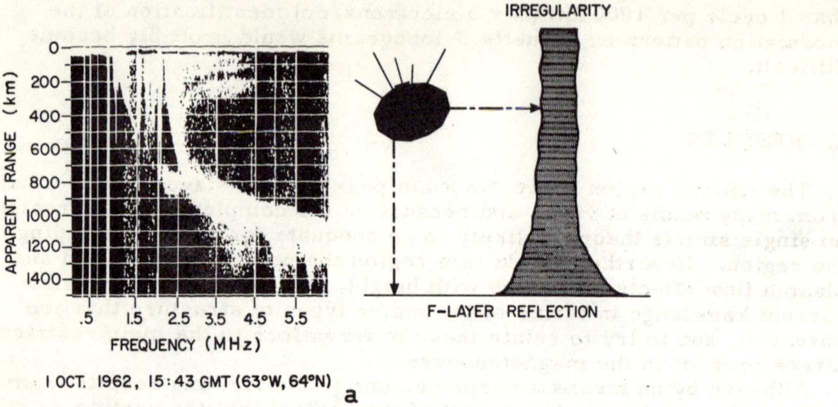

Figure 3a. Alouette 1 ionogram with a spread trace between 200 and 300 km apparent range and 2.5 and 3.5 MHz resulting from echoes scattered from a high-latitude field-aligned irregularity (Petrie, 1963).

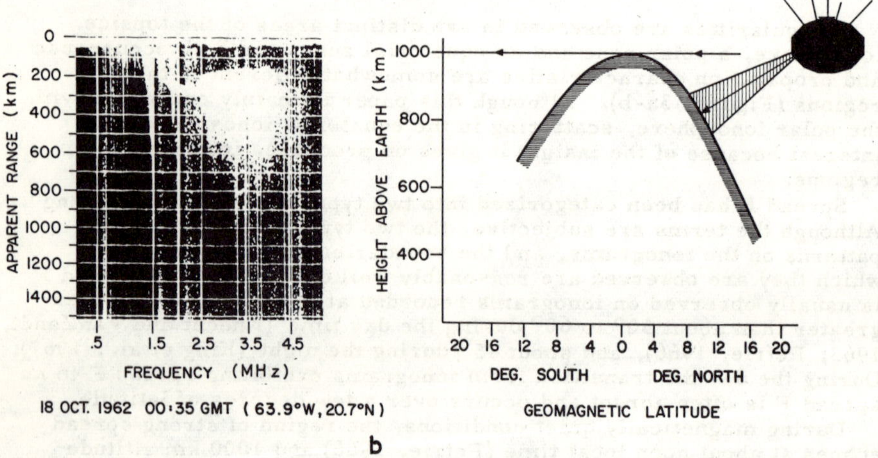

Figure 3b. Alouette 1 ionogram with a spread trace between 2.5 and 3.5 MHz resulting from echoes scattered from a low-latitude field-aligned irregularity (Lockwood and Petrie, 1963).

along the magnetic field lines, and the striking propagation effects which are observed.

Lockwood and Petrie (1963) identified echoes scattered from irregularities aligned along the equatorial magnetic field, forming a shell of ionization extending many km in the E-W direction. Alouette 1 at 1000 km altitude (Figure 3b) usually passed over this region, but Alouette 2 and ISIS-1, having perigees several hundred km lower, occasionally pass through this shell of irregular ionization. Figure 4 shows such an example from ISIS-1.

In Figure 4 the ISIS-1 satellite was just outside of the sheet, resulting in reflections from the sheet which exhibit a relatively sharp upper boundary that is located just a few km from zero range over a frequency range of several MHz. On the ionogram of Figure 5, recorded

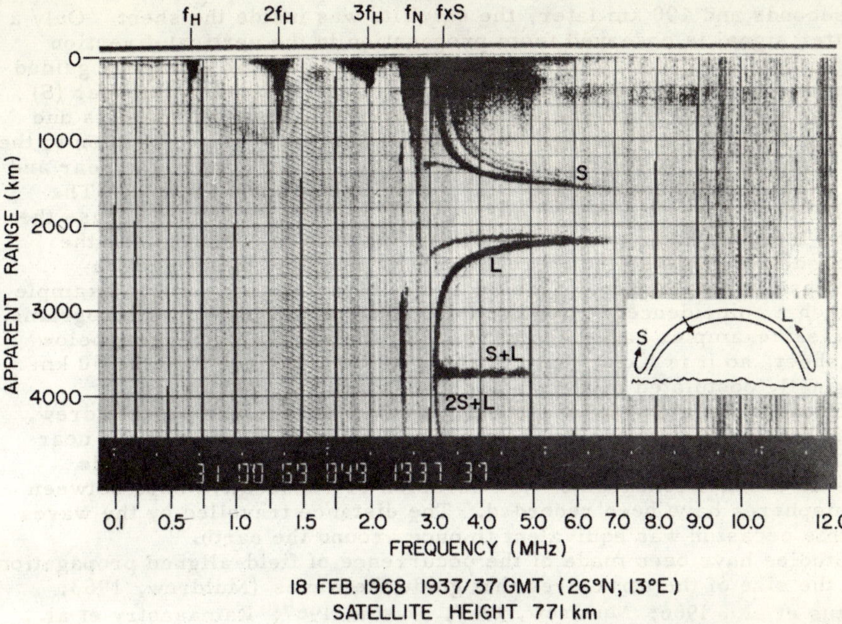

Figure 4. An ISIS-1 ionogram recorded when the satellite was just outside a region of equatorial field-aligned irregularities.

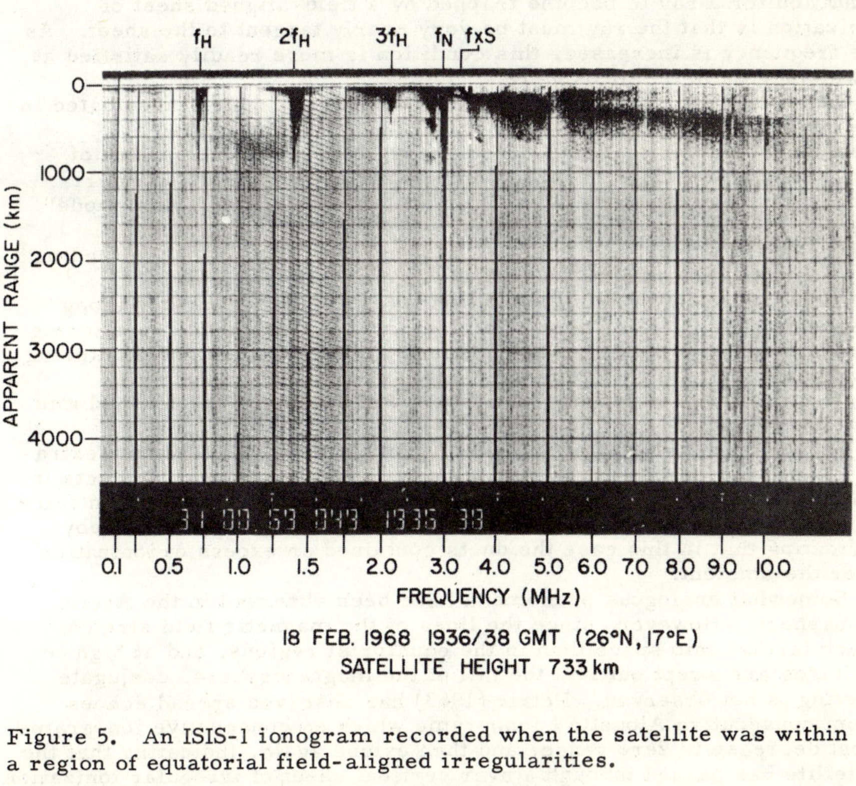

Figure 5. An ISIS-1 ionogram recorded when the satellite was within a region of equatorial field-aligned irregularities.

60 seconds and 400 km later, the satellite was inside the sheet. Only a scatter signal is observed from propagation in the vertical direction (e.g., 5 to 9 MHz at about 500 km range); most of the energy is guided along the field, within the sheet, to the reflection level in the near (S) and in the far (L) hemisphere. The remarkable ε shaped trace is due (Muldrew, Private communication) to the one "hop" of propagation to the far hemisphere (L) for the upper part of the ε, two hops to the near and one to the far hemisphere (2S + L) for the bottom part of the ε. The center bar of the ε is due to an S + L hop, and is horizontal because the S and L traces happen to be nearly mirror images. That is, as the frequency increases, the increase in path length is balanced by a decrease in retardation. This record is by no means a unique example of such a coincidence; it has been observed on several other ionograms. No clear examples have been obtained of the satellite emerging below the sheet, so it is likely that the sheet is much thicker than the 50 km originally postulated by Petrie and Lockwood.

Propagation along field-aligned ionization irregularities (Muldrew, 1963) (ducting), between the satellite and reflection points in the near and far hemispheres, is a common feature of mid and low latitude topside sounder ionograms. On occasion 10 consecutive hops between hemispheres have been recorded. The distance travelled by the waves on this occasion was equivalent to once around the earth.

Studies have been made of the occurrence of field-aligned propagation, and the size of the structures that guide the waves (Muldrew, 1963; Loftus et al., 1966; Muldrew, 1967; Dyson, 1967; Ramasastry et al., 1968). Several authors have used ray-tracing techniques to demonstrate the mechanism of radiowave propagation back and forth along irregularities of ionization aligned along the magnetic field. One condition for a ray to become trapped by a field-aligned sheet of ionization is that the ray must be very nearly tangent to the sheet. As the frequency is increased, this condition is more readily satisfied at lower heights in the ionosphere.

Scattering and ducting by irregularities are intimately associated in the equatorial region. Pitteway and Cohen (1961) suggested a combination of vertical and ducted propagation as an explanation of "frequency spreading" observed on equatorial bottomside sounder ionograms, and Muldrew (1963) showed that this "combination mode" was a common form of spread F on topside sounder equatorial ionograms. Calvert and Cohen (1961) proposed that "range spreading" on equatorial bottomside sounder ionograms was due to aspect sensitive backscatter from ionization irregularities elongated along the field. Dyson (1967) showed a close correlation between scattering and ducting, with ducting more common at low latitudes and scattering more common at high latitudes. Loftus et al. (1966) found that scattering occurred only in the presence of ducting in the low and mid latitude regions.

Figure 5 is particularly interesting since it shows about four extraordinary wave traces which are evidence of propagation along ducts in which the average ionization was significantly different; the main trace is the strongest signal occurring at the lowest frequencies, thereby indicating that in this case the ducts contained an excess of ionization over the ambient.

Somewhat analogous phenomena have been observed in the Arctic ionosphere. However, since the lines of the magnetic field stretch much farther into space than in the equatorial regions, and at high latitudes are swept out into the tail of the magnetosphere, conjugate ducting is not observed. Petrie (1963) has observed spread echoes from consecutive Alouette 1 ionograms which on consecutive ionograms first decrease to zero range, and then extend again, indicating that the satellite has passed through a near vertical sheet of irregular ionization at 1000 km height.

Figure 3a shows an example of reflections from such a sheet. In this case the sheet was magnetic field-aligned and about 100 - 200 km thick transverse to the field. The deviations in electron density are sufficiently large to scatter signals 2 MHz or more above fxS. In the ionogram of Figure 3a, backscattered signals of this type appear between 200 - 300 km apparent range and about 2.5 to 4.5 MHz on the frequency scale.

Calvert and Schmid (1964) proposed backscatter from long, thin field-aligned irregularities as the dominant mechanism for producing spread F. Calvert and Warnok (1969) suggest that the disappearance of ducted echoes at high latitudes may be due to their being masked by scatter echoes. However, there is an alternate explanation for spread F, based on scatter from a rough surface, that does not require either field-aligned irregularities or ducting (Gallet, 1955; Renau, 1960).

2.3 Precipitation Boundaries at High Latitudes

The ambient density distributions are also governed by the presence of the field, not only by forcing diffusive equilibrium, if and when it exists, to occur along the field rather than in a vertical direction, but also by providing a major restriction to the free flow and distribution of ionization that is not present where the field lines are open into the large volume of the magnetospheric tail (Nishida, 1966; Bauer, 1969). The boundary between these two regions, those of open and of closed magnetic field lines, is the boundary of major significance in the high ionosphere, and is probably of greater importance in the low ionosphere than has thus far been recognized.

Particles of energy > 35 keV. McDiarmid and Burrows (1968) defined several boundaries found in the energetic particle data from the Alouette 2 satellite. They found that for $Kp < 2^-$ between December 1965 and July 1966, the "background" boundary for 35 keV particles, which they regarded as the limit of the closed field lines, varied between about 70° geomagnetic at 2100 LMT and 77° at 0800 hours, whereas the "smooth" boundary, which they interpreted as the limit of stably-trapped electrons, varied between about 67° at midnight and 70° at 1100 LMT. The low latitude boundary of the region of maximum frequency range for spread F echoes, observed by Petrie (1966) from Alouette 1 ionograms, is in good agreement with the "background" boundary of 35 keV particles.

Hartz and Brice (1967) combined a wide variety of upper atmosphere phenomena that they considered to be linked in some way to the influx of auroral particles, into an idealized representation of two main zones of auroral particle precipitation in the northern hemisphere (Figure 6). They considered the outer zone to be associated with a particle influx (> 20 keV electrons) that was steady and widespread, and which they described as a "drizzle", and the inner zone to be associated with intense and spatially limited precipitation (a few keV electrons) which they called, after O'Brien (1964) "splash". The splash zone should occur very close to the boundary between the open and closed field lines. It should be noted that the McDiarmid and Burrows (1968) and Petrie (1966) data are included in the data used by Hartz and Brice. In Figure 6 a dashed curve representing the most recent determination by McDiarmid and Wilson (1968) from the Alouette 2 particle data, of the boundary of the closed field region, is superimposed on the original Hartz-Brice diagram. This curve will be used for comparison later in this paper.

Low energy electrons (E < 1 keV). The precipitation of low energy electrons (E < 1 keV) is of particular importance in the topside

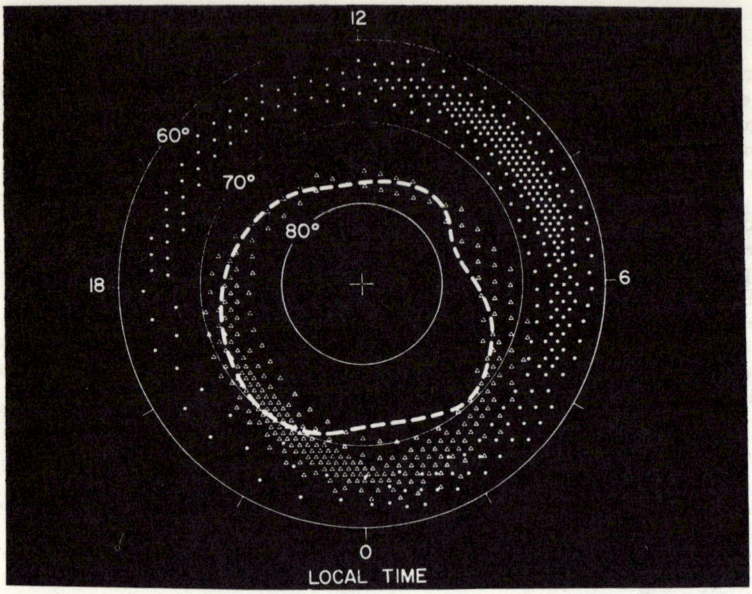

Figure 6. An idealized representation of the two main zones of auroral particle precipitation (northern hemisphere). Average intensity of influx is indicated approximately by the density of the symbols. Coordinates are geomagnetic latitude and geomagnetic time. "Splash" precipitation is represented by triangles, "drizzle" precipitation by dots (Hartz and Brice, 1967). Added to the Hartz-Brice diagram is an estimate of the boundary of the closed field region (dashed line) as given by McDiarmid and Wilson (1968), from particle measurements.

ionosphere because such particles have their maximum ionization effects above about 150 km (above about 250 km for 500 eV electrons). Until very recently, few measurements had been made of near thermal particle fluxes. Since other papers in these proceedings deal specifically with the present state of knowledge of near thermal particle fluxes, no attempt will be made here to review the subject. However, as background to this paper, three general features of their distributions are outlined below (Frank, 1968; Hones, 1968; Maehlum, 1968).

(i) In the tail of the magnetosphere low energy electrons are found near the neutral sheet and are distributed across the tail from dawn to dusk, corresponding to an extension of the sheet, called the "Plasma Sheet". They are more numerous on the dawn than on the dusk side, with maximum population at 0400 and 1900 magnetic local time.

(ii) In very general terms, the spectrum of low energy electrons softens with increasing geomagnetic latitude and the less energetic the particle, the higher the latitude at which it is precipitated.

(iii) The high latitude limit of observations of electron fluxes of energy 280 eV $< E <$ 2.7 keV is near 70° geomagnetic on the night side and near 80° at noon. At high latitudes there are occasional intermittent "islands" of such electrons.

The boundary given in (iii) is well outside the region of stably-trapped 35 keV electrons given by McDiarmid and Wilson (1968), but is close to their limit of the closed field lines (cf. Figure 6).

Maehlum (1968) found that in the polar region, electrons of 50 eV to

1 keV occurred at about 73° geomagnetic latitude at night and 83° during the day, with an absence of such electrons (i. e., a "polar hole") at higher latitudes. He found that these electrons showed a strong GMT control, with maximum latitude occurring at 1800 to 2000 GMT and minimum latitude at 0800 to 1200 GMT. He suggests that the variation of low energy flux might be due to "wobbling" of the geomagnetic axis with respect to the solar wind direction, since the maximum latitude occurs when the angle between the geomagnetic axis and the sun-earth line is a minimum.

2.4 Plasmapause and Troughs

The presence of a plasmapause, i. e., a sharp drop in electron number density at a radial distance of 2 to 6 earth radii above the equatorial region, was established by Carpenter (1963) from whistler measurement, and by Gringauz (1963) from particle measurement. Reber and Ellis (1956) from cosmic noise, and Warren (1963) from topside sounding data reported significant depletions in the electron density at the peak of the F-layer near 50° geomagnetic latitude. Muldrew (1965) showed that "troughs" in the electron density at the peak of the F-layer, are a persistent feature of the nighttime and mid and high latitude ionosphere. The troughs are aligned in the magnetic east-west direction, and are usually a few degrees wide in north-south direction. The electron density in the main (the most southerly, or mid-latitude) trough is typically one-fourth that of the background density. The average location of the main trough (Jelly and Petrie, 1969) during November 1962 to March 1963 is shown in Figure 7. The trough is generally masked by day time production; during magnetic storms it shifts toward the equator by about 2° per unit Kp (Muldrew,

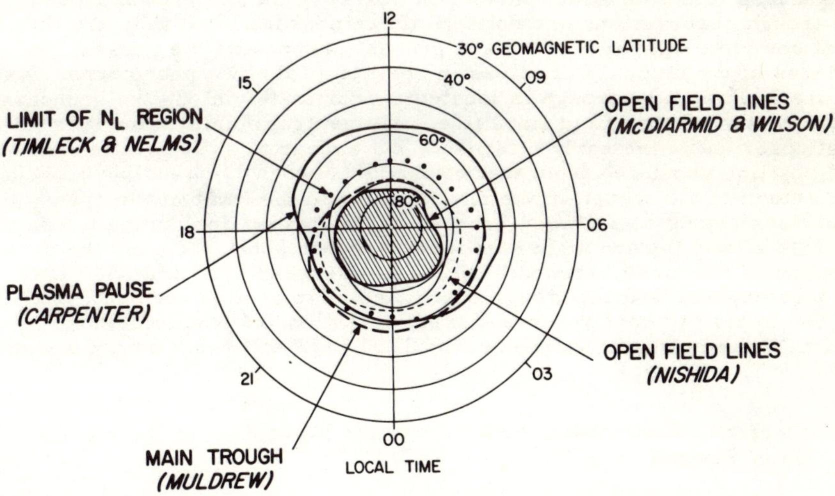

Figure 7. A local time vs. geomagnetic latitude plot of the location of the main trough (Muldrew), the equatorward limit of the region of electron density less than 100 electrons/cc (Timleck and Nelms, 1969), the plasmapause (Carpenter, 1966), and experimental and theoretical determinations of the limit of the region of open field lines (McDiarmid and Wilson, 1968; and Nishida, 1967). All curves are projected to a height of 300 km.

1965; Liszka, 1967; Rycroft and Thomas). Sharp et al. (1964) and Sharp (1966) reported abrupt changes in ion concentration measurements, made on low altitude satellites, that are probably related to the troughs.

Muldrew (1965) suggested that the main trough was an extension along magnetic field lines, of the whistler "knee". Subsequent work has further confirmed a close correlation between the two (Rycroft and Thomas; Thomas and Andrews, 1968), but there are important differences. First, the decrease of N at the whistler knee refers to a decrease observed primarily in the radial direction at the equator, but at great distances from the earth, whereas the troughs are observed in N_{max} F2, which occurs at practically constant heights above (and close to) the earth. Second, it is not clear whether or not the electron density beyond the plasmapause increases significantly, whereas north of the main trough, the density at the peak of the F-layer increases, often to values as great as those south of the trough, and other troughs are observed at higher latitudes. Third, the troughs are primarily nighttime phenomena, and are generally masked during the day, whereas the plasmapause can be observed at all times of day. Fourth, the troughs occur in the height range of maximum production of ionization, whereas nearly all of the ionization present near the plasmapause must have been produced elsewhere, and have been diffused or otherwise transported over large distances. Fifth, although the magnetic L-shell along which the plasmapause and the trough are located agrees well during the night, during morning and afternoon the main trough is observed at higher L-shells than the plasmapause (Figure 7).

In Figure 7, the plasmapause is from Carpenter's observations for July 1963 (Carpenter, 1966), the trough from Muldrew's northern hemisphere winter 1962-63 data (Muldrew, 1965), and the two limits of open field lines are an estimate from the energetic particle data of McDiarmid and Wilson (1968), and the theoretical limit given by Nishida (1967).

Because of the different periods of observations the plasmapause and trough observations are not strictly comparable, but they are the most complete data available, and probably represent the gross features of the diurnal variations in position of the two phenomena. As Figure 7 shows, the trough is located approximately along the boundary of the region of open field lines (i.e., the neutral points) when it is first observed in the early afternoon, but appear to shift equatorward to a position about 10° from that boundary between 1500 and 2000 LMT. The return of the trough in the late morning to the limit of the open field lines cannot be followed because of its obscuration during the day; it shifts slowly toward higher latitudes between about 0100 and the time it becomes obscured, around 0700 LMT. Between 1900 and 0500 LMT the plasmapause and the trough positions are in good agreement, but the two appear to have inverse diurnal variations of position; the plasmapause moves poleward around 1700 LMT and equatorward about 0500 LMT.

2.5 Electron Density Distribution Over the Polar Region During Quiet Periods

In general spatial and temporal variations of the distribution of electron density above the peak of the F-layer are much more regular in the latitude region below the main trough than in the higher latitude region, for both quiet and disturbed conditions (Nelms, 1966) (Figure 8). Nevertheless, several studies of the distribution in the high latitude regions have succeeded in isolating some persistent structures.

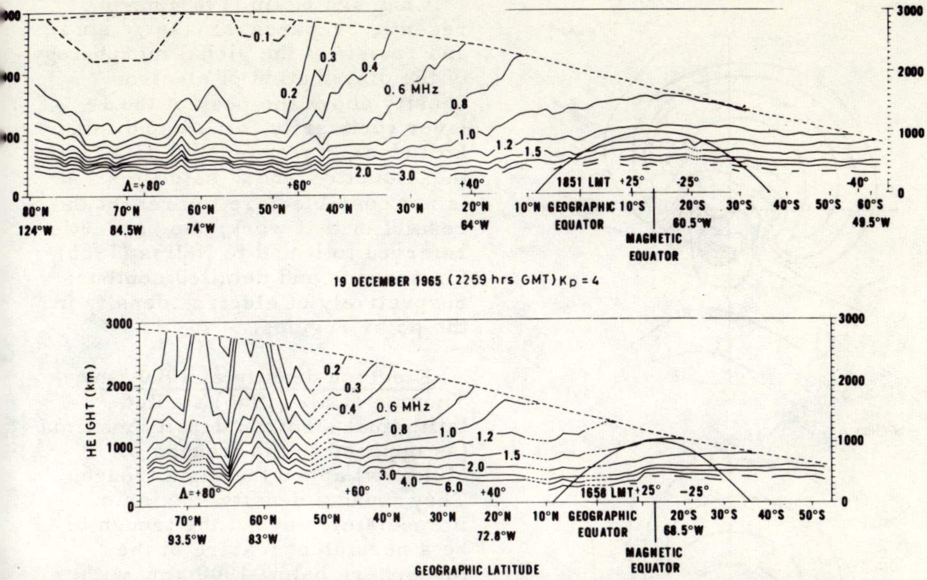

Figure 8. The contours of constant plasma frequency (electron density) as a function of height and geographic latitude for two passes in which Alouette 2 was near apogee over the north polar region (Nelms and Lockwood, 1967).

N(h) distributions below 1000 km. Nishida (1967) produced "average" electron density contours on a geomagnetic latitude vs local time plot for quiet intervals of the autumn (northern hemisphere) 1962-63, as shown in Figure 9. For this study he used a "mesh" 4° in longitude and 2 hrs in local time. The general features of the main trough and the decreased complexity at the higher heights can be seen from this figure. Unfortunately, some of the detail is lost in these figures because the N(h) data used (the DRTE N(h) data publications) tend to avoid some of the important but complex structures in the topside ionosphere, while emphasizing other phenomena. In particular, the data probably favour profiles in which the density at the satellite is substantial, and neglect some data in which the electron densities are very small. Nevertheless, the diagrams provide a useful picture of the polar quiet ionosphere. Figure 10 shows the variability of the data, defined by Nishida as the standard deviation divided by the average, for each mesh point, expressed as a percentage. It is of interest that the greatest variability occurred at 1000 km, and with the exception of the 0800 LT graph, occurred in the region of the splash zone (Figure 6).

Nishida (1966; 1967) explained the plasmapause and the trough regions in terms of an "outward flow" model in which plasma is removed by the combined action of magnetospheric convection and plasma escape from the tail of the magnetosphere. In his model, ionization will be removed by upward flow along field lines that are open into the magnetospheric tail (i.e., within the cross-hatched region of Figure 7). However, all field lines poleward of the plasmapause become open (i.e., are convected through the cross-hatched region of Figure 7) at some time during the night, and in the process can cast off ionization in proportion to the time they spend in the "open" region.

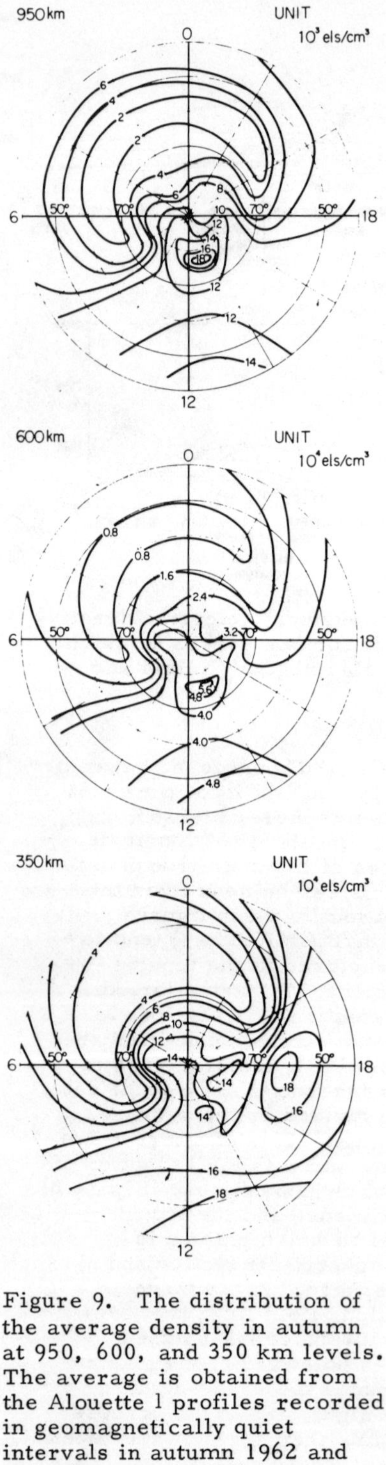

Figure 9. The distribution of the average density in autumn at 950, 600, and 350 km levels. The average is obtained from the Alouette 1 profiles recorded in geomagnetically quiet intervals in autumn 1962 and 1963 (Nishida, 1967).

Chan and Colin (1969) have recently prepared a massive study and review of the global morphology of the distribution of electron density above the peak of the F-layer (primarily for low and mid latitudes), based on N(h) data from the Alouette 1 and 2 satellites. It is not possible here to present the results of that work; the reader is referred to it and to Nelms (1966) for average and detailed contours respectively of electron density in the polar regions.

<u>Electron density at 1000 km.</u> Thomas et al. (1966) studied latitudinal electron density maxima (as opposed to the troughs of Muldrew) above the main trough. They found a density maximum immediately beyond the trough to be a persistent feature of the ionosphere below 1000 km, with maximum to trough density ratios of 25 to 1 a common occurrence. Although the diurnal variation of this maxima was not established, at about 1500 LMT it occurred at about 77° geomagnetic latitude, the location of the boundary of the open field lines. This maxima was related by Thomas et al. to that found at the peak of the F-layer by Duncan (1962) from ground-based ionosonde data. Duncan showed that polar enhancements of foF2 were largely GMT controlled, with maximum ionization occurring at 1800 GMT in the north and 0600 GMT in the south, the time at which the angle between the geomagnetic axis and the sun-earth line is a minimum. He suggested that the enhancements were due to particle precipitation; Thomas et al. (1966) suggested that the maximum they observed was due to upward diffusion of ionization from the precipitation production region. A similar GMT control of low energy electrons (50-1000 eV) was found by Maehlum (1968), who noted the possibility of a relation between the ionospheric maxima and the peak of low energy particle precipitation.

Sato and Colin (1969) showed that at 1000 km high latitude ionization maxima (Figure 11)

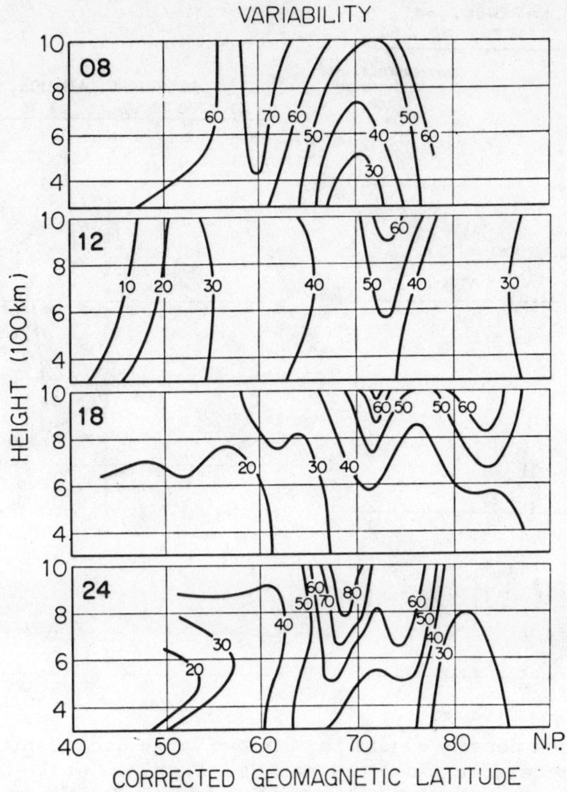

Figure 10. Meridianal distribution of the variability in 0800, 1200, 1800, and 2400 meridians from Alouette 1 data. Variability is derived by dividing, in each mesh point, standard deviation by average, and is expressed in the unit of per cent (Nishida, 1967).

generally occurred more frequently and had greater amplitude during the day than during the night, especially in winter, and that the amplitude of the maxima was greater in summer than in winter. They showed that the maxima did not occur uniformly over the high latitude region, but in three zones (Figure 12), with the maxima in a particular zone related, but independent of those in the other zones. In zone I and III, the maxima occur most frequently during periods of high geomagnetic activity. In zone I the maxima occur most often at dawn and dusk, in zone II most frequently in daytime regardless of geomagnetic activity or season, and in zone III mainly at night. The boundaries of the three zones vary, depending upon magnetic local time, season and magnetic activity.

Zone II corresponds approximately to the auroral oval, i. e., to the zone of splash precipitation given by Hartz and Brice (Figure 6). The maxima previously found by Thomas et al. (1966) occur within zone II. Sato and Colin concluded that the electron density maxima were due to low energy electrons ($E < 1$ keV) precipitated from three sources: (i) from electron islands outside the region of trapping into zone I, (ii) from an unstable region of trapping, close to the trapping boundary, into zone II, and (iii) from regions of stable trapping into zone III.

N(h) distributions below 3000 km. The foregoing studies were, in general, for altitudes below 1000 km; Nelms and Lockwood (1967)

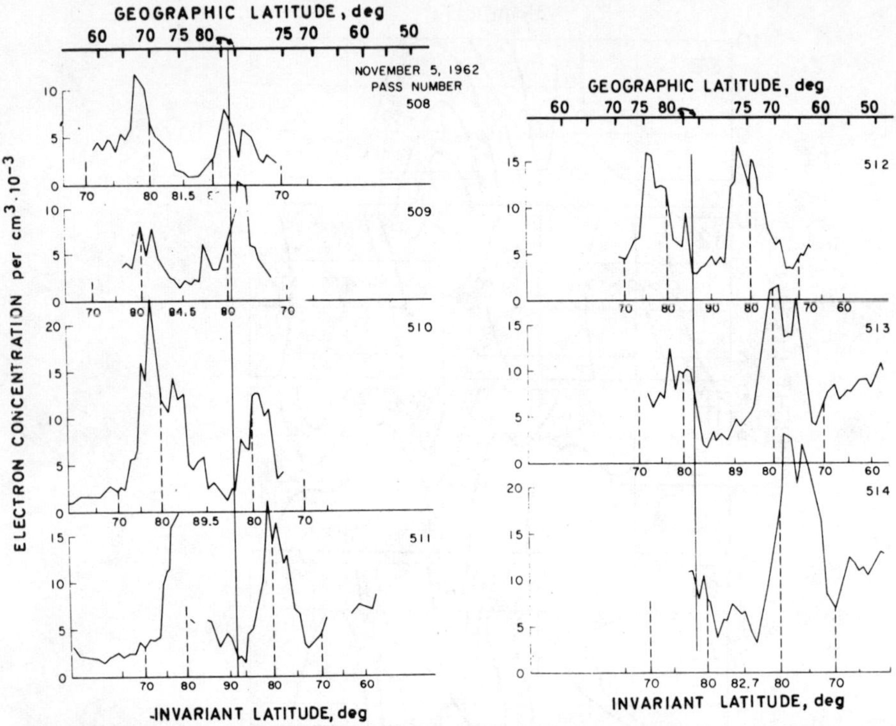

Figure 11. Electron density variation with two large enhancement peaks for seven successive Alouette 1 passes. Numbers at the top and bottom represent geographic latitudes and numbers under the reference levels indicate invariant latitudes (Sato and Colin, 1969).

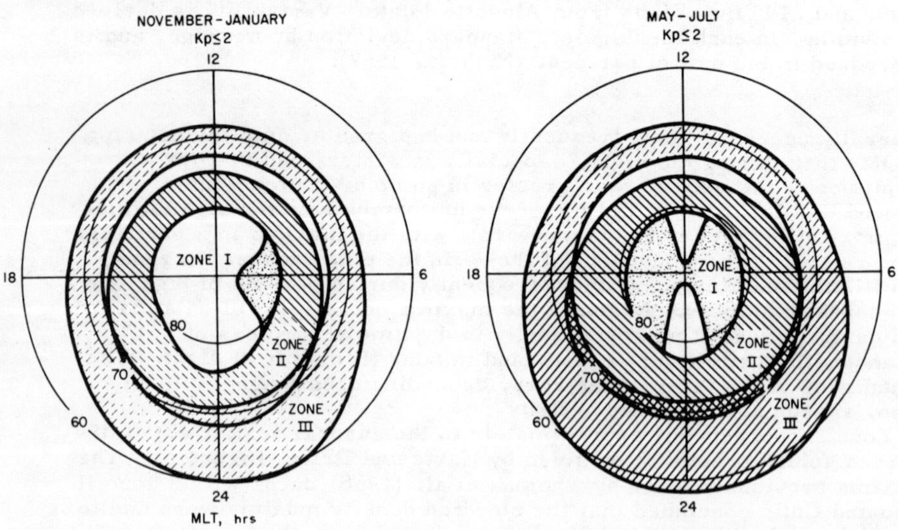

Figure 12. Schematic representation of three zones, Zone I, II, and III, where peaks of electron density enhancements appear on disturbed days in winter (a) and summer (b). This zoning is only qualitative. Dotted shading areas show highly speculative zones, owing to rare occurrence of the enhancement in those zones (Sato and Colin, 1969).

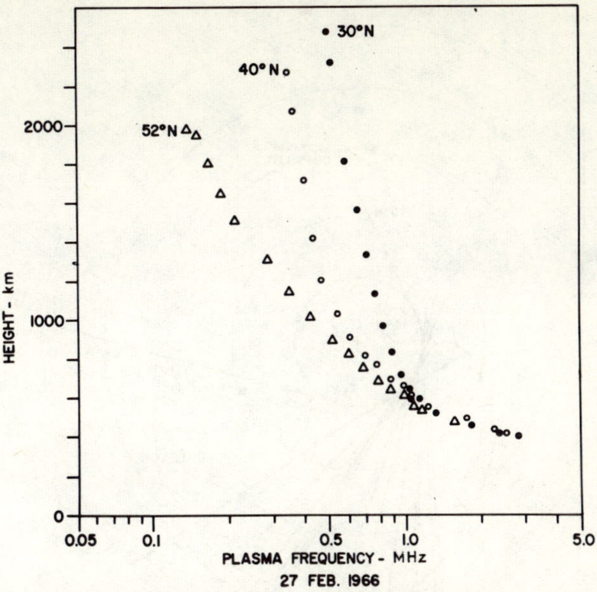

Figure 13. The N(h) profiles calculate from Alouette 2 ionograms recorded near the 75 °W meridian and 00 LMT (Nelms and Lockwood, 1967).

showed that at heights near 2000 km, N(h) profiles taken while approaching (and close to) the low latitude edge of the trough showed successive decreases of scale height near 2000 km, whereas the distribution near 500 km remained approximately the same from record to record (Figure 13). They also showed that at heights above 2400 km, the electron density was less than 300 to 500 electrons/cc over a large area of the polar region (Figure 14). The southern boundary of this low density region was found to agree in general with the location of the main trough. (It will be noted that Figure 14 is a polar plot of geographic coordinates, and is not directly comparable with the geomagnetic latitude vs local time plots of Figures 6 and 7.) They interpreted the decrease in scale height and the presence of the extensive low density region at high heights as equivalent to a shift of the level at which the H^+ concentration became comparable with the O^+ concentration to great heights in the vicinity of the trough and at latitudes greater than the trough.

Electron density at the height of the Alouette 2 satellite (500 to 3000 km). The range of values of number density reported in the Nelms-Lockwood paper was set by the lower frequency limit of the Alouette 2 sounder; Hagg (1967), using the resonance beat method outlined in Section 3, showed that electron densities in the low density region were often less than 30 electrons/cc and frequently were as low as 8 to 15 electrons/cc. The interpretation that the very low electron densities are due to movement of hydrogen ions to greater heights is supported by the data of Hoffman (1969), who found that on several occasions when Hagg observed very low electron densities, the predominant ion, even near 3000 km, was O^+. In addition, Hoffman has observed upward streaming of H^+ ions at certain times in the polar regions (Hoffman, 1968; 1969).

Hagg found that densities less than 30 electrons/cc were observed in

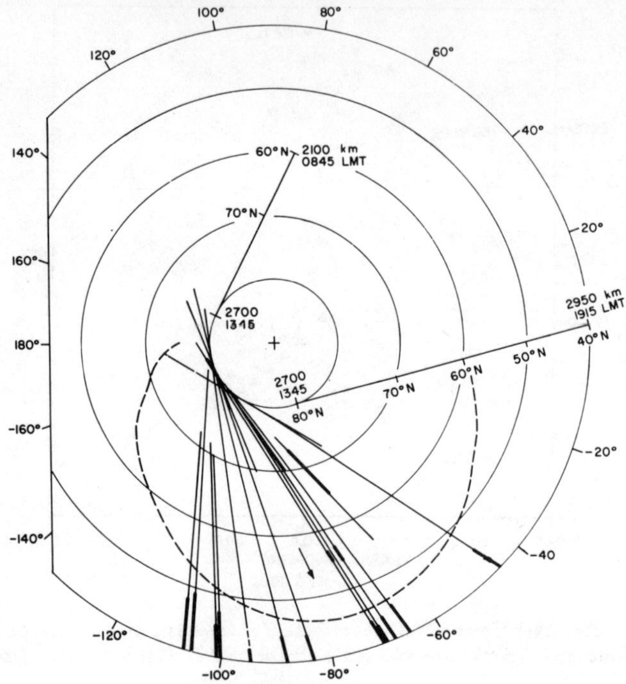

I TO 15 DECEMBER 1965

Figure 14. A polar plot of electron density at the height of the Alouette 2 satellite, on a coordinate system of <u>geographic</u> latitude and longitude. The heavy and light lines indicate densities of greater and less than 500 electrons/cc respectively for 18 passes taken between 1-15 December 1965. The planetary magnetic index Kp was less than 4 for the data shown. The local time and altitude of the satellite are constant at each latitude for this diagram, and are given along the additional axes. The location of the main trough in the density at the height of the peak of the F-layer, found by Muldrew on 24 October 1962 (Kp = 4) at comparable local times, is shown as a dashed line (Nelms and Lockwood, 1967).

the northern hemisphere throughout the region above 60° geomagnetic latitude at night and 65° geomagnetic during the day. Timleck and Nelms (1969) used the Hagg method to study the occurrence of densities less than 100 electrons/cc between December 1965 and February 1968 in both northern and southern hemispheres. They found that below 3000 km altitude the low densities were much more prevalent during winter than during summer; during the summer the low densities were observed only near the region of the main trough and only near midnight.

Figure 15 is a scatter diagram from the Timleck-Nelms study, showing the occurrence of densities less than 100 electrons/cc as a function of geomagnetic latitude and local mean time for all of the northern hemisphere data collected between December 1966 and February 1968. Densities less than 100/cc occurred on about 12% of all ionograms recorded at geomagnetic latitudes above 60 °N during these winter periods. The figure only shows the occurrence of densities less than 100/cc, and does not show regions of higher electron density that occur within the region (Figures 11 and 12). The overall picture is, of course, a combination of regions of low electron density and regions of density maxima. The study has not been completed that

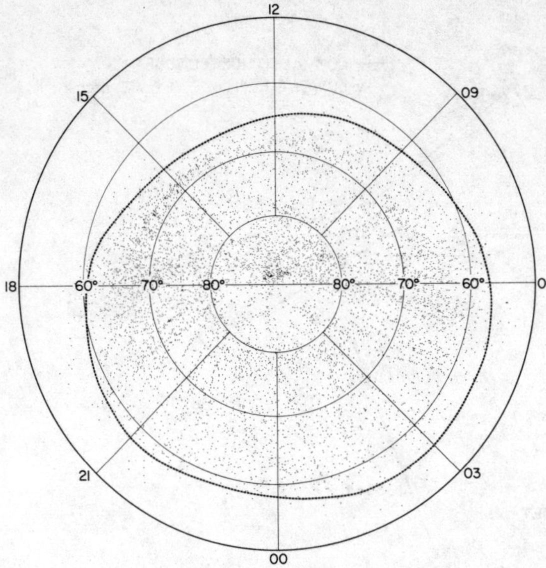

Figure 15. Occurrence of electron densities $\leq$ 100/cc, observed on Alouette 2 ionograms, plotted against geomagnetic latitude (northern hemisphere) and local time for Dec 66 to Feb 67 and Dec 67 to Feb 68. The broken line indicates the approximate boundary for undisturbed conditions (Timleck, Privat communication).

puts these two sets of data together, but the preliminary results indicate that electron densities less than 100/cc tend to occur on successive ionograms over extended areas of the polar region, separated by "islands" of higher densities.

About 4.5 % of the data of Figure 15 are for Kp $\geq$ 4. The dashed line in Figure 15 is a crude estimate of the low latitude boundary of the low density region for relatively quiet conditions. Although only an estimate, the real boundary is not likely to be significantly displaced from the one shown. The projection of the boundary down the field to the 300 km level, from an assumed height of 2800 km, is plotted on Figure 7.

The separation of altitude, LMT and latitude effects in the low density data is difficult because of the high inclination elliptical orbit of Alouette 2. However, Timleck and Nelms isolated a diurnal and sunspot cycle dependence of the minimum height at which electron density of 100/cc occurred, as shown in Figure 16. Because of the elliptical orbit and the fact that the low densities occur above about 60° geomagnetic latitude, the satellite was not in a position to establish the low height limit at times other than those shown in Figure 16.

The densities less than 100/cc occur at lower heights during the night than during the day, and there is a general tendency toward higher heights during more recent years. The exception, midnight to 0400 LMT, is thought (Timleck and Nelms, 1969) to be due to a seasonal effect overriding the diurnal effect; the 1965-66 data for these hours were collected during December whereas the 1967-68 data were collected during February. Two of the periods when no low electron

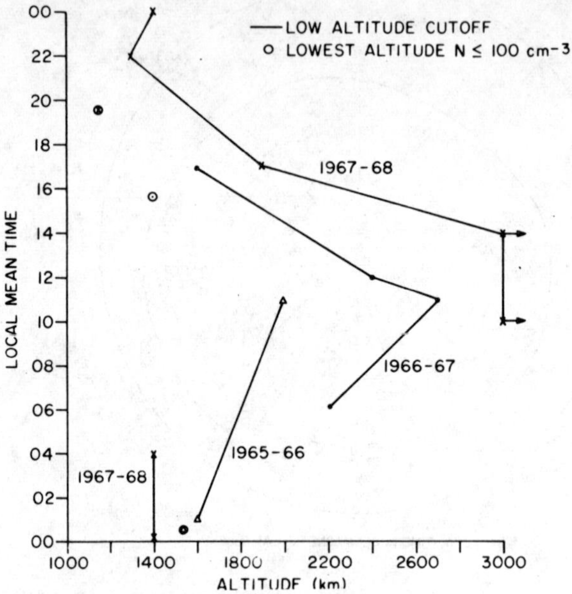

Figure 16. Low altitude cut-off of occurrence of electron densities $\leq 100/cc$, for the northern hemisphere winter from Alouette 2 data. The circled points indicate the lowest height at which low densities were observed. Data are not available to establish the low altitude cut-off at times other than those shown here (Timleck and Nelms, 1969).

densities were observed over the polar region, even though they were to be expected from immediately preceding and following data, were found to coincide with proton events observed by the IMP-F satellite.

Banks and Holzer (1968) interpreted the Hagg (1967) results, and Timleck and Nelms (1969) interpreted their results in terms of the removal of light ions by upward flow in regions where the field lines are swept back into the tail of the magnetosphere. Banks and Holzer showed that essentially all of the thermal H^+ and He^+ ions created in the polar ionosphere can be accelerated away from the earth into the tail of the magnetosphere, but that the upward velocity of the O^+ ions, because of their greater mass, will be small for heights below a few thousand kilometers. Timleck and Nelms showed that the electron densities observed in their regions of very low electron density were approximately what would be expected if the light ions were removed, leaving behind a relatively unaltered O^+ distribution. They explained the diurnal variation of the cut-off height for $N \leq 100/cc$ as due to daytime production (of O^+) raising the height at which $N = 100/cc$ occurs. The summertime absence of observations of $N \leq 100/cc$ below 3000 km, except for a few occurrences immediately above the main trough, was attributed by them to increased production during the time when much of the polar region is continually illuminated. They pointed out that the cut-off height would be expected to rise during times of disturbance or particle precipitation, either due to increased production or increased temperature.

It should be noted, however, that the main trough involves a large reduction in the density at the peak of the F-layer, which indicates that there is a reduction in the O^+ number density as well as removal of the lighter ions. This implies that a further process or a great enhancement of the outward flow is present in the trough to cause the reduction in the heavy ion (O^+) density.

2.6 Electron Density Distribution Over the Polar Region During Disturbed Periods

Although this paper concentrates on the high latitude topside ionosphere, it is useful to introduce the subject of the disturbed ionosphere with a brief discussion of the general global behaviour. The relevant literature on this complex subject is extensive, so the following brief discussion will be incomplete; the reader is referred for more detail to Nishida (1967), Sato and Chan (1969), Sato (1968), Ondoh (1967) and Warren (1969) from which the present authors have drawn much of the following summary.

In general, ionospheric storms can be considered to be due to enhanced influx of energetic particles (from thermal to MeV) from the solar wind near the earth, and may be due to direct effects of the particles such as ionization or heating of the upper atmosphere, or to indirect effects such as distortion of the geomagnetic field (the geomagnetic storm) or establishment of ring currents that cause ionization drifts.

At the sudden commencement of a magnetic storm, the front (day-side) of the magnetosphere becomes compressed (Bauer and Krishnamurthy, 1968), and this increases both the electron density and the magnetic field in the topside ionosphere. This initial phase of the storm may last up to 10 hours. At the beginning of the main phase of the magnetic storm the magnetic field is inflated by energetic particles (Cahill, 1966) that appear to be injected (and trapped) into orbits near the plasmapause at late afternoon and evening longitudes. The decay of this enhanced particle belt may take several days.

The ionospheric storm effects are observed as changes (increases or decreases) of electron number density in various height and latitude regions, changes in electron and ion temperature, neutral and in composition, recombination rates, and transport velocities. All of these may vary independently in different height, latitude, seasonal and diurnal regimes, and may also have different storm time behaviour. In spite of this great complexity, it appears that some useful general patterns are now beginning to emerge (Warren, 1969).

At mid latitudes the beginning of the ionospheric storm on the day side of the earth is associated with the beginning of the main phase of the magnetic storm (Lewis and McIntosh, 1953; Thomas and Venables, 1966), rather than with the sudden commencement. However, on the night side the full ionospheric storm does not occur until the following day (Ratcliffe and Weekes, 1960). Gledhill et al. (1967) found that ionospheric disturbances in the D-region invariably occurred when the flux level of precipitated electrons ($E > 40$ keV), measured in Alouette I, rose above $1.4 \cdot 10^4$ electrons/cm^2/sec, and that the storm lasted a time proportional to the time this critical flux was exceeded.

The major disturbances during an ionospheric storm are in the electron density in the D- and F-region, and in the ion and electron temperatures near the peak of the F-layer. Effects do not necessarily occur simultaneously in both regions. At high latitudes the electron density in the D-region is enhanced while at mid latitudes the total content (Titheridge and Andrews, 1967; Hibberd and Ross, 1967) and the density at the peak of the F-layer usually are first enhanced over about 24 hours (positive phase) and then depressed over a period of a few days (negative phase). However, the decrease in density at the peak precedes the decrease in total content by several hours (Hibberd and Ross, 1967), and occasionally, large systematic storm-time variations in topside total content occur that are not accompanied by significant or regular changes in the peak density or in the bottomside content (Titheridge and Andrews, 1967). At lower latitudes, the positive phase usually increases and the negative phase decreases until at the equator only the positive phase remains (Appleton and

Piggott, 1952). In some particularly strong storms a negative phase does occur at the equator.

Plasma temperatures during storms. It has been shown by incoherent scatter measurements at mid latitudes that during the day the disturbed and quiet day electron temperatures in the F-region below about 400 km appear to be about the same (Evans, 1965; Rao, 1967), whereas above 400 km the electron temperature is slightly higher on disturbed days. During the night, the disturbed time electron temperature is 300° to 600 ^{o}K greater, and the ion temperature is slightly greater than in quiet times. The ion temperature below 400 km is greater, and above 400 km is slightly less on disturbed than on quiet days.

Willmore (1965), using Ariel 1 data, found that between 0400 and 1200 km altitude electron density enhancements, and electron temperature decreases of about 180 ^{o}K, were associated with geomagnetic storms, both at night and during the day. He attributed the temperature changes to increased collisional cooling due to the increased electron density.

Ondoh (1967), from Alouette 1 data, found that during mild storms the electron density increased above 500 km height and decreased below at all local times. He concluded that the increase was due to expansion of the neutral atmosphere, which would lift the whole ionosphere, causing higher densities at high heights (Garriott and Rishbeth, 1963). The decrease below 500 km he attributed to increased chemical losses. Like Willmore, he expected the enhanced electron density above 500 km to cause decreased temperature through additional collisional cooling. Decreased electron density in this region during severe storms should then allow the temperature to increase. Reddy et al. (1967) found a similar result at 640 km, by relating bottomside sounding data to probe data from Tiros 7, and also found that measured decreases in electron temperature were associated with the increases in density at 640 km. They also interpreted the storm-time enhancements as due to thermal expansion of the neutral gas lifting the whole ionosphere.

N(h) distributions during storms. Nishida (1967) investigated structures in the polar disturbed ionosphere, using Alouette 1 N(h) data. He constructed "deviation" contours, similar to his variability plots of Figure 10, using the difference between the storm and average quiet day data, expressed as a percentage of the quiet day data. An example is given in Figure 17, for the magnetic storm of 23 Sept 1963, for $Kp = 9^-$. Regions of storm-time density reduction are shaded. Below 66 ^{o}N geomagnetic at noon, and below 82 ^{o}N geomagnetic at midnight the storm time ionosphere is enhanced above and depleted below 450 km. A general storm time depletion at all heights over the polar region is evident. Nishida showed that storm-time changes enhanced the high latitude peaks, often to as much as five times their average value, and caused them to move several degrees south. He also found that the plasma was depleted at the pole and in the trough.

Nishida explained his results in terms of the number of field lines convected through the "open" tail region of the magnetosphere, and of heating of the magnetospheric plasma by the hot storm-time solar wind (i. e., the area and velocity of plasma exit). He pointed out that the plasma temperature increased toward the magnetopause (Serbu and Maier, 1966), the temperature of the solar wind increases with increasing Kp (Snyder et al., 1963; Neugebauer and Snyder, 1966), and more field lines are transported into the tail region (Snyder et al., 1963) during storm than during quiet times. All of these would tend to increase the escape of plasma from the polar regions through the magnetospheric tail, leading to the observed reduction of electron density over the polar region. He also invoked lifting of the plasma due

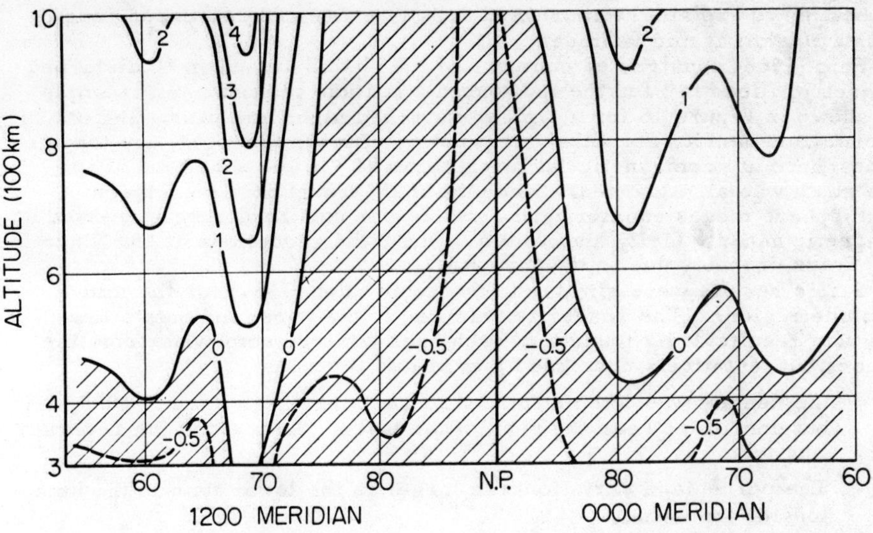

Figure 17. Alouette 1 distribution of the relative deviation S = (n - N)/N of the electron concentration from the average value for 0035-0053, September 23, 1963, Kp = 9⁻, where n and N are the disturbed and quiet day distributions respectively. The region where the density is less than the average quiet interval value is shown by hatching (Nishida, 1967).

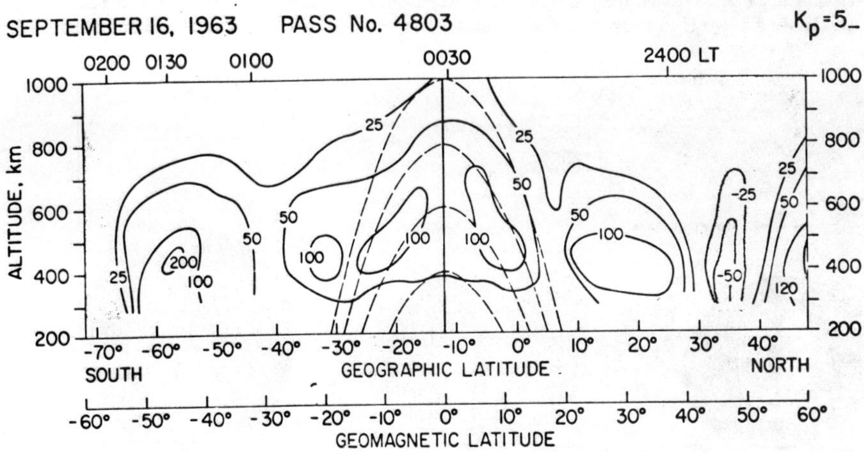

Figure 18. Alouette 1 profiles of electron density variations in the topside ionosphere during geomagnetic disturbances in the nighttime. Shaded parts represent reduction of the electron density and non-shaded parts indicate enhancement or no variation. Solid lines show isopleths of the enhancement or reduction rates in percentage (Sato, 1968).

to heating to explain the increased density above and decreased density below 500 km at mid latitudes.

Sato (1968) constructed contours of the Nishida type for 20 disturbed passes of Alouette 1 for the low and mid latitude regions. An example is shown in Figure 18 for a typical disturbed nighttime pass. Regions of storm time density reduction are shaded. During the night, the topside ionosphere at geomagnetic latitudes below 40° in the north and 50° in the south was all enhanced. Since the main trough in the northern hemisphere moves equatorward about 2° per unit Kp during increased magnetic activity (Jelly and Petrie, 1969), the reductions at 45 °N and 55 °S are probably due to the trough.

Sato's results were similar to those of Ondoh (1967) for the mid latitude region. (The reader is referred to the paper for Sato's low latitude results.) He found that each pass differed somewhat from the others but tentatively concluded that at mid latitudes

(i) the density was often enhanced in the upper region (near 1000 km) but decreased near the peak of the layer. Less often the reverse was true.

(ii) The percentage variation was larger in the lower than in the upper topside ionosphere.

(iii) The density was most often enhanced between 30° and 40° and reduced between 40° and 50° geomagnetic latitude. The enhancements occurred over wider ranges of latitudes and height in local summer and equinox than in winter.

He concluded that drift motion and ambipolar diffusion could not alone explain his mid latitude storm-time observations, and suggested that lifting of the ionosphere due to thermal expansion might be involved.

Sato and Chan (1969) extended the study of storm-time deviation contours of Sato (1968) to cover 24 Alouette 1 passes over the north polar region. They found that above 60° invariant latitude the deviations from quiet day showed a strong dependence on magnetic local time as well as on latitude and height. Figure 19 summarizes

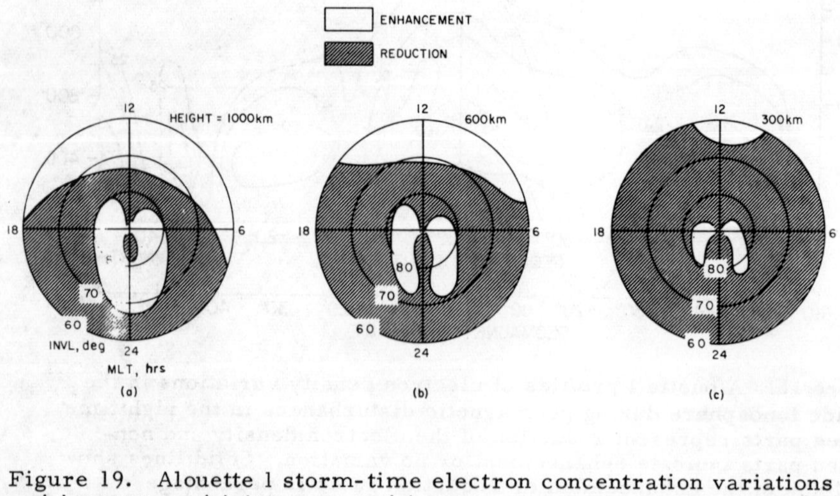

Figure 19. Alouette 1 storm-time electron concentration variations at (a) 1000 km, (b) 600 km and (c) 300 km against invariant latitude and magnetic time in polar coordinate. Shaded areas represent electron concentration reduction and white areas, enhancement (Sato and Chan, 1969).

the results of their study. Storm-time enhancements occur in two well defined regions above 80 °N near dawn and dusk at low heights, with the rest of the region above 60 °N, except for a small region near noon at 60 °N, a region of electron density reduction. At higher heights the regions of enhancement all expand, until at 1000 km the enhanced region covers a large portion of the day time region below 70° latitude and of the nighttime region above 80° latitude.

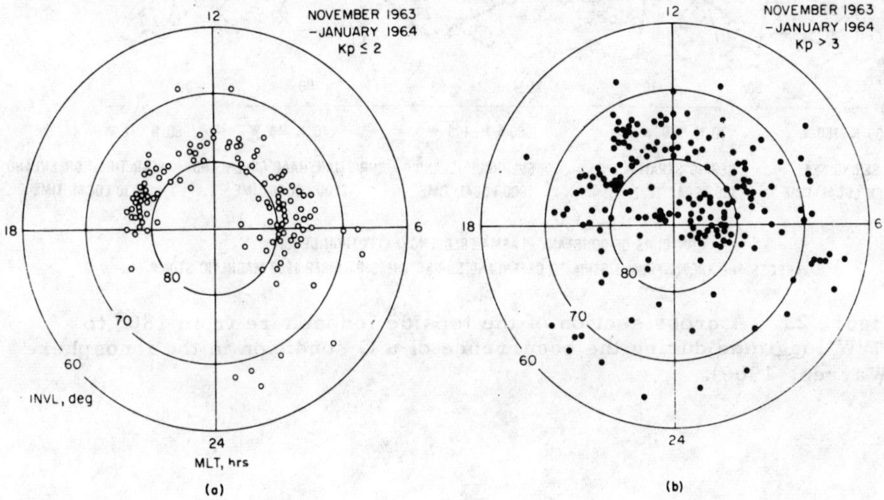

Figure 20. Distribution of electron concentration enhancement peaks against invariant latitude and magnetic local time in the polar region at a height of 1000 km (from Alouette 1) on (a) quiet and (b) disturbed days in winter (Sato and Colin, 1969).

The Nishida (1967) results (reduction over a large region near the geomagnetic pole) are explained by Sato and Chan as due to the fact that Nishida's data was only in the noon-midnight local time plane (both authors used the same Alouette 1 data, so it is expected that their results will agree wherever they overlap). Sato and Chan also relate their results to the measurements of abnormally large electron peaks at 1000 km (Duncan, 1962; Sato and Colin, 1969). The distribution of peaks in winter for low and high magnetic activity (Sato and Colin, 1969) are shown in Figure 20. On quiet days the peaks are confined to zone II (Figure 12) on the day side, whereas on disturbed days, they spread over the whole region above 70°, although still concentrated on the day side. Although the evidence for a movement of the peaks to lower latitudes is not overwhelming, Sato and Chan note the indication, and relate it to similar storm time behaviour of the auroral oval, that is thought to be due to the lowering of the precipitation latitudes of low energy electrons ($E < 1$ keV) during storms. They explain the presence during storms of peaks in the region above 80 °N as due to precipitation of low energy electrons from the neutral sheet into the polar region. Thus they interpret the stormtime behaviour of the topside polar ionosphere as due to

(i) lowering of the particle precipitation latitude of low energy electrons

(ii) enhancement of the precipitating electron flux intensity at high latitudes.

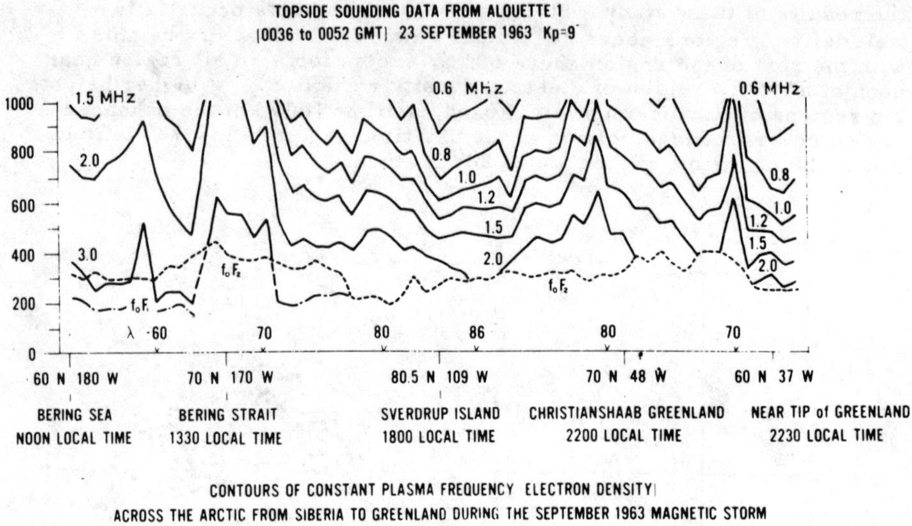

Figure 21. A cross section of the topside ionosphere from 180° to 37°W longitude during the occurrence of a G condition in the ionosphere (Warren, 1969).

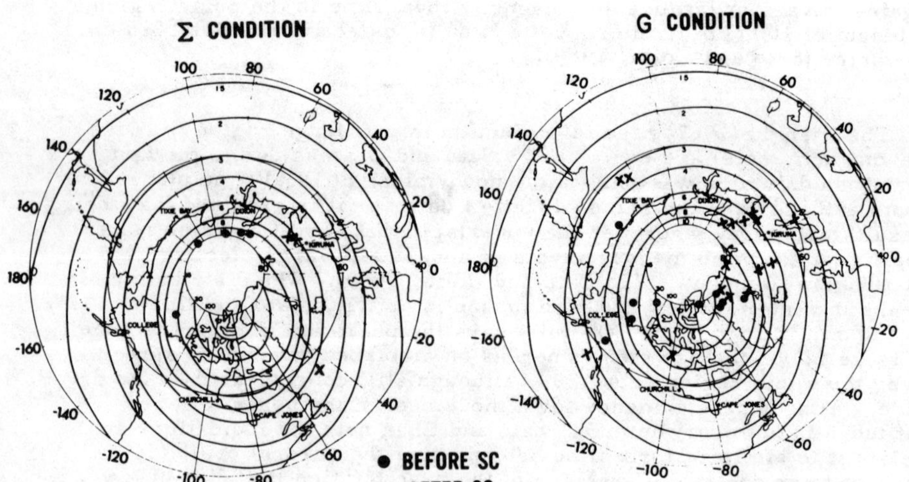

Figure 22. Geographical distribution after the solar proton flare of July 7, 1966, of the Σ condition and the G condition (Herzberg and Nelms (1969) using Alouette 1 and 2 data).

The effect of (i) is to produce a reduction region (from the absence of particle precipitation), while the effect of (ii) is to produce an enhancement region near the pole.

It will be noted that Figures 19 and 20 are in a sense contradictory; since the peaks move to lower latitudes during storms, as shown in Figure 20, there should be an enhancement region between 70° and 75° at 1500 to 1800 LMT and near 65° at 2000 to 2300 LMT. However, this amount of detail is probably not to be expected to survive the averaging processes used in producing Figure 19.

Several authors (Nelms and Warren, 1965; Herzberg and Nelms, 1969; Norton, 1969; and Warren, 1969) have studied the occurrence of large electron density decreases at the peak of the F2-layer in which the F1-layer became visible from the topside (the "G" condition). An example of an Alouette 1 pass during a severe storm, during which the G-condition was observed (Nelms and Warren, 1965; and Warren, 1969) is shown in Figure 21. Herzberg and Nelms (1969) did not find a systematic pattern to the occurrence of G-condition before and during the sudden commencement associated with the proton flare storm of 7 July 1966 (Figure 22). Norton (1969) concluded that the vertical drifts alone could not explain the G-condition data, and suggested that an increase of the loss rate by a factor of 16 over the quiet day values was required to fit the G-condition data. The loss rate can be increased by increaseing the N_2 or O_2 density or by increasing the rate coefficient for reactions of these molecules with O^+; Norton suggests that increased mixing might increase the neutral density sufficiently to explain the results, but favours rather an enhanced reaction (Thomas and Norton, 1966) between vibrationally excited N_2 and O^+. Laboratory measurements of the reaction rate (Schmeltekopf et al., 1967) indicate that at electron temperatures of 3000° to 4000°K the enhancement should be sufficient to explain the G-condition observations.

During the proton event of 7 July 1966, Herzberg and Nelms (1969) also observed a series of vertical bands across the normal ionospheric reflection traces on Alouette 2 ionograms, which they called a Σ-condition, and interpreted as due to narrow (~ 2 km) vertical regions of electron density depletion, or "mini-troughs". This interpretation was supported by the plasma probe on Alouette 2, which measure small scale electron density depletions at the time the Σ-condition was observed (Brace, Private communication). The Σ-condition occurred at L-values of about 11 (about 77° geomagnetic) before the sudden commencement, and about 6 (about 65° geomagnetic) afterward (Figure 22). For the very limited data available, it was found to occur in coincidence with sudden changes in the intensity of energetic particles of energy greater than 35 keV, and only in conjunction with proton events. The note added in proof to their paper, in which they reported a Σ-condition for which no proton flare was reported was later found to be premature. The associated proton flare was later reported from satellite observations.

Norton and Marovich (1969) obtained Alouette 1 and 2 topside sounder data in coincidence with visual observations of red arcs (Barbier, 1958) (enhanced 6300 Å airglow) over the central USA (L-values of ~ 3). The peak emission rate for red arcs occurs at heights of 350 to 450 km. They usually extend a few hundred km in altitude but a few thousand km in magnetic longitude. Red arcs move from high to mid latitudes during magnetic storms (F. E. Roach and J. R. Roach, 1963). Norton and Marovich found that a major depression in topside electron density occurred over the region of the arc, and from scale height measurements, inferred that the plasma temperature and the ion mass within and above the arc were considerably greater than equatorward of it. They concluded that the electron density reduction was the main (mid latitude) trough. Recent explanations of the red arc invoke collisional excitation of atomic oxygen by energetic electrons (Walker and Rees, 1968), which implies that energetic electrons may play an important role in the production of the main trough, at least during storm conditions when red arcs are observed.

3. DISCUSSION

Many of the phenomena of the high latitude high altitude ionosphere may be understood as the extension of all F-region heights of the processes and structure of the high magnetosphere. During the polar winter, the enhancements of electron density in the upper ionosphere, while apparently highly irregular in individual behaviour, occur in zones that are relatively well behaved, and that appear to be related to regions of particle precipitation. Observed reductions of electron density, on the other hand, appear to be related either directly, or through convection, to regions of open field lines in the magnetosphere. The two regions, of course overlap, so the overall picture tends to be complicated.

Axford and Hines (1961) proposed a model in which viscious drag from the solar wind caused large scale convection of the plasma in the magnetospheric cavity beyond L-shells of about 4 R_E, and used the resulting convection patterns to explain high latitude auroral phenomena. Their model has been extended by, among others, Nishida (1966; 1967) and Brice (1967), who use it to explain a number of ionospheric phenomena.

Both Nishida and Brice consider the plasmapause to be the boundary of the plasma that co-rotates with the earth, and the plasma beyond the plasmapause to be convected through the magnetosphere at some time during the diurnal rotation. However, a basic difference between the Nishida and the Brice models is that the former explained the decreased density in the region beyond the plasmapause as due to the convection of plasma of terrestrial origin out into the essentially open tail of the magnetosphere, whereas the latter considered the plasmapause as an equipotential boundary between the plasma of terrestrial origin (inside) and plasma that has migrated (or been convected) in from the solar wind. The two models are not necessarily incompatible; the inward flow model is supported by observations of energetic (above a few eV) particle data and, as shown by Brice, agrees with the observations of the ambient plasma, whereas the outward flow model appears to find its major support from the observations of the near-thermal plasma.

The height distribution of electron density in the high ionosphere is governed primarily by electron production (which occurs predominantly at low heights) and loss, temperature, ionic mass (gravitational forces), and the earth's magnetic field. In regions where the field lines are simply connected to the conjugate hemisphere, the field acts as a container to the ionization; light ions that have sufficient energy will not readily escape, but will simply travel along the field and may even reach the opposite hemisphere. Thus in time a general equilibrium is built up along the field, and an increase in production or temperature simply places more ionization at high heights in the "container", forcing different ion distributions to adjust themselves so that a new pressure balance is obtained. In the polar regions the field lines are convected into the tail of the magnetosphere at some time during the diurnal rotation, and the light ions can expand into a very large (essentially open) volume. Thus a given temperature of the ion gas will have a very different effect in the mid latitude and high latitude regions.

Bauer (1969), in a comprehensive study of diffusive equilibrium as applied to the topside ionosphere shows that diffusive equilibrium, which is a consequence of a Maxwellian velocity distribution of particles and therefore is dependent on collisions, may break down at

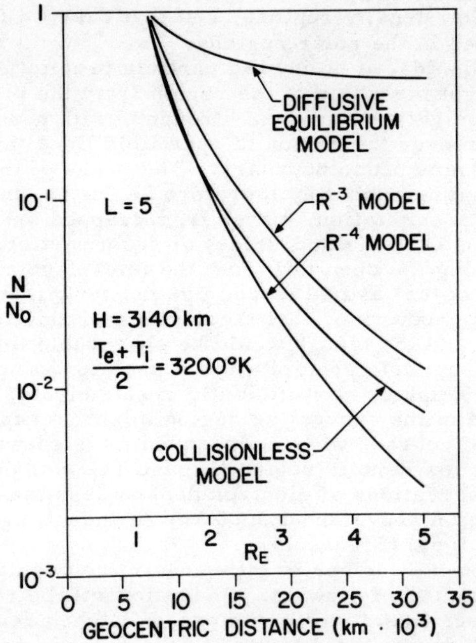

Figure 23. Plasma density distributions along a magnetic field line, whose apex in the equatorial plane is at $L = 5\ R_E$, are shown for a diffusive equilibrium, and a collisionless or ion-exosphere model, as well as for R^{-3} and R^{-4} power law models (Bauer, 1969).

latitudes beyond the trough (beyond the plasmapause). In these regions, where the local electron density can be very low, a "collisionless" or ion-exosphere distribution is required, as derived by Eviatar et al. (1964). Figure 23 compares the diffusive equilibrium and the collisionless models, and illustrates the very rapid decrease of density with altitude that would be expected in the collisionless region. Experimental data from knee whistlers are found to correspond to the R^{-4} distribution outside the plasmapause (Angerami and Carpenter, 1966), i. e., to approximate the collisionless distribution.

Banks and Holzer (1968) have shown that essentially all of the H^+ and He^+ ions created in the polar ionosphere can be accelerated away from the earth into the tail of the magnetosphere, but because of its greater mass, the upward velocity of O^+ will be small for heights below a few thousand kilometers. The process of removal of the light ions over the polar regions (the "polar wind" of Axford, 1968) is a dynamic one, and during periods of ionization production may well be obscured.

The picture presented by the "outward flow" models is, of course oversimplified, just as a picture that relies entirely upon particle precipitation to explain the structures observed in polar ionosphere is probably oversimplified. A combination of the two, however, may provide a reasonably accurate model. We know that there is precipitation of particles from the tail region and from the neutral points, that, these particles will produce ionization, and that low energy (0. 1 to 1. 0 keV) component of this particle flux will produce ionization in the upper ionosphere. Also, we know that there are extensive areas of very low electron densities at a few thousand km above the polar regions, that there are troughs that extend to the peak of the F-layer, that very few light ions are found, even at 3000 km height in these very low density regions, and that outward flow of light ions has been observed in the polar regions.

As suggested by Nishida, if no inward particle precipitation were present during the night the whole of the region from the plasmapause boundary to the pole might be one large "trough", with perhaps a greater depletion of ionization over the region of open field lines than between that region and the plasmapause boundary. The peaks of ionization at latitudes above the main trough may therefore be due to ionization produced by particle precipitation, either from trapped orbits (the "drizzle" precipitation of Hartz and Brice) or from particles that have entered the region along the open tail from the interplanetary medium. Also, one might expect that as one moved upward away from the main region of ionospheric production, that the daytime production would have less apparent effect, and the trough would be observable throughout the day. Since energetic particle precipitation caused ionization predominantly at low heights, at sufficiently great heights, the large trough over the whole of the convective region might be expected to be revealed. This is, of course, what is observed as is shown by the Alouette high latitude low density region (Figure 14) and, even more strikingly, by the observations of electron density less than 100/cc over much of the polar region above about 2000 km altitude (Hagg, 1967; Timleck and Nelms, 1969) (Figure 15).

The extent to which the regions of influence of the two models overlap can be appreciated from the following. Ionization will be removed along field lines that are open (i. e., within the cross-hatched region of Figure 7), some time during the day, and in the process can dump ionization in proportion to the time they spend in the open region. On the day side of the earth the boundary of the region of open field lines is also the locus of the neutral points, and along this boundary particles can directly penetrate the magnetosphere from interplanetary space. Field lines within the cross-hatched region are swept back in the nightside tail, and are also open and accessible to particles from the interplanetary medium. On the night side of the earth, the boundary of the region of open field lines is the locus of the neutral or plasma sheet, and corresponds to the regions in which energetic particles from the magnetospheric tail can obtain direct access.

During the polar summer when solar production is relatively continuous, the electron density to well above the peak of the F-layer is substantial, and much of the removal process is swamped by the high production under these conditions, the region is probably also under the influence of the neutral air winds.

The main trough appears to be at the locus of the lowest latitude region along which ionization can be readily removed from the peak of the F-layer. If this removal were by upward acceleration of ionization, as suggested by the outward flow models, production of ionization during the day probably would be faster than the removal process, which would explain why the trough does not appear as a well defined structure during the day. In the afternoon, as production decreases, the trough at the peak of the layer appears, first along those field lines that are open into the tail of the magnetosphere, and later at lower latitudes, down to the limit of the convective region (i. e., the limit of the plasmapause) during most of the night. However, the trough is a major depression in the density of the peak of the F-layer, and represents a large depletion in the O^+ ion density. Therefore a process in addition to the outward flow of light ions is needed to explain the trough. The correlation between the Red Arc observations and the trough is of interest; it may be that the trough result from heating of the plasma by energetic electron precipitation, causing either increased recombination due to the enhanced reaction between vibrationally excited N_2 and O^+ suggested by Norton (1969), or an enhanced outward flow that is sufficient to remove significant amounts of O^+ ionization.

Thus the model that we appear to require over the polar regions is one in which there is simultaneously a continual upward flow of near thermal ions (primarily H^+ and He^+) and electrons along the field, to be convected or dumped directly out the tail of the magnetosphere, and a (probably more sporadic) inward flow of low energy energetic particles from the solar wind into the tail of the magnetosphere (or through the neutral points) and down along the field into the ionosphere. As well, there will, of course be the flow of energetic particles into trapped orbits and precipitation from these trapped "belts" into the lower ionosphere.

It appears that the crucial measurements to be made at the present time are high altitude measurements of the flow of low energy (thermal to a few keV) particles, in both the upward and downward directions, along with the distribution and temperature of the ambient plasma.

4. SUMMARY

The study of the global variations of the topside ionosphere, as revealed by the topside sounder satellites, is by no means complete, but some useful patterns are beginning to emerge. In brief, the global structure of the high latitude ionosphere at a given height at or above the peak of the F-layer is characterized by areas of enhanced and areas of reduced electron density. At least some of these peaks and troughs of electron density occur in well defined zones that can be related to the configuration of the magnetic field lines in the magnetosphere and to the precipitation of energetic particles.

There appear to be two major sources of ionization in the high altitude, high latitude ionosphere; production by solar ultraviolet and by low energy particles. Since the magnetic field near the earth is nearly vertical at high latitudes, horizontal transport of ionization seems unlikely, although a flux of thermal particles down the field from the magnetospheric tail may be part of the low energy particle flux, and indeed may be the major source of thermal ionization at the pole during the winter night. Solar production produces the maximum effect at F1-layer heights, but is often the predominant source of ionization at higher heights (at least up to 1000 km) due to the diffusive redistribution of the ionization along the field.

Loss of ionization below about 250 km probably is due to chemical recombination but at higher heights transport along the field is important. The effect of this transport is either to move ionization down into a region of increased recombination, or to move it upward into the great volume of the magnetospheric tail. The latter appears to be a predominant effect at least at certain times of day.

During the day (which arbitrarily will be taken as solar zenith angles of less than about 80°, regardless of the hour) and near the peak of the F-layer, solar production is usually great enough to mask both the effects of particle production and of loss of ionization through upward flow, but during the night when solar production is absent, both of these effects are revealed. At heights above 1500 to 2000 km all three effects are often observable during the day. However, near the summer solstice, when the polar F-region is under continual illumination, the effects of solar production predominate to much higher altitudes. In fact, near the summer solstice, average electron densities in the polar region, at least between the peak of the F-layer and 1000 km altitude, have little diurnal variation, whereas electron densities near the winter solstice have a pronounced diurnal variation (Chan and Colin, 1969).

During the winter an extensive depletion of electron density occurs over the polar region from about 60° geomagnetic latitude on the night side to about 75° geomagnetic latitude on the day side of the earth, that is, approximately between the location (mapped down along magnetic field lines) of the plasmapause on the night and day sides of the earth. On the night side, the boundary of the region of depletion agrees well with the location of the plasmapause, as determined from whistler measurements, but on the day side of the earth the two do not appear to be in good agreement, with the depletion boundary at several degrees higher latitude than the plasmapause. Superimposed upon this region of depletion are enhancements of electron density that appear to be related to zones of particle precipitation. The overall effect of the combination of the two is to produce a series of peaks of electron density in regions of particle precipitation, and a series of rather narrow troughs located between adjacent regions of particle precipitation. The main (or mid latitude) trough is the most equatorward trough, and is located between the plasmapause and the region of particle precipitation that coincides approximately with the inner of the two main zones of auroral particle precipitation, given by Hartz and Brice (1967), i.e., with the zone of low energy (< 1 keV) particle precipitation found by Maehlum (1968).

Poleward of the highest latitude zone of particle precipitation electron densities are very low during the winter, but near the peak of the F-layer are higher than would be expected if no source of ionization (either production or transport) were available to the region.

The region of depletion is thought to be related to the area over which magnetic field lines are convected through the tail of the magnetosphere at some time during the diurnal rotation; the reduced electron densities may be due to the flow of the lighter ions upward along the field and the eventual loss of them through the open tail or through the boundary of the magnetosphere.

At altitudes of 2000 to 3000 km, electron densities of less than 100/cc (occasionally as low as 8/cc) are a frequent occurrence over the region between about 60° geomagnetic latitude on the night side and 70° geomagnetic latitude on the day side of the earth, that is, over the region of depletion. (At lower latitudes, the electron densities are typically a few thousand electrons/cc at these altitudes.) At these heights there are also zones of electron density enhancement within the region of the low densities, but the location of these zones has not yet been determined.

During the summer the extensive region of depleted electron density is not observed below 3000 km altitude, but at 3000 km electron densities of less than 100 electrons/cc are observed over the region of the main trough.

During magnetic storms, the high latitude ionosphere is highly variable, but in general the regions of enhancements of electron density broaden, often covering much of the polar region and occasionally spreading to lower latitudes as well. The main trough moves equatorward by about 2° per unit increase of Kp.

Acknowledgements. The authors are indebted to P. L. Timleck for the use of Figure 15 and to D. B. Muldrew for many helpful suggestions, particularly with reference to Figure 7, which is along the lines of a similar figure from an unpublished paper by Muldrew. The authors have drawn liberally from review papers by K. L. Chan and L. Colin, D. H. Jelly and L. E. Petrie, D. B. Muldrew, and E. S. Warren, and wish to express their gratitude to those authors for access to prepublication copies of the papers.

REFERENCES

Angerami, J. J. and D. L. Carpenter, Whistler studies of the plasmapause in the magnetosphere 2; electron density and total tube electron content near the knee in magnetospheric ionization, J. Geophys. Res., 71, 711, 1966.

Appleton, E. V. and W. R. Piggott, The morphology of storms in the F2-layer of the ionosphere, J. Atmospheric Terrest. Phys., 2, 236, 1952.

Axford, W. I., The polar wind and the terrestrial helium budget, J. Geophys. Res., 73, 21, 1968.

Axford, W. I. and C. O. Hines, A unifying theory of high-latitude geophysical phenomena and geomagnetic storms, Can. J. Phys., 39, 1433, 1961.

Banks, P. M. and T. E. Holzer, The polar wind, J. Geophys. Res., 73, 6855, 1968.

Barbier, D., L'Arc auroral stable, Ann. Geophys., 14, 344, 1958.

Bauer, S. J., Diffusive equilibrium in the topside ionosphere, Proc. IEEE, 57, 1114, 1969.

Bauer, S. J. and B. V. Krishnamurthy, Topside ionosphere behaviour during a magnetic storm, Planetary Space Sci., 16, 653, 1968.

Behannon, K. W. and N. F. Ness, Magnetic storms in the earth's magnetic tail, J. Geophys. Res., 71, 2327, 1966.

Brace, L. H., Private communication.

Brice, N. M., Bulk motion of the magnetosphere, J. Geophys. Res., 72, 5193, 1967.

Cahill, L. J., Inflation of the inner magnetosphere during a magnetic storm, J. Geophys. Res., 71, 4505, 1966.

Calvert, W. and R. Cohen, The interpretation and synthesis of certain spread F configurations appearing on equatorial ionograms, J. Geophys. Res., 66, 3125, 1961.

Calvert, W. and G. B. Goe, Plasma resonances in the upper ionosphere, J. Geophys. Res., 68, 6113, 1963.

Calvert, W. and C. W. Schmid, Spread F observations by the Alouette topside sounder satellite, J. Geophys. Res., 69, 1839, 1964.

Calvert, W. and J. M. Warnock, Ionospheric irregularities observed by topside sounders, Proc. IEEE, 57, 1019, 1969.

Carpenter, D. L., Whistler evidence of a "knee" in the magnetospheric ionization density profile, J. Geophys. Res., 68, 1675, 1963.

Carpenter, D. L., Whistler studies of the plasmapause in the magnetosphere, J. Geophys. Res., 71, 693, 1966.

Chan, K. L. and L. Colin, Global electron density distributions from topside soundings, Proc. IEEE, 57, 990, 1969.

Duncan, R. A., Universal-time control of the arctic and antarctic F-region, J. Geophys. Res., 67, 1823, 1962.

Dyson, P. L., Magnetic field-aligned irregularities at mid-geomagnetic latitudes, J. Atmospheric Terrest. Phys., 29, 857, 1967.

Evans, J. V., Mid-latitudes ionospheric temperatures on magnetically quiet and disturbed days, J. Geophys. Res., 70, 2726, 1965.

Eviatar, A., A.M. Lenchek, and S.F. Singer, Distribution of density in an ion-exosphere of a non-rotating planet, Phys. of Fluids, 7, 1775, 1964.

Frank, L.A., Recent observations of low-energy charged particles in the earth's magnetosphere, in Physics of the Magnetosphere, Reidel Pub. Company, p. 271, 1968.

Franklin, C.A. and M. Maclean, The design of swept frequency topside sounders, Proc. IEEE, 57, 897, 1969.

Gallet, R.J., Aerodynamical mechanisms producing electron density fluctuations in turbulent ionized layers, Proc. IRE, 43, NLO, 1240, 1955.

Garriott, O.K. and H. Rishbeth, Effects of temperature changes on the electron density profile in the F2-layer, Planetary Space Sci., 11, 587, 1963.

Gledhill, J.A., D.G. Torr, and M.R. Torr, Ionospheric disturbance and electron precipitation from the outer radiation belt, J. Geophys. Res., 72, 209, 1967.

Gringauz, K.I., The structure of the ionized gas envelope of earth from direct measurements in the USSR of local charged particle concentration, Planetary Space Sci., 11, 281, 1963.

Hagg, E.L., Electron densities of 8-100 electrons cm^{-3} deduced from Alouette II high latitude ionograms, Can. J. Phys., 45, 27, 1967.

Hagg, E.L., Private communication.

Hagg, E.L., E.J. Hewens, and G.L. Nelms, Interpretation of topside sounder ionograms, Proc. IEEE, 57, 949, 1969.

Hartz, T.R. and N.M. Brice, The general pattern of auroral particle precipitation, Planetary Space Sci., 15, 301, 1967.

Herzberg, L. and G.L. Nelms, Ionospheric conditions following the proton flare of 7 July 1966 as deduced from topside soundings, Annals of IQSY, 3, 572, 1969.

Hibberd, F.H. and W.J. Ross, Variations in total electron content and other ionospheric parameters associated with magnetic storms, J. Geophys. Res., 72, 5331, 1967.

Hoffman, J., Ion composition measurements in the polar region from the Explorer XXXI satellite, Trans. Am. Geophys. Union, 49, 253, 1968.

Hoffman, J., The ion-mass spectrometer on Explorer XXXI satellite, Proc. IEEE, 57, 1063, 1969.

Hones, E.W., Review and interpretation of particle measurements made by the VELA satellites in the magnetosphere, in Physics of the Magnetosphere, Reidel Pub. Company, p. 392, 1968.

Jelly, D.H. and L.E. Petrie, High-latitude ionosphere, Proc. IEEE, 57, 1005, 1969.

King, J.W., A.J. Legg, and P.A. Smith, Observations of spread ionization in the topside of the ionosphere, J. Atmospheric Terrest. Phys., 29, 1373, 1967.

Knecht, R.W. and T.E. VanZandt, Some early results from the ionosphere topside sounder satellite, Nature, 197, 641, 1963.

Lewis, R.P.W. and D.H. McIntosh, Geomagnetic and ionospheric relationships, J. Atmospheric Terrest. Phys., 4, 44, 1953.

Liszka, L., The high-latitude trough in ionospheric electron content, J. Atmospheric Terrest. Phys., 29, 1243, 1967.

Lockwood, G. E. K., Plasma and cyclotron spike phenomena observed in topside ionograms, Can. J. Phys., 41, 190, 1963.

Lockwood, G. E. K. and L. E. Petrie, Low latitude field aligned ionization observed by the Alouette topside sounder, Planetary Space Sci., 11, 327, 1963.

Loftus, B. T., T. E. VanZandt, and W. Calvert, Observations of conjugate ducting by the fixed-frequency topside-sounder satellite, Ann. Geophys., 22, 530, 1966.

Maehlum, B. N., Universal-time control of the low-energy electron fluxes in the polar regions, J. Geophys. Res., 73, 3459, 1968.

McDiarmid, I. B. and J. R. Burrows, Local time asymmetries in the high-latitude boundary of the outer radiation zone for the different electron energies, Can. J. Phys., 46, 49, 1968.

McDiarmid, I. B. and M. D. Wilson, Dependence of the high-latitude electron (E > 35 keV) boundary on the orientation of the geomagnetic axis, J. Geophys. Res., 73, 7237, 1968.

Muldrew, D. B., Radio propagation along magnetic field-aligned sheets of ionization observed by the Alouette topside sounder, J. Geophys. Res., 68, 5355, 1963.

Muldrew, D. B., F-layer ionization troughs deduced from Alouette data, J. Geophys. Res., 70, 2635, 1965.

Muldrew, D. B., Medium frequency conjugate echoes observed on topside-sounder data, Can. J. Phys., 45, 3935, 1967.

Muldrew, D. B., Non-vertical propagation and delayed-echo generation observed by the topside sounders, Proc. IEEE, 57, 1097, 1969.

Muldrew, D. B., Private communication.

Nelms, G. L., Seasonal diurnal variations of the distribution of electron density in the topside of the ionosphere, in Electron Density Profiles in the Ionosphere and Exosphere, edited by J. Frihagen, North-Holland Publishing Company, Amsterdam, 1966.

Nelms, G. L. and G. E. K. Lockwood, Early results from the topside sounder in the Alouette II satellite, Space Res. VII, North-Holland Publishing Company, Amsterdam, p. 604, 1967.

Nelms, G. L. and E. S. Warren, Some irregular variations of the electron density in the topside of the ionosphere, Space Res. V, North-Holland Publishing Company, Amsterdam, p. 637, 1965.

Neugebauer, M. and C. W. Snyder, Mariner 2 observations of the solar wind, 1, average properties, J. Geophys. Res., 71, 4469, 1966.

Nishida, A., Formation of a plasmapause or magnetospheric plasma knee by combined action of magnetospheric convection and plasma escape from the tail, J. Geophys. Res., 71, 5669, 1966.

Nishida, A., Average structure and storm-time change of the polar topside ionosphere at sunspot minimum, J. Geophys. Res., 72, 6051, 1967.

Norton, R. B., The mid-latitude F-region during a severe ionospheric storm, Proc. IEEE, 57, 1147, 1969.

Norton, R. B. and E. Marovich, Alouette observation taken during a mid-latitude red arc, Proc. IEEE, 57, 1158, 1969.

O'Brien, B. J., High-latitude geophysical studies with satellite Injun 3: precipitation of electrons into the atmosphere, J. Geophys. Res., 69, 13, 1964.

Ondoh, T., Morphology of disturbed topside ionosphere for 1962-1964, J. Radio Res. Labs., Japan, 14, 267, 1967.

Petrie, L. E., Topside spread echoes, Can. J. Phys., 41, 194, 1963.

Petrie, L. E., Preliminary results on mid and high latitude topside spread F, in Spread F and Its Effects Upon Radio Wave Propagation and Communications, edited by P. Newman, Technivision Press, Maidenhead, England, p. 67, 1966.

Pitteway, M. L. V. and R. Cohen, A waveguide interpretation of "temperate-latitude spread F" on equatorial ionograms, J. Geophys. Res., 66, 3141, 1961.

Ramasastry, J., J. Walsh, and J. R. Herman, Research on field-aligned propagation of HF radiowaves using Alouette 2 topside sounder data and digital ray-tracing techniques, Proc. AGARD/EPC XIVth Symposium, Scatter Propagation of Radio Waves, Oslo, 1968.

Rao, P. B., Electron concentrations and electron and ion temperatures in the F-region for magnetically quiet and disturbed conditions, J. Geophys. Res., 72, 5331, 1967.

Ratcliffe, J. A. and K. Weekes, The ionosphere, in Physics of the Upper Atmosphere, edited by J. A. Ratcliffe, Academic Press, London, 1960.

Reber, G. and G. R. Ellis, Cosmic radio-frequency radiation near one megacycle, J. Geophys. Res., 61, 1, 1956.

Reddy, B. M., L. H. Brace, and J. A. Findlay, The ionosphere at 640 km on quiet and disturbed days, J. Geophys. Res., 72, 2709, 1967.

Renau, G., Theory of spread F based on aspect sensitive backscatter echoes, J. Geophys. Res., 65, 2269, 1960.

Roach, F. E. and J. R. Roach, Stable 6300 Å auroral arcs in mid-latitudes, Planetary Space Sci., 11, 523, 1963.

Rycroft, M. J. and J. O. Thomas, The magnetospheric plasmapause and the electron density trough at the Alouette I orbit, submitted to Planetary Space Sci.

Sato, T., Electron concentration variations in the topside ionosphere between 60 °N and 60 °S geomagnetic latitude associated with geomagnetic disturbances, J. Geophys. Res., 73, 6225, 1968.

Sato, T. and K. L. Chan, Storm time variations of electron concentration in the polar topside ionosphere, J. Geophys. Res., In press, May 1969.

Sato, T. and L. Colin, Morphology of electron density enhancement at a height of 1000 km at polar latitudes, J. Geophys. Res., In press, May 1969.

Schmeltekopf, A. L., F. C. Fehsenfeld, G. I. Gilman, and E. E. Ferguson, Reaction of atomic oxygen ions with vibrationally excited nitrogen molecules, Planetary Space Sci., 15, 401, 1967.

Serbu, G. P. and E. J. R. Maier, Low-energy electrons measured in IMP 2, J. Geophys. Res., 71, 3755, 1966.

Sharp, G. W., Mid latitude trough in the night ionosphere, J. Geophys. Res., 71, 1345, 1966.

Sharp, G. W., T. J. Crowther, and C. W. Gilbreth, Some recent ion concentration measurements obtained from a polar-orbiting satellite, (abstract), Trans. Am. Geophys. Union, 45, 87, 1964.

Snyder, C. W., M. Neugebauer, and U. R. Rao, The solar wind velocity and its correlation with cosmic-ray variations and with solar and geomagnetic activity, J. Geophys. Res., 68, 6361, 1963.

Thomas, J. O. and M. K. Andrews, Transpolar exospheric plasma, J. Geophys. Res., 73, 7407, 1968.

Thomas, J. O., M. J. Rycroft, L. Colin, and K. L. Chan, The topside ionosphere, II, in Electron Density Profiles in Ionosphere and Exosphere, edited by J. Frihagen, North-Holland Publishing Company, Amsterdam, p. 322, 1966.

Thomas, L. and R. B. Norton, Possible importance of internal excitation in ion-molecule reactions in the F-region, J. Geophys. Res., 71, 227, 1966.

Thomas, L. and F. H. Venables, The onset of the F-region disturbances at middle latitudes during magnetic storms, J. Atmospheric Terrest. Phys., 28, 599, 1966.

Timleck, P. L. and G. L. Nelms, Electron densities less than 100 cm^{-3} in the topside ionosphere, Proc. IEEE, 57, 1164, 1969.

Timleck, P. L., Private communication.

Titheridge, J. E. and M. K. Andrews, Changes in the topside ionosphere during a large magnetic storm, Planetary Space Sci., 15, 1157, 1967.

Walker, James C. G. and M. H. Rees, Excitation of stable auroral red arcs at sub-auroral latitudes, In press, 1968.

Warren, E. S., Some preliminary results of sounding of the topside of the ionosphere by radio pulses from a satellite, Nature, 197, 636, 1963.

Warren, E. S., The topside ionosphere during geomagnetic storms, Proc. IEEE, 57, 1029, 1969.

Willmore, A. P., Electron temperature of the upper F-region, Proc. Roy, Soc., London, A286, 537, 1965.

IRREGULARITIES IN THE ELECTRON DENSITY OF THE POLAR IONOSPHERE

Jon Frihagen

Norwegian Defence Research Establishment
2007 Kjeller, Norway

Abstract. The large scale morphology of spread F and satellite scintillation over the polar cap is considered. Both phenomena are caused by F region irregularities and appear consistently above about 80° geomagnetic latitude. In addition an auroral oval maximum in occurrence or strength of satellite scintillation is evident and spread F shows an auroral zone midnight maximum.

The irregularities might be caused by direct ionization of precipitating particles with rms spatial variations of the order 10^4 - 10^8 cm^{-2} sec^{-1} of 1 keV electrons, depending on the time variations. Although published studies of high latitude low energy particle precipitation can not yet directly support this view, it seems difficult to ignore the possibility that the observed variations in particle flux does take part in producing ionospheric irregularities.

1. INTRODUCTION

One of the characteristic features of the polar ionosphere is the presence of irregular variations in electron density. The scale size of these irregularities range from several hundred kilometers down to irregularities of centimeters or less caused by the randome thermal motion of the ionospheric particles.

The large variations are conveniently studied by groundbased and satellite borne ionosondes, the smallest by radar methods. In this paper we will only include results of studies of irregularities of an intermediate range of typically hundreds of meters that give rise to spread F and scintillation of extraterrestrial radio sources.

I will first review some quite clear results on the occurrence of topside and bottomside spread F in Section 2. I will then pass on to review what is known about the extent of scintillation in the polar area in Section 3 and review briefly some production mechanisms that have been proposed to explain the presence of irregularities in Section 4.

2. HIGH LATITUDE SPREAD F

Penndorf (1962; 1966) has studied the occurrence of spread F over both polar caps during IGY. The main features over both polar caps are:

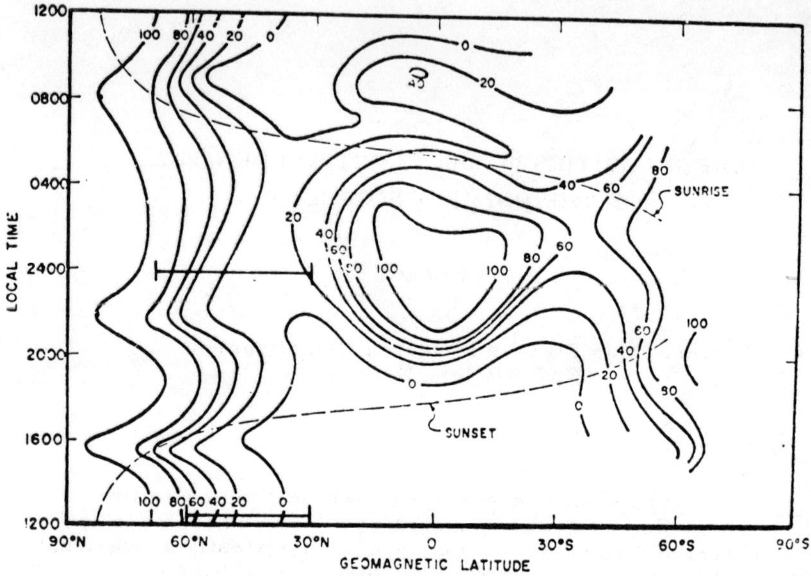

Figure 1. The occurrence probability of topside spread F (Calvert & Schmid, 1964).

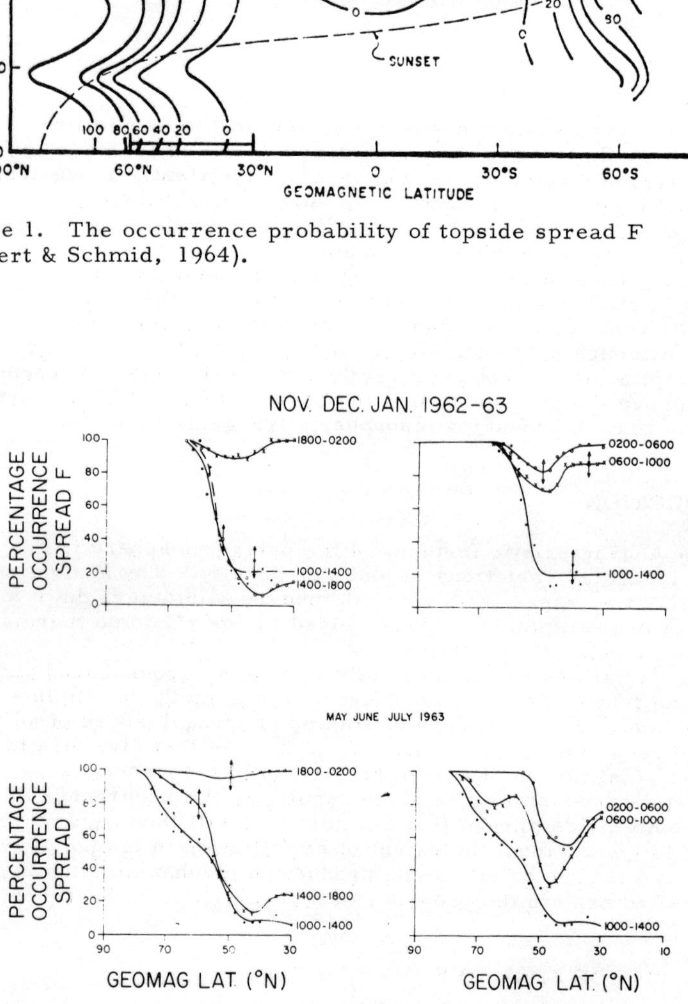

Figure 2. The diurnal variation of the occurrence of spread F for winter 1962 - 1963 and summer 1963 (Petrie, 1966).

(i) A fixed maximum in occurrence probability centered in the general vicinity of the magnetic poles, of roughly 10° in diameter. In these regions there is little diurnal variation and the probability of observing spread F on bottomside soundings generally exceeds 80%.

(ii) There is a travelling maximum, near local magnetic midnight travelling round the earth, close to the classical auroral zone.

This general picture seems to apply to both polar caps and to all seasons. There is, however, a seasonal variation in that the occurrence probability of spread F is less in local summer than in local winter.

The diurnal variation is then such that at auroral latitudes and below there is a midnight maximum in the occurrence probability of spread F, while at higher latitudes there is little diurnal variation. These observations lead, of course, to a general statistical relation between the electron density and the occurrence of spread F, in that small night time electron densities are accompanied by large probability for spread F occurrence. I am not aware of any more detailed study of these quantities which may indicate a physical relationship between them.

Figure 1 shows the occurrence probability of aspect sensitive topside spread F after Calvert & Schmid (1964).

Again we see that spread F occurs with near 100% probability at all times of day at high latitudes, with a steep gradiant towards the equator. Around auroral latitudes there seems to be a rather complicated diurnal variation. On this figure there is no evident separation of a midnight maximum from the fixed polar maximum. Partly this may be caused by the sampling of observations which probably, in this case, is largely from over the American continent, where according to Penndorf the maxima tend to merge.

Petrie (1966) has also studied the occurrence of spread F on topside ionograms, and Figure 2 again shows the high occurrence probability over the polar cap and the sharp equator-ward boundary and how this moves equatorward at night. Petrie also studied the occurrence of zero range spread echoes. In Figure 3 the locations of the satellite (at 1000 km above ground) are shown at which the spread maximized on a number of passes. The locus of these points is in the general vicinity of the auroral zone.

The above observations indicate that the irregularities giving rise to spread F can extend from above 1000 km height and, at night, down to below the F-layer peak in the auroral zone. Inside this region there can be a zone where irregularities are only present near, but above, the F-layer peak and still closer to the magnetic poles a region where spread F occurs on both sides of the F-layer peak.

It might be interesting to speculate over whether the presence of irregularities at 1000 km height as well as throughout the F-layer indicate the presence of ducts through the ionosphere or if only a series of elongated "blobs" are present along a field line. The evidence on this is at present inconclusive. Ducted echoes on topside ionograms (Calvert & Schmid, 1964) can only be distinguished in the absence of strong spread F, but there is a fairly high occurrence probability of ducted echoes equatorward of the auroral zone.

It might also be appropriate to speculate a little on the reliability of spread F as an indication of the presence of ionospheric irregularities. It is clear that system parameters can influence the data considerably. Thus one will always see an increase in spread when receiver gain is increased and conversely by reducing the gain, or for bottomside sounders, when absorbtion increases, spread F will decrease or even disappear. For these reasons one cannot state that the absence of spread F indicates absence of irregularities. On the other hand increased spread, specially on topside ionograms where absorption

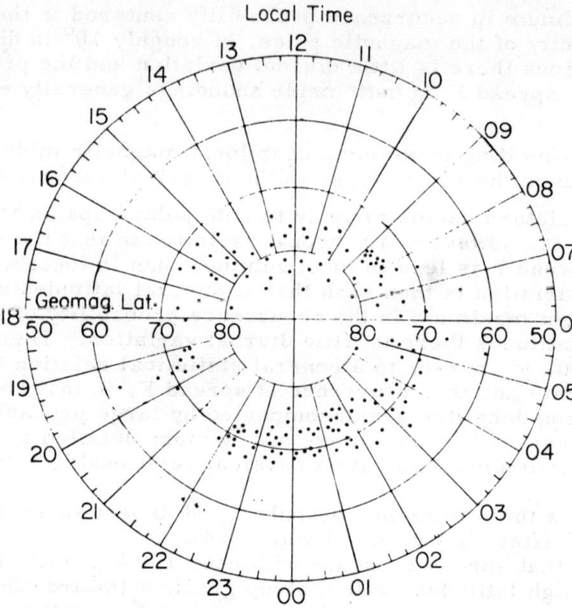

Figure 3. The distribution of the maximum frequency range of the spread echoes at the satellite height (Petrie, 1966).

and change in environment etc are small, indicates increased presence of irregularities.

Partly for this reason we will not embark on commenting upon correlations between the presence of spread F and other geophysical parameters that are not direct observations of irregularities.

3. SCINTILLATION OF EXTRATERRESTRIAL RADIO SOURCES

Briggs (1964) found the presence of radio star scintillation, observed from England, to be closely correlated with the presence of spread F. Further north (Owren et al., 1966) this correlation is not very strong or may even be negative. For the reasons given above we shall not take this too seriously. Furthermore Calvert & Schmid found that north of 30° geomagnetic latitude, topside spread F is not Kp dependent whereas we shall see below that high latitude scintillation is. This discrepancy may be partly due to spread F being scaled either as being present or not while for scintillation studies the degree of scintillation is noted. There is, however, a possibility that spread F and scintillation give measures of quite different phenomena in the ionosphere.

It is well known that scintillation of extraterrestrial sources, at latitudes below the auroral zone, is a nighttime phenomenon showing increasing scintillation with increasing Kp (Aarons et al., 1963; Liszka, 1963a).

Very few studies have been made of scintillation over the polar caps. Titheridge and Stuart (1968; 1969) have published results from the southern polar cap, and Frihagen (1969) has published part of his results of studies of satellite scintillation as observed from Spitsbergen.

3.1 Morphology of Scintillation Producing Irregularities

Titheridge and Stuart (1968) report a maximum in the vicinity of the magnetic poles and an erratic decrease, with several small maxima and minima, towards lower latitudes. In particular there is a tongue of irregularities extending equatorwards between 165° and 175°E in winter only, and a region of low scintillation activity between the geomagnetic and dip poles in summer and winter.

I will now turn to some of my own data which are from observations of a 40 MHz beacon on Explorer 22 made at Spitsbergen (Frihagen, 1969). First a point of clarification. I will present a number of figures showing the distribution of irregularities. In all these the position of irregularities have been plotted under the assumption that they are at 400 km height. If the actual height is lower the figures should be radially contracted towards the observing point, if it is higher the figures should be delated. A justification for the selection of 400 km will be given below.

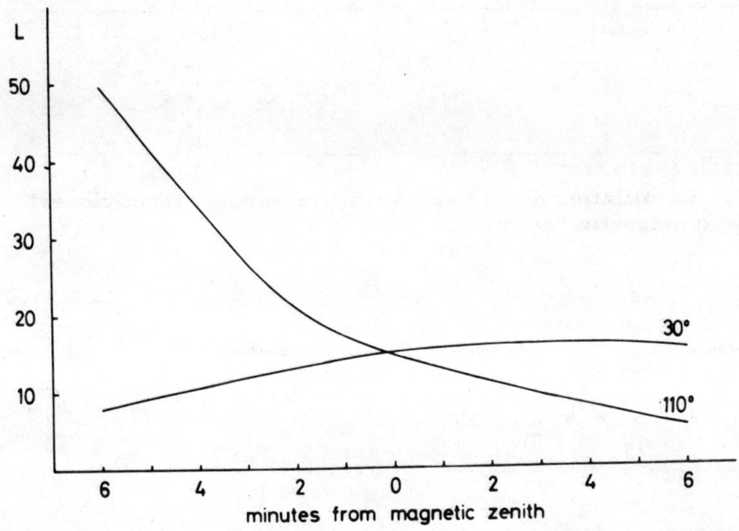

Figure 4. The L value of the subionospheric point for a northbound (30°) and southbound (110°) pass coming close to magnetic zenith at the Spitsbergen site.

The paths of the subionospheric points of two passes that come very close to the magnetic zenith of the observing station are shown in Figure 4. It may be seen that the northbound pass follows very closely curves of constant L (and geomagnetic latitude) while the southbound pass crosses these curves nearly perpendicularly. For such passes the scintillation depth was scaled for every minute, and in Figure 5 we see how this turned out for the northbound and southbound passes, for day and night, and for different Kp.

There are three main features:

(i) Apart from a hump near magnetic zenith, which is to be expected from field aligned irregularities, the northbound passes show little systematic variation with time during the passes, while the scintillation level for the southbound passes fall off. This is of

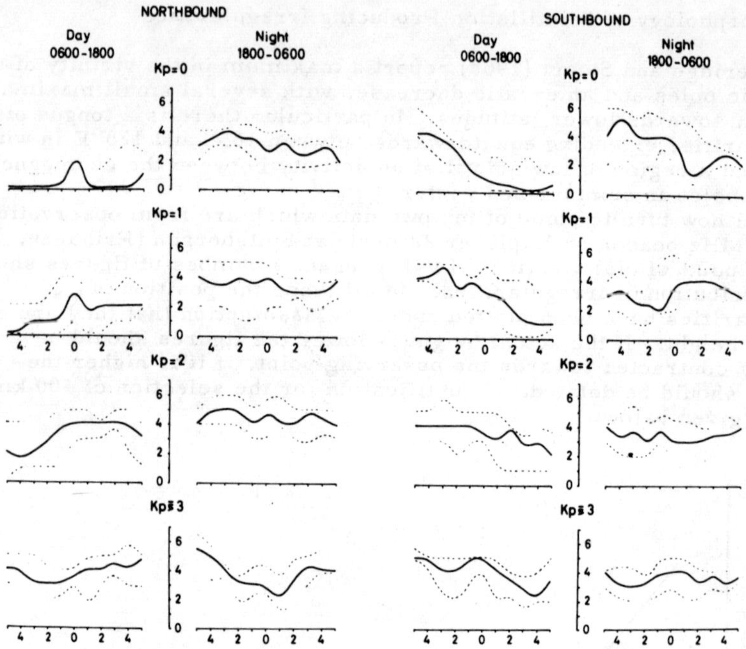

Figure 5. Scintillation depth versus time in minutes from closest approach to magnetic zenith.

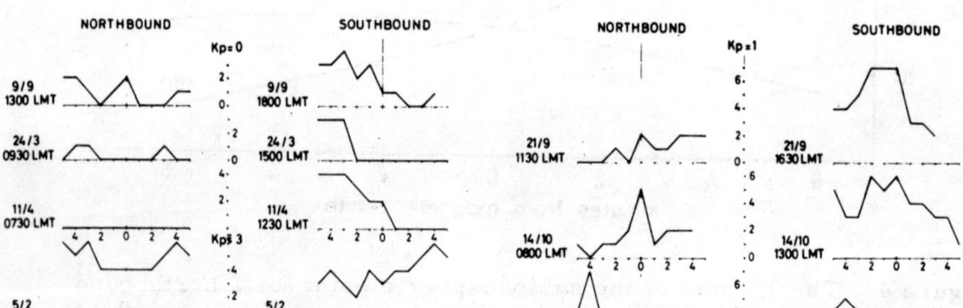

Figure 6. Scintillation depth versus time in minutes from closest approach to magnetic zenith for a number of passes.

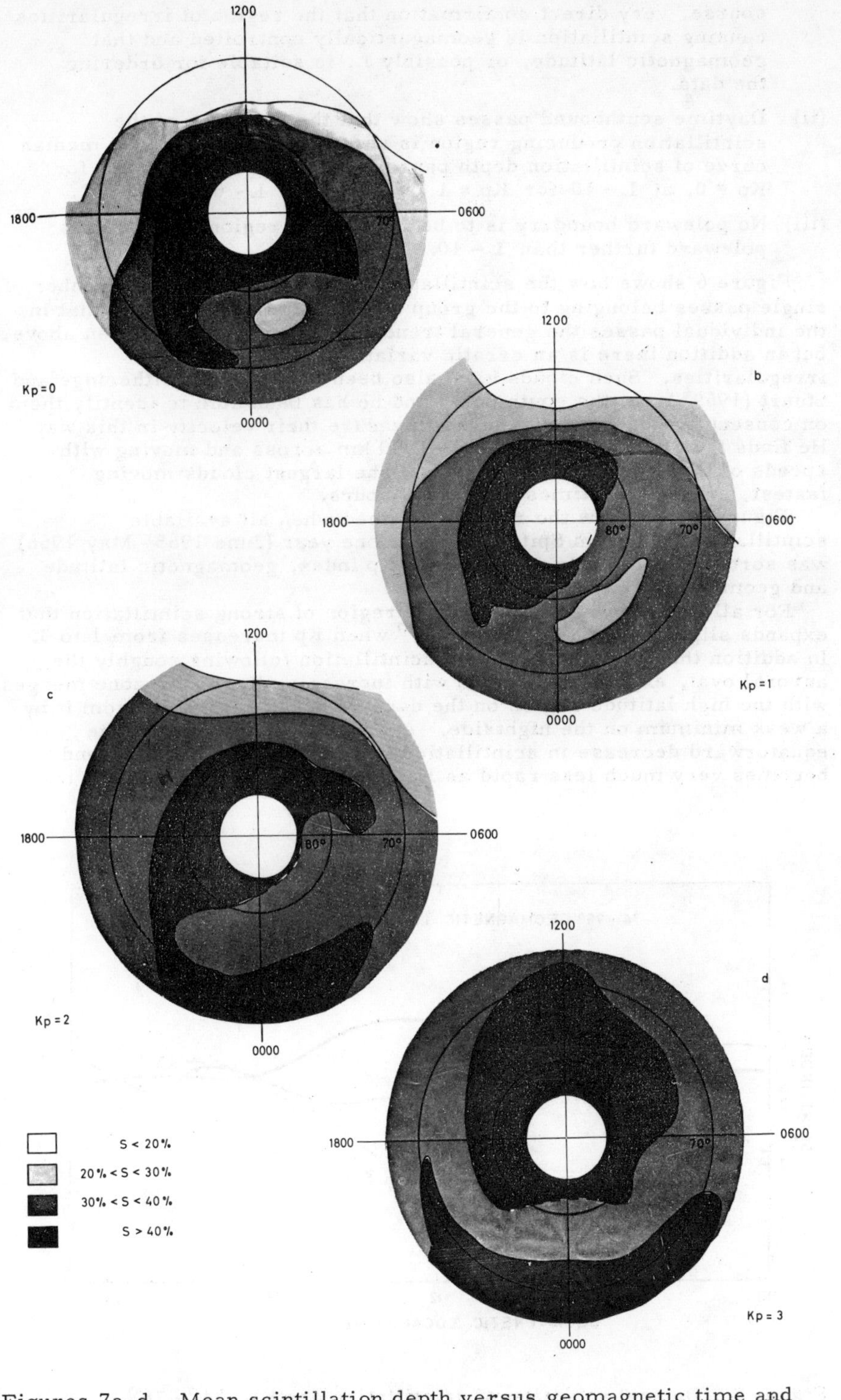

Figures 7a-d. Mean scintillation depth versus geomagnetic time and latitude for different Kp.

course, very direct confirmation that the region of irregularities causing scintillation is geomagnetically controlled and that geomagnetic latitude, or possibly L, is suitable for ordering the data.

(ii) Daytime southbound passes show that the boundary of the scintillation producing region is very Kp dependent. The median curve of scintillation depth passes the 20% level at L ~ 18 for Kp = 0, at L ~ 10 for Kp = 1, and south of L ~ 6 for Kp $\geq$ 2.

(iii) No poleward boundary is to be seen. The region then extends poleward further than L ~ 40.

Figure 6 shows how the scintillation depth varied during a number of single passes belonging to the group discussed above. We see that in the individual passes the general trend corresponds to that given above, but in addition there is an erratic variation due to clouds of irregularities. Such clouds have also been observed by Titheridge and Stuart (1969) from the south pole, and he has been able to identify them on consecutive passes and roughly measure their velocity in this way. He finds them to be typically 100 - 1000 km across and moving with speeds of the order 20 - 100 m/sec. The largest clouds moving fastest, and with lifetimes of several hours.

Figures 7a-d show the results obtained when all available scintillation data from Spitsbergen for one year (June 1965 - May 1966) was sorted according to geomagnetic Kp index, geomagnetic latitude and geomagnetic local time.

For all Kp there is a high latitude region of strong scintillation that expands slightly, say from 80° to 77° when Kp increases from 1 to 3. In addition there is a zone of high scintillation following roughly the auroral oval, expanding in width with increasing Kp. This zone merges with the high latitude region on the dayside but is separated from it by a weak minimum on the nightside. Around geomagnetic noon the equatorward decrease in scintillation is very rapid for Kp = 1, and becomes very much less rapid as Kp increases. Figure 8 shows the

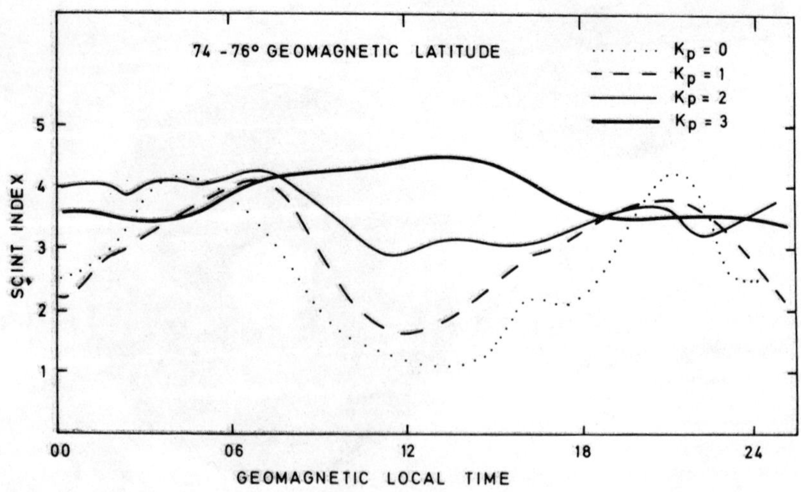

Figure 8. Diurnal variation of scintillation depth at 74° - 76° geomagnetic latitude.

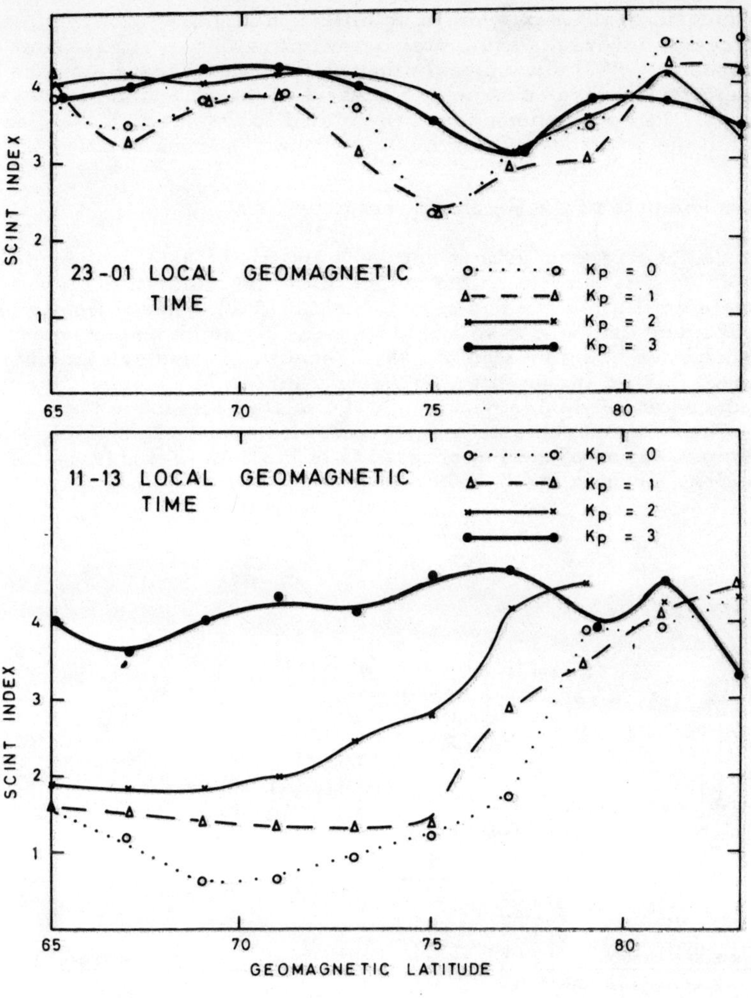

Figure 9. Latitudinal variation of scintillation depth round geomagnetic noon and midnight.

diurnal variation of scintillation depth at 74° - 76° geomagnetic latitude for different Kp. Figure 9 shows the latitudinal variation at geomagnetic noon and midnight. The diurnal variation maximizes at 0700 and 2100 for Kp = 0. 1 with troughs around midday and midnight. For Kp $\geq$ 2 the diurnal variation is less, due to filling in of the minima, which we see from the latitudinal variation is caused by a poleward movement and some reduction in depth of the trough between the high latitude and "auroral" zone maxima.

If this general picture of scintillation applies also to other locations it should be noted that it agrees very well with Penndorf's (1962) picture of a fixed high latitude maximum and a moving midnight maximum. Groundbased stations observing overhead spread F might well miss the possibility of the midnight region being connected, in a rather narrow zone, to the high latitude maximum.

The "auroral oval" maximum in scintillation might well be related to the presence of aurorae. Radio star scintillation observations from Alaska (Benson, 1960) show clearly that scintillation always appears if not before at least when aurorae are present in the ray path. However, a number of other phenomena are also related to the auroral oval, and may explain the presence of scintillation producing irregularities there.

3.2 Measurements of the Height of Irregularities.

By spaced receiver measurements of scintillating satellites one can, in principle, easily determine the height of the irregularities giving rise to the scintillation (Frihagen and Trøim, 1960). The majority of all measurements show that scintillation is an F region phenomenon, the irregularities being located at, say, between 250 and 450 km above the ground. Basler and deWitt (1962) found auroral zone scintillation to be produced at heights varying from 100 to 1000 km above the ground. These measurements are, however, not entirely reliable due to oversimplification of the interpretation of the data (Frihagen, 1969). Some Kjeller, Tromsø and Spitsbergen results are given in Figure 10.

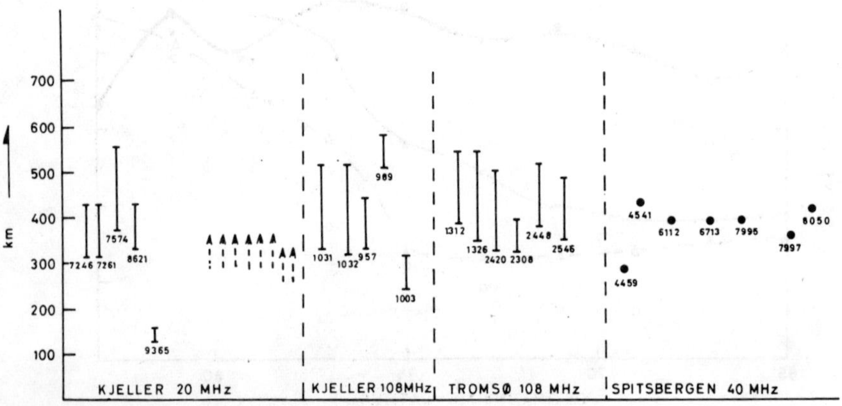

Figure 10. Observed heights of irregularities.

It seems that at Spitsbergen the irregularities that produce the dominating scintillation are situated higher in the ionosphere than at lower latitudes. Furthermore, we are currently studying a number of cases where the maximum cross-correlation between signal levels at spaced receivers (spacing ~ 150 m) at Spitsbergen is very low, indicating very thick active regions. This very rarely happens in our auroral zone or lower latitude observations.

3.3 The Size and Shape of Irregularities

It is rather well established that irregularities producing scintillation are field aligned cylinders at auroral and lower latitudes. Over the polar cap little is yet known. A number of our observations from Spitsbergen indicate field alignment but the transverse structure is uncertain and is under study. The correlation distance of the diffraction pattern on the ground is of the order of a few hundred to a few thousand meters, at Spitsbergen. These numbers are not yet corrected for ellipticity of the pattern. If field aligned cylinders are the cause of scintillation, there may have a cross-section of the order 100 m and might be several km long.

3.4 Strength of the Irregularities

A simple but essentially correct way of estimating the strength of the irregularities, that is the percentage variation in electron density in the ionosphere, may be done in the following way.

Simple diffraction theory shows (Ratcliffe, 1956) that the amplitude fading depth is determined by the place variation imposed on an incoming wave by variations in refractive index. To produce deep scintillation an rms phase variation of 1 radian is sufficiant.

The phase path of a ray passing through a region with refractive index n is given by

$$\phi = \frac{2\pi}{\lambda} \int n \, ds \qquad (1)$$

The deviation from the mean is given by

$$\Delta\phi = \frac{2\pi}{\lambda} \int \Delta n \, ds \qquad (2)$$

In our case the refractive index is given with sufficiant accuracy by

$$n^2 = 1 - \frac{81 \, N}{f^2} \qquad (3)$$

where N is electron density (cm^{-3}) and f is radio frequency (kHz).

At f = 40 000 kHz

$$\Delta n = \frac{1}{2n} \frac{81 \, \Delta N}{f^2} \sim 2.5 \cdot 10^{-8} \, \Delta N \qquad (4)$$

where ΔN is the deviation from the mean electron density.

$$\Delta\phi = 2 \cdot 10^{-10} \int \Delta N \, ds = 2 \cdot 10^{-10} \, \Delta n_t \qquad (5)$$

where Δn_t is the deviation from the mean electron content in the slab.

For rms variations of $5 \cdot 10^9$ cm^{-2} in electron content rms variations of $\Delta\phi$ of 1 radian are produced giving fairly deep scintillation If the irregularities extend over 100 km along the ray path and the mean electron density there is 10^5 cm^{-3}, we find that 0.2% variations in electron density are sufficiant, while if the electron density is 10^4 cm^{-3}, 2% variations are necessary to give 1 radian rms phase variations.

One should note that irregular variations in electron content of ~ $10^9 - 10^{10}$ cm^{-2} will give rise to scintillation regardless of what the surrounding or background electron content is.

4. PRODUCTION OF IONOSPHERIC IRREGULARITIES

To produce irregularities in electron density one or more of three possibilities exist:

(i) Redistribution of existing ionization (by waves, diffusion etc).

(ii) Variations in ionization loss rates (by for instance temperature variations).

(iii) Variations in production rates (density variations, particles etc).

Present day techniques of measuring ionospheric plasma density are hardly able to measure variations in density accurately enough to distinguish between these possibilities. For the high latitude region of interest here it is tempting to associate the production of irregularities with precipitating electrons.

If one assumes the loss coefficient $\beta \sim 10^{-4}$ sec^{-1}, and that every incident 1000 eV electron to produce 30 electron-ion pairs, and steady state conditions, a flux of 10^6 cm^{-2} sec^{-1} of 1 keV particles will maintain the midnight midwinter solar minimum F-region at about 75° geomagnetic latitude (Bratteng and Frihagen, 1969). The observed fluxes are orders of magnitude larger than this and show rapid and strong variations (Hoffman, 1969; Burch, 1969). Still assuming steady state conditions a flux of 10^4 cm^{-2} sec^{-1} will maintain irregularities of sufficient strength, and 1 second bursts of 10^8 cm^{-2} sec^{-1} at 1 keV will give irregularities that decay in the order of 1 hour. That such variations exist in the precipitation seems extremely plausible according to the data shown by Hoffman. It is, however, not known whether the data shown are examples of consistantly appearing phenomena in this respect. When comparing his figures 8 and 9 with Figures 7a-d in this paper there are a number of similarities and dissimilarities between the occurrence of burst regions and of scintillation. One important difference is the apparent lack of bursts above 80° latitude. This may in part be explained by the definition of bursts requiring peak fluxes larger than $6 \cdot 10^8$ cm^{-2} sec^{-1} (assuming isotropy) which is far more than we may require. Furthermore whereas his instruments have to be the right place at the right time to see a burst, the ionosphere will maintain irregularities for the order of hours - store the memory of the burst - untill it may be observed. It is perhaps more difficult to explain the lower boundary of the burst region around midday. My only suggestion is that the bursts near the boarder may be broader, produce too large irregularities to be observed as scintillation, than further into the burst region.

Further evidence that this production mechanism may produce irregularities of reasonable scale size may be found by extrapolating from the higher energy particles producing aurora, which of course may be observed as narrow rays of the order 100 meters across, and Davis (1967) has shown that even homogeneous arcs may be composites of such rays. Whereas electrons thus can precipitate in narrow beams, this is not the case for protons that below about 350 km rapidly spread out in broad arcs due to charge exchange. Particles may of course act in other ways to produce irregularities. One possibility is the streaming instability forming waves modulating the existing ionospheric plasma. A proper theory for this (at least 2. order) is difficult and, I believe, has not been done. Other hydromagnetic or acoustic waves may accompany or be produced by particle interaction with the ionosphere. The very mechanism that forms the precipitating particles into rays may be important. Heating of the plasma as a consequence of particle precipitation may produce temperature differences (that can be maintained according to Bowhill (1966)) and let the plasma on "neighbouring", field lines be distributed according to their own temperatures, thus creating irregularities.

Other theories include streaming instabilities in the equatorial and auroral electrojets giving rise to field aligned waves travelling with new ion-acoustic velocity (Farley, 1963). The theory only applies to the E-region. Crossed electron density gradients and electric fields give rise to instabilities which at least may be important in the auroral zone and near the equator (Reid, 1968). These theories regrettably lack higher than 1. order terms and do thus not give results in terms of preferred wavelengths and amplitudes. I will refer to other papers, for example that of Singleton (1966) for a discussion of other possible production mechanisms that are generally not specially related to high latitudes.

5. CONCLUDING REMARKS

The differences in the occurrence of spread F and scintillation at high latitudes require explanation. Specially one might investigate whether they are produced by irregularities of different amplitudes and/or wavelengths that are selectively produced.

In general terms both phenomena are produced in the F-region. They both appear consistantly above about 80° geomagnetic latitude. Scintillation also has a relatively weak maximum near the auroral oval, whereas groundbased spread F maximizes at midnight in the auroral zone. The Kp dependence of spread F seems to be small, whereas the low latitude boarder of daytime scintillation moves rapidly southwards with increasing Kp. Scintillation is caused by field aligned variations in electron density of the order 1% over lateral scales of the order 100 m. If observed variations in electron precipitation are at least partly spatial, soft electrons must produce F-region irregularities. It is certainly not yet well enough known whether these are of the correct size and amplitude, and a number of apparant differences in morphology are not satisfactory explained. Other mechanisms may certainly explain part of or all of the observations. If, however, particle precipitation is the cause of irregularities in the ionosphere, their study may give important contributions towards the understanding of the behaviour of particle precipitation in time and space.

REFERENCES

Aarons, J., J.P. Mullen and Sunanda Basu, Geomagnetic control of satellite scintillations, J. Geophys. Res., 68, 3159, 1963.

Basler, R.P. and R.N. deWitt, The heights of auroral irregularities in the auroral zone, J. Geophys. Res., 67, 587, 1962.

Benson, R.F., Effect of line of sight aurorae on radio star scintillation, J. Geophys. Res., 65, 1981, 1960.

Bowhill, S.A., Origin of field aligned irregularities in the F2-layer, in Spread F and its Effects upon Radiowave Propagation and Communications, Agardograph 95, edited by P. Newman, Technivision, 1966.

Bratteng, O.M. and J. Frihagen, The electron content in the polar F-region, Progress Report no 8, The Auroral Observatory, Tromsø, Norway, 1969.

Briggs, B.H., Observations of radio star scintillation and spread F echoes over a solar cycle, J. Atmospheric Terrest. Phys., 26, 1, 1964.

Burch, J.L., Satellite measurements of low energy electrons precipitated at high latitudes, in The Polar Ionosphere and Magnetospheric Processes, edited by G. Skovli, Gordon and Breach, New York, 1969.

Calvert, W. and C.W. Schmid, Spread F observations by the Alouette topside sounder satellite, J. Geophys. Res., 69, 1839, 1964.

Davis, T.N., Cinematographic observations of fast auroral variations, in Aurora and Airglow, edited by B.M. McCormac, Reinhold, New York, p. 113, 1967.

Farley, D.T., A plasma instability resulting in field aligned irregularities in the ionosphere, J. Geophys. Res., 68, 6083, 1963.

Frihagen, J. and J. Trøim, Scintillations of the 20 MHz signal from the earth satellite 1958 δII, J. Atmospheric Terrest. Phys., 18, 75, 1960.

Frihagen, J., Satellite scintillation at high latitudes and its possible relation to precipitation of soft particles, J. Atmospheric Terrest. Phys., 31, 81, 1969.

Herman, J.R., A charged particle production mechanism for spread F irregularities, in Spread F and its Effects upon Radiowave Propagation and Communications, Agardograph 95, edited by P. Newman, Technivision, 1966.

Hoffman, R.A., OGO-4 satellite measurements of low energy - high latitude electron precipitation, in The Polar Ionosphere and Magnetospheric Processes, edited by G. Skovli, Gordon and Breach, New York, 1969.

Liszka, L., Satellite scintillation observed in the auroral zone, Arkiv för Geofysik, 4, 211, 1963a.

Liszka, L., A study of ionospheric irregularities using satellite transmissions at 54 Mc/s, Arkiv för Geofysik, 4, 227, 1963b.

Owren, L., E.J. Fremow, and R.D. Hunsucker, Radio star scintillation and spread F in the auroral zone, in Spread F and its Effects upon Radiowave Propagation and Communications, Agardograph 95, edited by P. Newman, Technivision, 1966.

Penndorf, R., Geographic distribution of spread F in the arctic, J. Geophys. Res., 67, 2279, 1962.

Penndorf, R., Frequency of spread F occurrence over antarctica, in Spread F and its Effects upon Radiowave Propagation and Communications, Agardograph 95, edited by P. Newman, Technivision, 1966.

Petrie, L.E., Preliminary results on mid and high latitude topside spread F, in Spread F and its Effects upon Radiowave Propagation and Communications, Agardograph 95, edited by P. Newman, Technivision, 1966.

Ratcliffe, J.A., Some aspects of diffraction theory and their applications to the ionosphere, Rep. Progr. Phys., 19, 188, 1956.

Reid, G., The formation of small scale irregularities in the ionosphere, J. Geophys. Res., 73, 1627, 1968.

Singleton, D.G., The production of spread F irregularities, in Spread F and its Effects upon Radiowave Propagation and Communications, Agardograph 95, edited by P. Newman, Technivision, 1966.

Titheridge, J.E. and G.F. Stuart, The distribution of irregularities in the antarctic ionosphere I, J. Atmospheric Terrest. Phys., 30, 85, 1968.

Titheridge, J.E. and G.F. Stuart, Irregular regions in the antarctic ionosphere, J. Geophys. Res., 74, 1308, 1969.

REGULAR AND IRREGULAR DIURNAL VARIATIONS OF ELECTRON NUMBER DENSITY IN THE LOWER IONOSPHERE OVER RESOLUTE BAY

J.S. Belrose, L.W. Hewitt, and R. Bunker

National Radio Propagation Laboratory
Communications Research Centre
Department of Communications
Shirley Bay, Ottawa, Canada

Abstract. Observations of vertical incidence partial reflections at 2.66 MHz have been carried out at Resolute Bay, Cornwallis Island, N.W.T., near the centre of the polar cap (geographic latitude 74.7 °N, magnetic invariant latitude 84 °N), since 1963. By measuring the amplitude ratios of the extraordinary to ordinary backscattered waves, information is obtained about the electron number densities in the lower ionosphere.

Previous results have revealed regular diurnal and seasonal changes in the electron densities; and have shown that when the experiment is operated continuously, small solar cosmic ray (SCR) events, which would not otherwise be detected by ground-based methods, can be clearly recognized.

During August to October, 1968 the experiment was operated continuously for 24 hours a day, and digital results were recorded on an incremental tape recorder in a format compatible with computer analysis. An electron number density-height profile was deduced for each hour during the period. From this sequence of profiles, it was possible to study diurnal variations during days when the sun never sets, during days which had normal sunrise-sunset transitions, and during days of almost complete darkness. Superposed on these regular diurnal variations are several minor enhancements in electron number density, which appear similar to the enhancements which were previously reported in association with the small SCR event of 17 July, 1966; as well as the enhancement associated with the major event of 31 October, 1968.

The paper discusses the regular and irregular variations in electron density, and relates the disturbance variations with proton events detected by satellites.

1. INTRODUCTION

The Communications Research Centre (formerly Defence Research Telecommunications Establishment) has for a number of years been studying the polar D-region employing several techniques: (i) the partial reflection experiment; (ii) propagation of long radio waves over a number of high latitude paths; and (iii) measurement of the absorption of galactic radio noise at a frequency of 30 MHz (riometer experiment). This paper is concerned with the partial reflection experiment, employing a frequency of 2.66 MHz, at Resolute Bay, Cornwallis Island, N. W. T., which has been operating since September 1963.

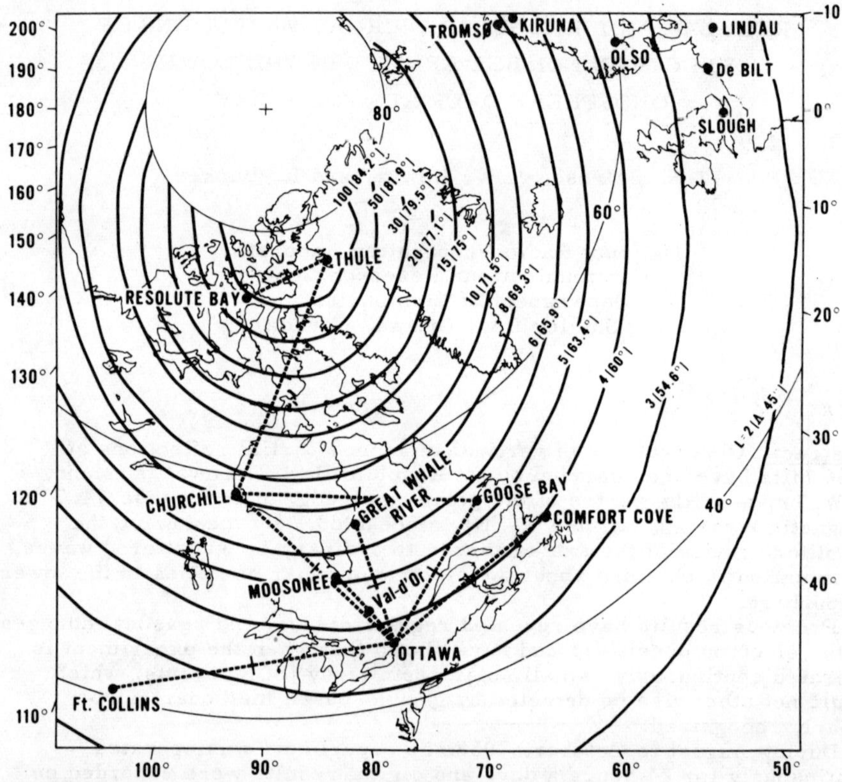

Figure 1. Northern polar map showing location of Resolute Bay. Geographic and magnetic invariant (at 100 km) latitudes are shown.

The location of Resolute Bay (Figure 1) is unique, in regards to observations by this type of experiment, not only because of the high magnetic latitude of the station (magnetic invariant latitude $\Lambda \sim 84^{\circ}$), but also because of the high geographic latitude (74.7 $^{\circ}$N). The high magnetic latitude makes it possible to study the polar D-region, essentially unperturbed by energetic particle precipitation associated with auroral and auroral absorption phenomena. The high geographic latitude places constraints on the observation of diurnal changes. In mid-winter the sun never rises (the solar zenith angle $\chi > 98^{\circ}$); in mid-summer the sun never sets ($\chi < 83^{\circ}$); only at the equinoxes can the transition from night ($\chi > 102^{\circ}$) to day ($\chi < 85^{\circ}$) be observed.

The partial reflection experiment is one of the best ground-based methods to study synoptically the D-region. By measuring the amplitudes of the extraordinary to ordinary backscattered waves, employing pulsed transmissions, information can be obtained about the electron number densities in the lower ionosphere.

Previous results (Belrose et al., 1964; 1966; 1967), some of which will be summarized in this paper, have revealed regular diurnal and seasonal changes in electron densities, and have shown that when the experiment is operated continuously small solar cosmic ray (SCR) disturbances can be recognized which would otherwise be undetected by ground-based methods (Hewitt, 1969). During August to October, 1968 the experiment was operated as continuously as possible for 24 hours a day, and digital results were recorded on an incremental tape recorder in a format compatible with computer analysis. An electron density-height profile was deduced for each hour during the period. From this sequence of data, it was possible to study diurnal variations during days when the sun never sets, during days which had normal sunrise and sunset transitions, and during days of almost complete darkness. Superimposed on these regular diurnal variations are several minor enhancements in electron density, which will be related to proton events detected by satellites; as well as the large enhancement associated with the major SCR event of 31 October, 1968. After a brief resume of previous work, these data obtained during the summer and fall of 1968 will be described.

2. PREVIOUS WORK

Observational data at Resolute Bay have been analyzed to investigate regular seasonal and diurnal changes. The curves in Figure 2 show the amplitude ratio of the extraordinary and ordinary wave amplitudes,

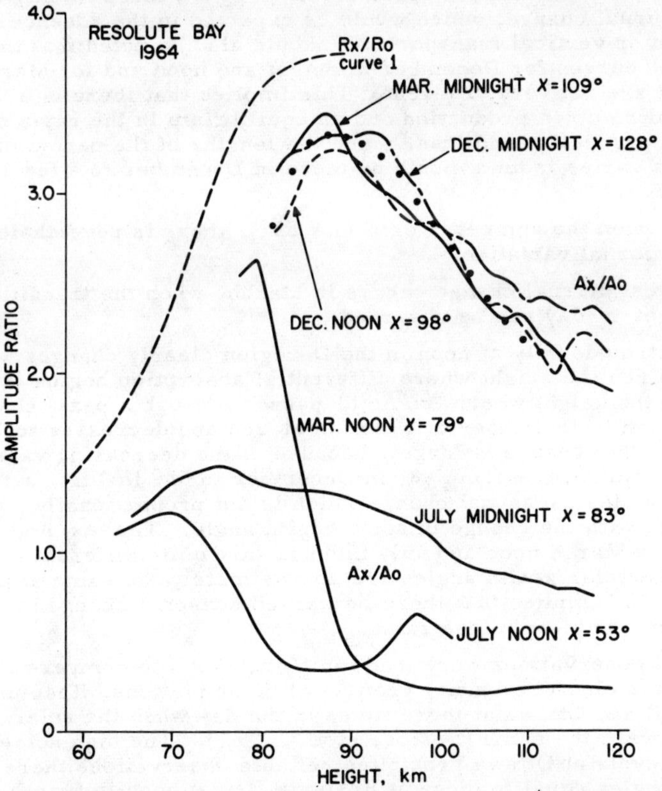

Figure 2. Seasonal variation in average ratio Ax/Ao against height at 2.66 MHz for midnight and noon at Resolute Bay in 1964. The dot curve shows results for December noon 1963.

Ax/Ao, measured as a function of height, for local noon and midnight in December, March and July 1964. Each curve represents the average of from 4 to 7 days. The various values of χ, the solar zenith angle, are marked. The dashed curve labelled Rx/Ro is the calculated reflection coefficient ratio, which depends on the collision frequency, gyro-frequency and wave frequency, and represents the amplitude ratio that would be measured in the absence of differential absorption. These curves have been calculated on the pressure-versus-height dependence given by the CIRA 1961 standard atmosphere (Belrose and Burke, 1964). A limitation of the experiment at Resolute Bay is the almost complete absence of detectable reflections from heights much below the height where Ax/Ao(h) is a maximum, and so the collision frequency cannot be estimated from the data. Previously the fact that the Rx/Ro(h) curve appears to be displaced to the left of the Ax/Ao(h) curves at low height (Figure 2) was taken to mean that the collision frequency over Resolute Bay was high (Belrose et al., 1966), but this is now believed to be at least partially incorrect. The displacement of the curves is now thought to result from underlying ionization or from (possibly) unreliable data from the lowest heights.

While the electron density-height, N(h) profiles have been deduced from the data given in Figure 2 (Belrose et al., 1967), these results will not be reproduced here; since the observed Ax/Ao(h) curves themselves show very clearly some interesting features of diurnal and seasonal changes, which for present purposes is all that is wanted:

(i) In December when the sun never rises ($\chi > 98^{\circ}$) there is only a small diurnal change, which would be expected in the absence of meridian or vertical transport. It should also be noted that the Ax/Ao(h) curves for December midnight and noon and for March midnight are not very different. This implies that there is a nocturnal electron production and an equilibrium in the rates of electron production and loss, since the lengths of the period of darkness varies from a month or more in December to a few hours in March.

(ii) In July, when the sun never sets ($\chi < 83^{\circ}$), there is nevertheless a clear diurnal variation.

(iii) The largest diurnal change occurs in March, when the transition from night-to-day can be observed.

(iv) The electron density at noon in the D-region clearly changes with season since the height where differential absorption begins (here taken as the height where Ax/Ao(h) passes through a peak) changes from 88 km in December to 79 km in March and decreases to 71 km in July. This change is largely because of the decreasing values of solar zenith angle from 98° in December to 53° in July, but there is a clear seasonal change which is not proportionally connected with the change in solar zenith angle. The Ax/Ao(h) curves for March noon and July midnight are quite different even though the solar zenith angles are approximately the same at these times. This implies that there is marked seasonal change in ionization or electron loss rates.

A series of observations were made in March 1965 to compare average daytime electron density profiles at three stations, Resolute Bay, Churchill and Ottawa at those times in the day when the solar zenith angles were the same (Belrose et al., 1967). The high noise and interference levels at Ottawa prohibited reliable observations there at solar zenith angles equal to those at Resolute Bay at noon in March ($\chi = 80^{\circ}$), and therefore the dependence on latitude had to be investigated by intercomparison between pairs of stations rather than between all three stations simultaneously. In Figure 3 the average Ax/Ao(h) curve

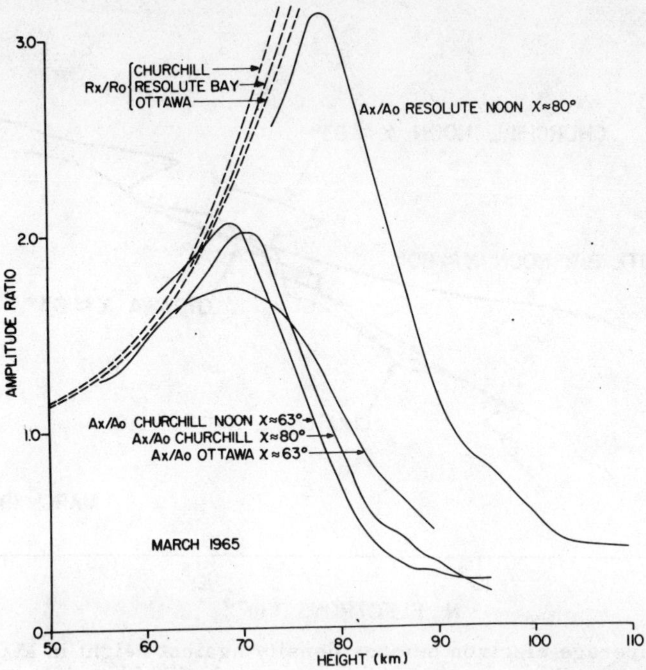

Figure 3. Average amplitude ratio Ax/Ao against height at 2.66 MHz in March 1965, for Resolute Bay noon ($\chi = 80^o$) compared with Churchill in the morning at the same solar zenith angle, and Churchill noon ($\chi = 63^o$) compared with Ottawa in the morning at the same solar zenith angle.

for Resolute Bay at noon ($\chi = 80^o$) is compared with the curve for Churchill in the morning at the same solar zenith angle; and the curve for Churchill at noon ($\chi = 63^o$) is compared with the curve for Ottawa in the morning at the same solar zenith angle. The electron densities deduced from these data are given in Figure 4. A study of this figure reveals that (i) the electron densities over Resolute Bay at heights below 80 km in March are very much smaller than those taken in the morning at the same solar zenith angle at Churchill; (ii) the electron density profiles over Churchill at noon and Ottawa in the morning at the same solar angle are not very different; and (iii) the electron density profile over Churchill does not change much for a change of χ from 80° to 63° (a larger change has been observed at Ottawa and at Resolute Bay).

The greatest differences between the profiles for Resolute Bay at noon in March and the profiles measured at lower latitudes are at low heights in the D-region. For heights less than 70 km the generally accepted source of ionization is galactic cosmic rays, the so-called CR-layer or C-layer, and so at these heights, assuming that the electron loss rate is not strongly dependent on latitude, the electron densities over Resolute Bay should have been larger, if anything, than those for lower latitudes because the cosmic ray flux is latitude dependent. The observational data are quite opposite to this expectation. However, a major limitation of the observational data at Resolute Bay is the absence of detectable echoes at low heights (< 70 km), and therefore acceptance of this observational fact must depend on a demonstration that the small electron densities are due to absence of ionization and not absence of detectable echo returns. In the following discussion we argue

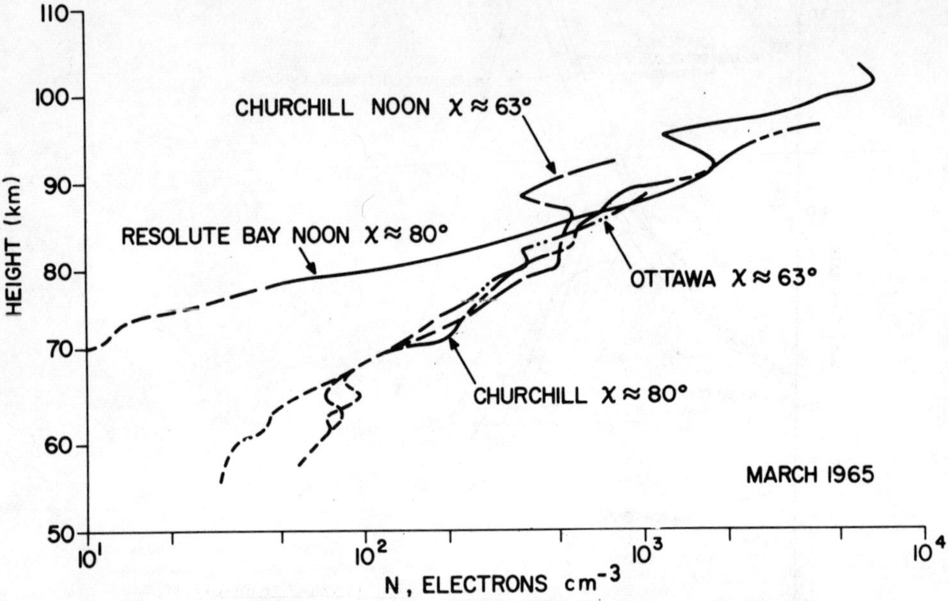

Figure 4. Average electron number density against height in March 1964 for Resolute Bay noon ($\chi = 80^{0}$) compared with Churchill in the morning at the same solar zenith angle, and Churchill noon ($\chi = 63^{0}$) compared with Ottawa in the morning at the same solar zenith angle.

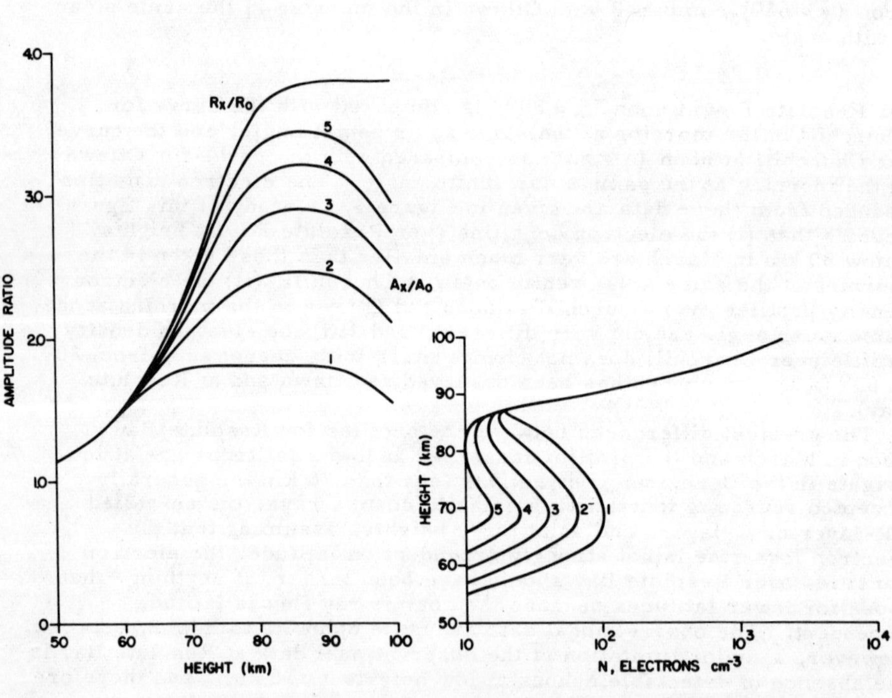

Figure 5. Model electron number density against height profiles and calculated Ax/Ao against height for 2.66 MHz at Resolute Bay, showing the change in ratio with changes in electron density in the underlying layer.

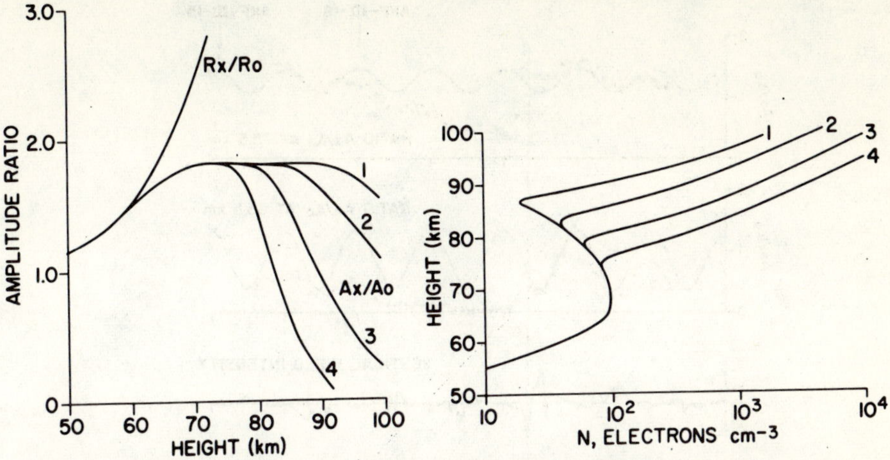

Figure 6. Model electron number density against height profiles and calculated Ax/Ao against height for 2.66 MHz at Resolute Bay, show the change in ratio with change in ionization in the overlying layer.

that the absence of ionization at low heights is real, and we further conclude that this situation is peculiar to March equinox.

In order to gain an insight into the presence, or absence of low lying ionization, ionization at heights below that from which detectable reflections are found, a series of model N(h) profiles were constructed, and Ax/Ao(h) was calculated for these model profiles. The various curves in Figures 5 and 6 show two sets of such model N(h) profiles and the calculated Ax/Ao(h) curves. It must be stressed that these are model profiles, for illustrative purposes only, and they have not been adjusted to agree with observational data; however the N(h) profiles in Figure 5 illustrate the general features of the seasonal change in ionization at midnight, and the profiles in Figure 6 illustrate the general features of diurnal change in summer (e.g., compare the calculated Ax/Ao(h) curves in Figures 5 and 6 with the observed data shown in Figure 2).

It is clear from these calculated curves that the magnitude of $(Ax/Ao)_{max}$ sets a limit on the amount of ionization present at low heights in the D-region. For $(Ax/Ao)_{max} \sim 3$, which is found in March (Figure 3), there can be little ionization at low heights, as is shown by Figure 5; whereas if $(Ax/Ao)_{max} \sim 1.5$, as it is for example at noon and midnight in July (Figure 3), the ionization at low heights can be quite substantial as is shown by Figure 6.

This concludes the review of regular diurnal and seasonal changes. We now consider departures from these regular diurnal changes. In July, 1966 observations were made (almost) continuously during a 10 day period to investigate in more detail the diurnal and inter-diurnal variations. The results in Figure 7 show the diurnal variation in the Ax/Ao ratio at two heights in the D-region, 75.5 and 85.5 km. The dashed curves show the computed ratio Ax/Ao based on 1964 noon and midnight values, given in Figure 2, and assuming a $\cos\chi$ dependence. The observations on 14 and 15 July 1966 agree with the assumed $\cos\chi$ dependence. The vertical dashed line through 1600 local time (2200 UT) on 16 July indicates the last partial reflection measurement which appeared to follow the $\cos\chi$ dependence. At 1900 local time (0100 UT on 17 July), the measured ratios were considerably smaller than the computed ratios indicating that an abnormal increase had occurred in the D-region ionization (the amplitude ratio at a fixed height is inversely proportional to the average electron density below the

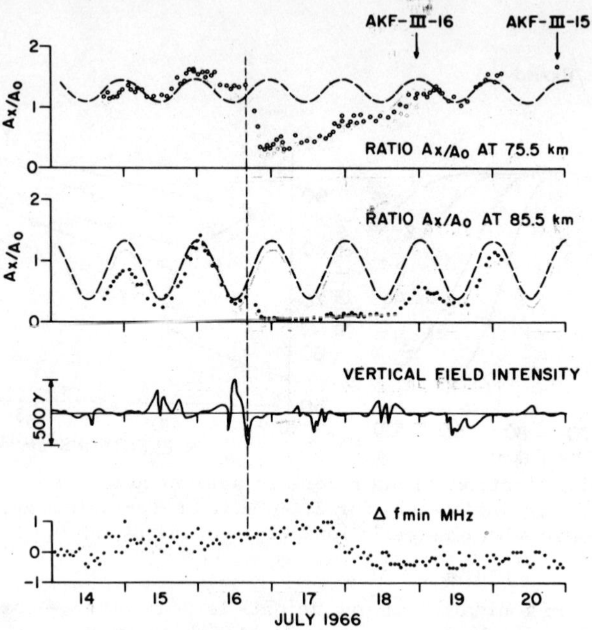

Figure 7. Variation of the amplitude ratio Ax/Ao at 75.5 and 85.5 km at Resolute Bay during 14 - 20 July 1966 (Central Standard Time). Also shown is the vertical intensity of magnetic field and the ionosonde parameter Δfmin.

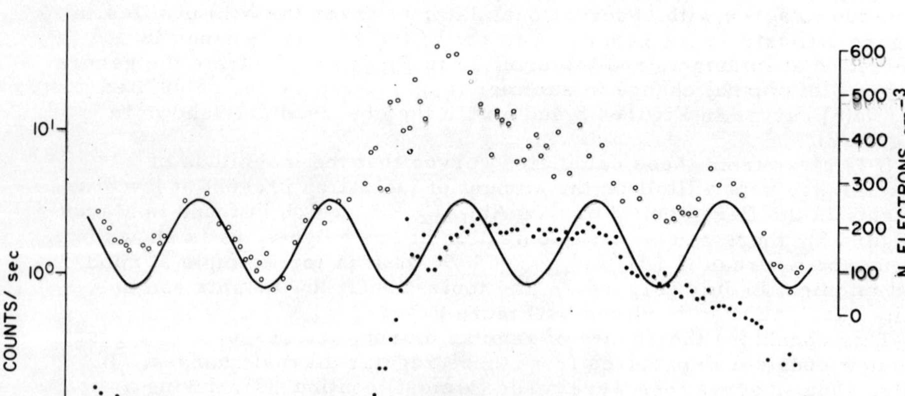

Figure 8. Variation of the average electron number density in the height range 71.5 - 79.5 km superimposed with the proton counting rate from Explorer 33 during 14 - 20 July 1966 (Central Standard Time).

reference height). The sharp drop in the ratio and the slow recovery over the next two days was taken as evidence that a SCR event was in progress, though it had not been detected by other ground-based techniques (e.g., transpolar VLF propagation). This identification, plus the fact that the event was of small magnitude, was subsequently verified by proton detectors in the Explorer 33 satellite (Hewitt, 1969). The intercomparison between ground-based and satellite data is shown in Figure 8, where the average electron density in the height range 71.5 to 79.5 km is displayed, along with the proton counting rates ($0.31 \leq Ep \leq 10$ MeV) from Explorer 33 during the interval 14 - 20 July 1966. The onset times indicated by the two parameters agree, within the time resolution of the measurements, and the duration of the event is approximately 3 days.

3. NEW OBSERVATIONAL RESULTS

During August to October, 1968 the partial reflection experiment was operated continuously for 24 hours a day, and digital results were recorded on an incremental tape recorder in a format compatible with computer analysis. This permitted a detailed study of diurnal and inter-diurnal variations for the first time over an extended period (previous data was all hand scaled from A-scan film recordings, a time consuming task which effectively limited the amount of data that could be scaled). An electron density-height profile was deduced for each hour during the period.

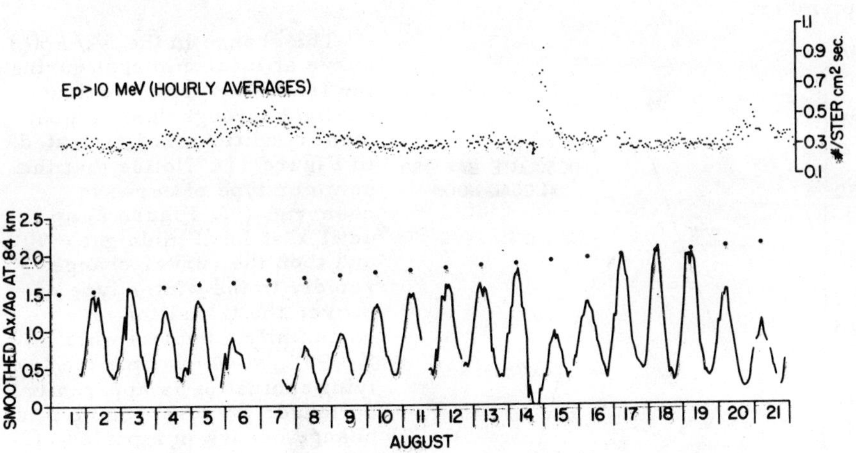

Figure 9. Variation of the amplitude ratio Ax/Ao at 84 km for 2.66 MHz observations at Resolute Bay during 1 - 21 August 1968 superimposed with the proton counting rate from Explorer 34 (IMP F) (Universal Time).

The continuous curve in the lower half of Figure 9 shows the diurnal change of the Ax/Ao ratio at 84 km. The bold dots are the estimated ratios which should have been measured at midnight, based on the long sequence of data and assuming a cos χ dependence. The increasing magnitude of the diurnal change in the Ax/Ao ratio with day of the month can be clearly seen, with the greatest change being observed in

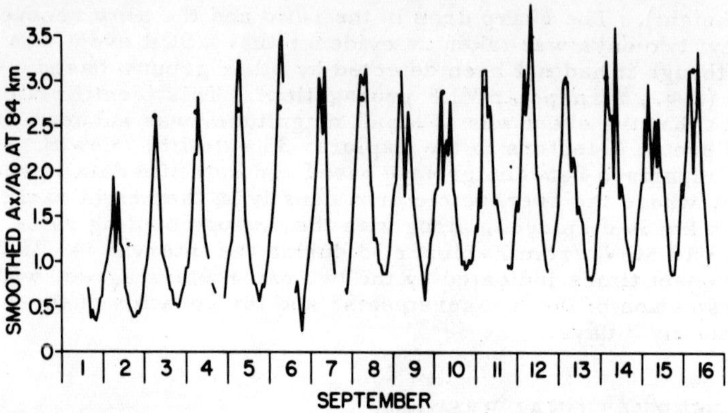

Figure 10. Variation of the amplitude ratio Ax/Ao at 84 km for 2.66 MHz observations at Resolute Bay during 1 - 16 September 1968 (Universal Time).

September, as shown in Figure 10. The diurnal change again becomes smaller and the minimum value of the Ax/Ao ratio increases as the season advances and the minimum value of χ at noon increases. This can be seen for the undisturbed interval 23 - 25 October 1968, in Figure 12; it must be noticed however that the reference height in this figure is 88 km rather than 84 km and so the values of the ratios in Figure 12 cannot be directly compared with those in Figures 9 and 10. The diurnal change in the Ax/Ao ratio is very much more irregular in October ($\chi_{min} = 86.4^o$ on 24 October) than it was during August and September.

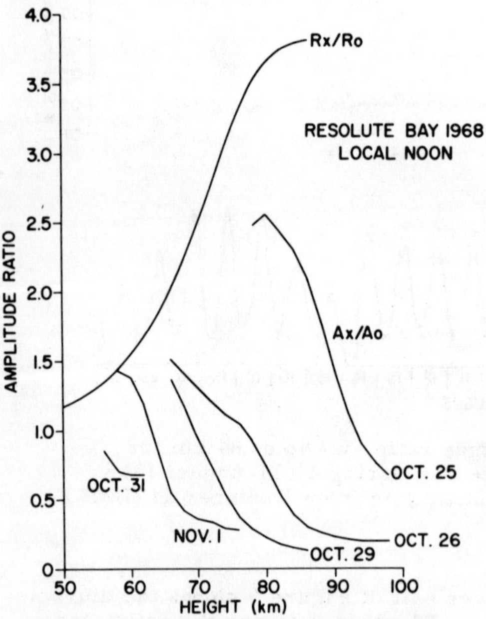

Figure 11. Amplitude ratio Ax/Ao against height at 2.66 MHz for days as marked in August and September, 1968 for Resolute Bay midnight.

The change in the Ax/Ao(h) curve at local midnight during the transition from continuous daylight through dusk to near night conditions is illustrated in Figure 11. Notice that the summer type of curve is observed (cf. Figure 2) up until χ at local midnight $\approx 90^o$, and then the curves change rapidly to the winter type of curve; the transition is essentially complete when $\chi \approx 98^o$, at which time the low lying ionization is apparently no longer present. Since this change occurs in a period of < 4 weeks it is presumably due almost entirely to the change in the solar zenith angle at the station, and illustrates that because of the high geographic latitude of Resolute Bay the diurnal change over dusk can be studied by observing the change with season at a fixed local time.

In the absence of disturbances the diurnal variations are remarkably regular, especially during summer months when $\chi < 90^{\circ}$. Because of this regularity small changes in the diurnal pattern are readily detected, and the data for August 1968 clearly show several minor disturbances that appear to be similar to the "event" associated with the small SCR event of 17 July 1966. The anomalies may be seen in Figure 9 during 6 - 13 August, 14 - 16 August and 20 - 21 August (equipment breakdown prevented the observation of the decay of this latter disturbance). They appear as a reduction in the Ax/Ao ratio, particularly the ratios at local midnight, with an onset that is either abrupt or more gradual, and the disturbance progressively disappears over a period of 2 - 7 days. These events were identified as SCR events at this meeting, and this has subsequently been verified by the satellite proton monitor on Explorer 34 (IMP F). Superimposed on the partial reflection data in Figure 9 is the observed counting rate for protons in the energy range $E_p \geq 10$ MeV (Williams, private communication). It is seen that all three events are associated with small increases in the proton counting rates; in fact the smaller than normal Ax/Ao ratios at local midnight on the nights of the 3rd and 4th of August, 1968 are quite clearly associated with a very small, but statistically significant increase in proton counting rates during this same period.

Since the 14 August 1968 event began abruptly in both satellite-proton and ground-based partial-reflection data, it is interesting to compare onset times. Hourly data only were available at the time of writing this paper. The ground-based partial reflection data indicated enhanced electron densities beginning abruptly at 1400 UT ± 1/4 hour, with a further abrupt and more drastic increase beginning at 1600 UT ± 1/4 hour. An importance one bright flare began at 1327 UT on 14 August 1968. The size of the event and its accompanying radio noise burst were strong. The earlier ionization increase was obviously due to solar X-ray flare (SXR) induced ionization, and before this event had decayed the SCR event began. The satellite proton data indicated an abrupt increase in counting rates at 1600 UT ± 1 hour, and therefore the onsets of the event at satellite and ground heights were the same within the time resolution of the data (one hour).

Partial reflection Ax/Ao data for the period 23 October to 1 November, 1968 are given in Figures 12 and 13. This period was one of the most active in the current solar cycle. The period 23 - 25 October was undisturbed, particularly on 25 October when the D-region was only weakly ionized at local noon. An importance one flare with weak X-ray emission began at 0150 UT on 26 October. This was followed by

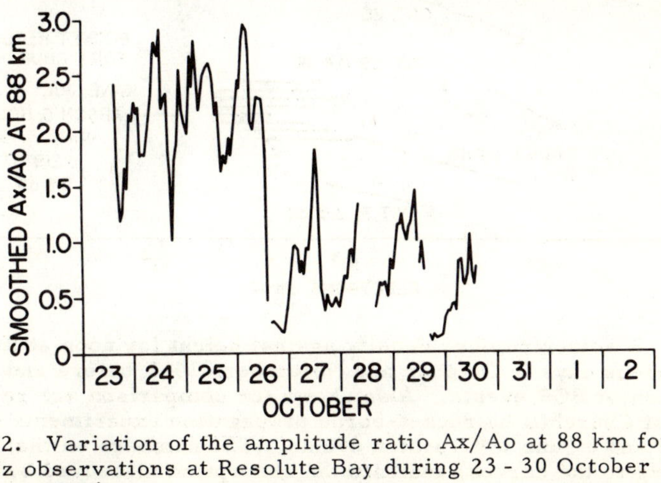

Figure 12. Variation of the amplitude ratio Ax/Ao at 88 km for 2.66 MHz observations at Resolute Bay during 23 - 30 October 1968 (Universal Time).

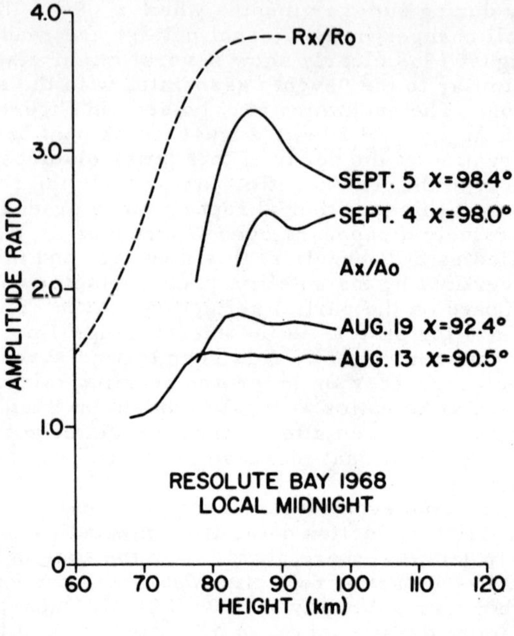

Figure 13. Amplitude ratio Ax/Ao against height at 2.66 MHz for noon at Resolute Bay for various days in October and November, 1968 before and during small and major SCR proton events.

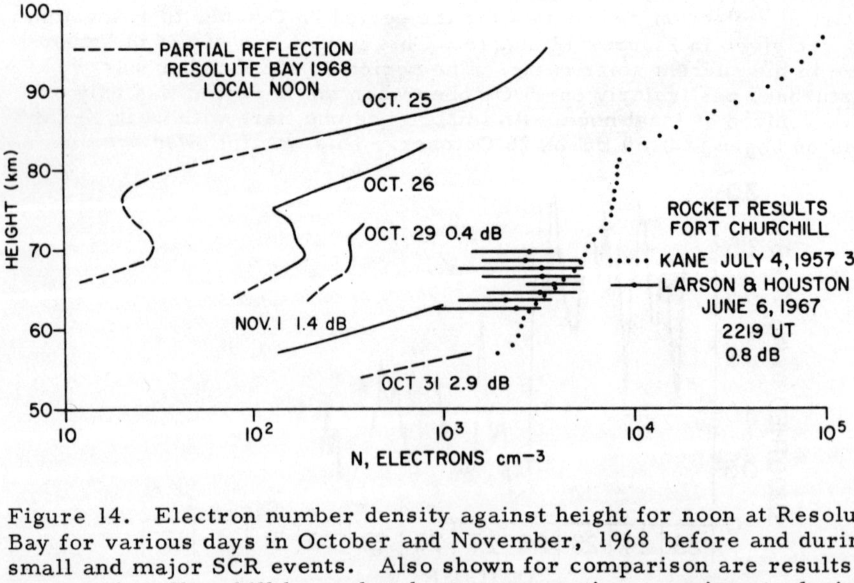

Figure 14. Electron number density against height for noon at Resolute Bay for various days in October and November, 1968 before and during small and major SCR events. Also shown for comparison are results measured at Churchill by rocket-borne propagation experiments during 4 July 1957 and 6 June 1967 proton events. The numbers on the curves (or referring to the curves) indicate the absorption measured by a 30 MHz riometer.

importance two bright flares at 1231 UT 27 October and 1218 UT 29 October. Both produced moderate to strong X-ray emissions. A sudden commencement magnetic storm began at 0909 UT 29 October. An importance three bright flare began at 2343 on 30 October, a great magnetic storm commenced at 0200 UT on 31 October and protons were detected by VELA satellite at approximately 1300 UT on 31 October 1968 (Space Disturbance Forecasts, ESSA Research Laboratories, Boulder, Colorado). Because the SCR event began when a magnetic storm was in progress, large SCR D-region ionization enhancements were observed as far south as Ottawa on 31 October 1968. The most marked feature of the Ax/Ao time record of Figure 12 is the sudden decrease in ratio at 1400 UT ± 1/4 hour on 26 October. The disturbance continued for several days, and became more pronounced, or perhaps a second disturbance began, on 29 October. On the 31 October and 1 and 2 November, the Ax/Ao ratios at 88 km were unmeasurably small at 88 km, consistent with a large increase in the D-region electron densities during the main disturbance, as is evident in Figures 13 and 14. It appears that the disturbances on 26 October and 29 October were SCR events, and were precursors of the 31 October 1968 event, though detailed satellite particle data are not yet available to permit a complete study of these disturbances.

The electron density profiles for midday on 25 October, a day of small electron density in the D-region, and 26, 29, 31 October and 1 November 1968 are given in Figure 14. The magnitude of the SCR disturbance can best be judged by the amount of polar cap absorption (PCA) measured by a 30 MHz riometer, which was 0.4 dB and 2.9 dB on 29 and 31 October, and 1.4 dB on 1 November, 1968. These N(h) results are compared with SCR D-region profiles measured by rocket experiments for events of 4 July 1957 (Kane, 1961) and 6 June 1967 (Larson and Houston, 1969). The magnitude of the PCA riometer absorption was 3 and 0.8 dB respectively. While a much better comparison needs to be made between profiles obtained at different latitudes, and at the same location by different methods, there does not seem to be any marked difference between the Resolute Bay and Fr. Churchill results, insofar as differences can be judged from the available data.

4. CONCLUDING REMARKS

Regular diurnal and seasonal changes in electron density over Resolute Bay, revealed by partial reflection data previously published, as well as by new observations made during August to October, 1968 have been summarized. A puzzling feature of these data is the apparent lack of a C-layer over Resolute Bay at noon in March. The fact that the ionization density at these times at heights below 70 km is very small, is further substantiated by a consideration of the value of $(Ax/Ao)_{max}$.

In the absence of SCR events the diurnal variations of ionization in the D-region is very regular, even more regular than diurnal variations which have studied at lower latitudes. Because of the great regularity it is very easy to detect quite small SCR events, and it is quite clear from the results presented here that the partial reflection experiment at Resolute Bay is the most sensitive ground-based method yet employed for detection of such disturbances. The experiment has its greatest sensitivity during nighttime hours when normal ionization densities at the altitudes of interest are very low.

The fact that precursors can be detected days before a main event by ground-based equipments has important geophysical implications, since such observations might indicate the operation of acceleration processes in the flare region, as well as warn about the possible occurrence of a major event. It could also give information, together with satellite data, of the kinds of particles that produce the earliest phases of SCR events.

Acknowledgments. The authors are indebted to Dr. D. J. Williams for the Explorer 34 (IMP F) data. They wish also to thank those colleagues who developed, constructed and operated the digital recording equipment that made possible the continuous acquisition of data at Resolute Bay.

REFERENCES

Belrose, J. S. and M. J. Burke, Study of the lower ionosphere using partial reflections, 1. Experimental technique and methods of analysis, J. Geophys. Res., 69, 2799-2818, 1964.

Belrose, J. S., L. R. Bode and L. W. Hewitt, Physical properties of the polar winter mesosphere obtained from low frequency propagation and partial reflections, Radio Science J. of Res. NBS/USNC-URSI, 68D, 1319-1323, 1964.

Belrose, J. S., L. R. Bode and L. W. Hewitt, A preliminary investigation of diurnal and seasonal changes in electron number density over Resolute Bay as observed by partial reflections, in Electron Density Profiles in Ionosphere and Exosphere, edited by J. Frihagen, North-Holland Publishing Company, Amsterdam, pp. 37-47, 1966.

Belrose, J. S., I. A. Bourne and L. W. Hewitt, A preliminary investigation of diurnal and seasonal changes in electron distribution over Ottawa, Churchill and Resolute Bay, in Proc. of Conf. on Ground Based Radio Wave Propagation Studies of the Lower Ionosphere, Def. Res. Board (Canada), pp. 167-188, 1967.

Hewitt, L. W., Ionization increases associated with the small solar proton events of 5 February 1965 and 16 July 1966, Can. J. Phys., 47, 131-134, 1969.

Kane, J. A., Re-evaluation of ionospheric electron densities and collisional frequencies derived from rocket measurements of refractive index and attenuation, J. Atmospheric Terrest. Phys., 23, 338-347, 1961.

Larson, L. E. and R. E. Houston, A measurements of the effective recombination coefficient in the lower ionosphere, J. Geophys. Res., 74, 2402-2406, 1969.

OPTICAL EMISSION OVER THE POLAR CAP

B. P. Sandford

Air Force Cambridge Research Laboratories
L. G. Hanscom Field, Bedford, Massachusetts 01730, USA

Abstract. A review of the typical morphology and behavior, in contrast to infrequent or unusual storm events, of night sky optical emissions at latitudes greater than 50°, with emphasis on the polar cap region at latitudes above 70°. The general aspects which have a bearing on the production and maintenance of the polar ionosphere are stressed. The main topics of discussion are the discrete visual aurora (night-peak, day-peak and polar cap discrete aurora), diffuse auroral emissions (mantle aurora and 6300 Å polar cap, noon sector and auroral oval enhancements), proton excited emission and a brief mention of some other emissions (sodium, infrared, ultra-violet, airglow).

1. INTRODUCTION

In this chapter the general morphology and behavior of optical emission from the high latitude night sky is reviewed. We shall primarily discuss the behavior of night sky optical emission (excluding twilight behavior) in the polar cap region, here considered as the region at corrected geomagnetic latitudes greater than 70° and I shall briefly consider the area between latitudes 50° and 70° where much of the classical auroral phenomena is observed. Emphasis will be placed on those general aspects of the optical emissions which relate to the production and maintenance of the polar ionosphere.

Our knowledge of upper atmosphere optical emissions in the polar cap is somewhat limited. Until the advent of the IGY there were very few stations where systematic scientific observations of the night sky had been made. Since that time a few stations have continued reliable and well calibrated observations of the optical night sky emissions using such instruments as interference filter photometers, photographic spectrographs and all-sky cameras. The use of polar orbitting satellites has enabled the difficulty of access to be overcome to some extent. Research aircraft are now flying more frequently into this region and they are helping greatly to build up information where it would otherwise be impossible to obtain.

It is now well established that auroral phenomena have a close relationship with the earth's magnetic field. There are, however, a multitude of magnetic field models (geomagnetic or centered dipole, eccentric dipole, corrected geomagnetic, invariant, surface field intensity) all of which have been used in studies of auroral phenomena. At very high latitudes the differences between these magnetic field models is very significant, making it difficult to relate different studies without converting to a common model. In this review we shall use the corrected geomagnetic coordinates described by Hakura (1965) which take account of non-dipole components of the magnetic field. High latitude phenomena under magnetic control have been found by Fukushima (1965), Zhulina (1968) and many others to be well ordered in these coordinates. The inter-comparison of data taken in different years must be made with caution because there are significant variations related to effects arising from the 11 year solar cycle. Another possible source of confusion may also arise from significant Universal Time variations originating from the diurnal changes in the tilt of the geomagnetic axis with respect to the ecliptic. As the geomagnetic axis does not pass through the earth's center, asymmetries arise between the north and south polar regions which can cause difficulties in the comparison of data (cf. Matsushita, 1968). These factors make it difficult to present a coherent picture of the optical emissions over the polar cap.

While the discrete visible aurora is a more spectacular phenomenon which gives rise to localized patches of sporadic ionization, the steady and weak optical emissions related to persistent ionization and recombination processes are equally important in the overall problem of the maintenance of the polar ionosphere. Daylight and twilight conditions prevail during most of the summer hours at these high latitudes, limiting night sky observations to winter and equinoctal months. We have consequently almost no knowledge of the summer season behavior of optical emissions. During the winter months, however, it is possible to observe for up to 24 hours of the day.

The importance of optical effects in relation to the ionosphere have been described by Omholt (1960), Thomas (1967), King (1967), and others, and are illustrated by the following examples. It is generally accepted that in the night sky the intensity of the first negative bands of ionized nitrogen, with the brightest bands at 3914 and 4278 Å, is directly proportional to the production rate of ionization by corpuscular bombardment. The 6300 Å line of atomic oxygen can be an important indicator of the F-region recombination rate because it is excited by the dissociative recombination of O_2^+ ions. However, significant emission at this wavelength can also arise from thermal electron excitation and from collisions with secondary electrons produced by energetic particle precipitation. The ionospheric information that the high latitude optical emissions can provide will be briefly discussed after the general morphology of the emissions is reviewed.

2. STORM BEHAVIOR

As this conference is primarily concerned with the typical or normal behavior of the polar cap ionosphere as opposed to the spectacular and unusual events we should determine what is to be considered as typical behavior.

In the auroral oval, quiet auroral forms are observed at all times. Originating around local midnight an elementary polar magnetic substorm with its closely related auroral substorm occurs on the average about every four hours with a duration of about one hour (Davis and Sugiura, 1966). The quiet auroral forms become active with the onset of a substorm, expanding to both higher and lower latitudes (Akasofu, 1965; Feldstein and Starkov, 1967). These auroral substorms are so frequent that they must be included as typical behavior. The magnitude of a substorm can be measured by the size of the planetary magnetic K-index, Kp, and even more precisely by local magnetic activity or Davis and Sugiura's (1966) electrojet activity index. One can set an arbitrary limit on the substorm size, by setting an upper limit on the magnetic K-index (say K = 5), above which the activity is intense and infrequent enough to be considered as an unusually disturbed event. The frequency of auroral substorms is independent of the solar cycle (Lassen, 1967; 1969; Afonia and Feldstein, 1967; Sandford, 1968), and it appears that they are part of the magnetospheric substorm (Akasofu, 1968) which is an inherent property of the earth's magnetosphere.

In particular, the largest auroral substorms are frequently associated with a solar flare. These substorms follow a magnetic sudden commencement (SC) and are relatively rare in relation to the nightly storms so that they cannot be considered as typical events. In a study of such storms Weill (1963) observed, in the period from plus 15 to plus 45 minutes after the SC, a very active and brilliant aurora which developed from the auroral oval to within 10° of the corrected geomagnetic pole. The SC storms are, therefore, an important part of the polar cap storm phenomena.

Another solar flare initiated storm event is the polar cap absorption event (PCA) caused by solar protons and alpha particles which produce a distinctive polar-glow aurora described by Sandford (1963; 1967) and Weill (1962). In large PCA an extensive glow is generated over the whole polar cap down to a latitude of about 60°.

As the occurrence of discrete aurora at latitudes above 80° is negatively associated with magnetic activity (Lassen, 1961; Davis, 1963; Feldstein, 1966; Sandford, 1968) the polar region can be most disturbed at times when the auroral oval is least disturbed. At times of very quiet magnetic activity Fukushima (1965), Fukushima and Hirasawa (1964), and Matsushita (1968) find the magnetic activity in the polar cap can be quite disturbed. This condition must also be considered as a typical feature of the polar cap.

It is quite common in correlation studies to employ the planetary K-index, but as this is not derived from high latitude observatories, it is not always a good parameter for polar cap studies. This factor is clearly illustrated by Silverman et al. (1962), who find the correlation coefficient between Kp and the local magnetic K-index to be greater than 0.9 for stations at mid- and auroral latitudes, but is reduced to 0.7 for a polar cap station. Sharp et al (1966), Feldstein and Starkov (1967), and Sandford (1968), also find the local magnetic K-index to be a much better parameter than Kp and one would hope that those studying the polar cap will consider better indicators of polar activity than the Kp index.

Until we have a better knowledge of the polar cap it will be quite difficult to establish definite criteria for disturbed and quiet behavior.

3. GENERAL MORPHOLOGY

The optical emissions at high latitudes can be divided into a number of groupings based on spatial distribution, time variations, magnetic dependence, etc. and the relative positions of the regions of maximum intensity or occurrence of some of these phenomena are presented in Figure 1. Such a diagram can be used because it is generally accepted that the emissions are produced in regions which are fixed relative to the earth's magnetic field and the earth-sun line and under which pattern the earth rotates once each day. The general morphology and behavior of these emissions will now be discussed.

3.1 Discrete Visual Aurora

The discrete visual aurora is the optical emission which can be observed by the naked eye and has well defined structure such as arcs,

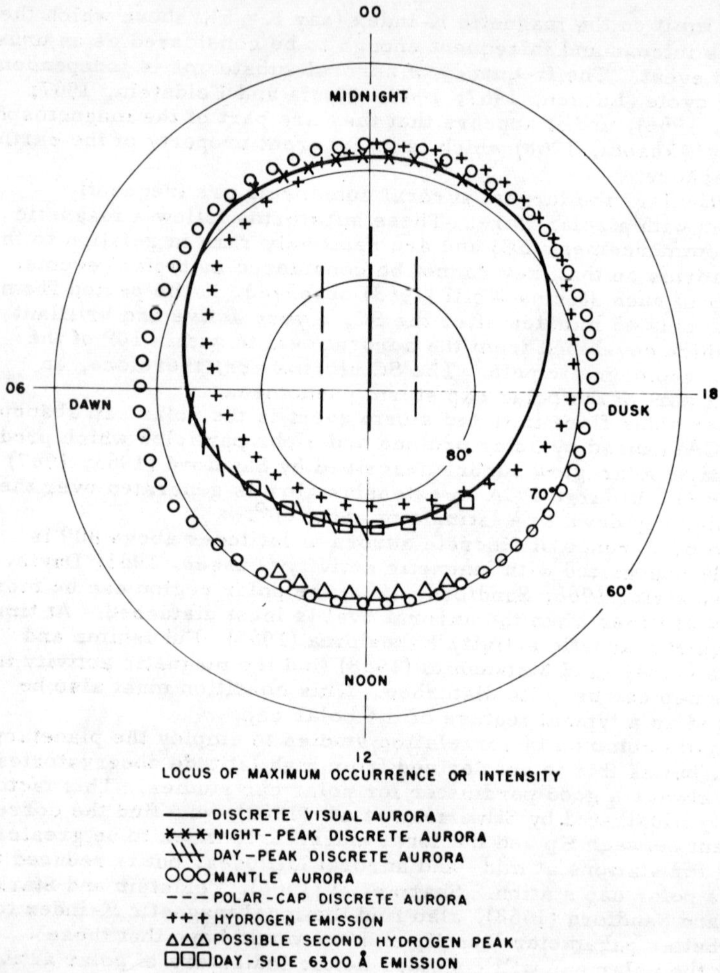

Figure 1. The relative positions in the high latitude night sky of the maximum occurrence or of maximum intensity of the classes of optical emissions discussed in this text. Coordinates are corrected geomagnetic latitude and time.

bands, rays, patches, etc. As there is no reason to arbitrarily use the visual limit, non-visual aurora with localized structure should also be included, but there is some evidence (Sandford, 1964; 1968) that the major proportion of non-visual emissions do not arise from discrete forms, but from types of diffuse emission like the mantle aurora. The discrete visual aurora has been studied most extensively (Chamberlain, 1961; Akasofu et al., 1966; McCormac, 1967) and some aspects that are of interest in relationship to the maintenance of the polar ionosphere will be briefly covered in the following. The discrete visual aurora occurs most frequently along the auroral oval, an annular zone about the corrected geomagnetic pole that lies near 68° latitude at midnight and 78° latitude at noon as shown in Figures 1 - 4. The occurrence of discrete aurora within the oval does not vary with the solar cycle, but the width of the oval is spread over a much larger latitude interval at solar maximum than at solar minimum (Feldstein and Starkov, 1967; Sandford, 1968).

The characteristics of the discrete aurora are sufficiently different in some regions that it is useful to establish a nomenclature to aid in the discussion. A distinct maximum of occurrence is centered in the oval near local midnight and activity related to this maximum we call the "night-peak" discrete aurora (Sandford, 1968). A second section of high occurrence with distinctive behavior occurs just before noon and the name "day-peak" discrete aurora is used to identify it. Over the polar cap at all latitudes higher than those of the poleward edge of the auroral oval there occurs a discrete aurora with characteristic behavior that is called the "polar cap" discrete aurora.

(i) Night-peak discrete aurora. The morphology of the night-peak discrete aurora has been described in a very systematic way by Akasofu (1965). In time of quiet magnetic activity, homogeneous arcs lie around the night side of the auroral oval. An auroral substorm

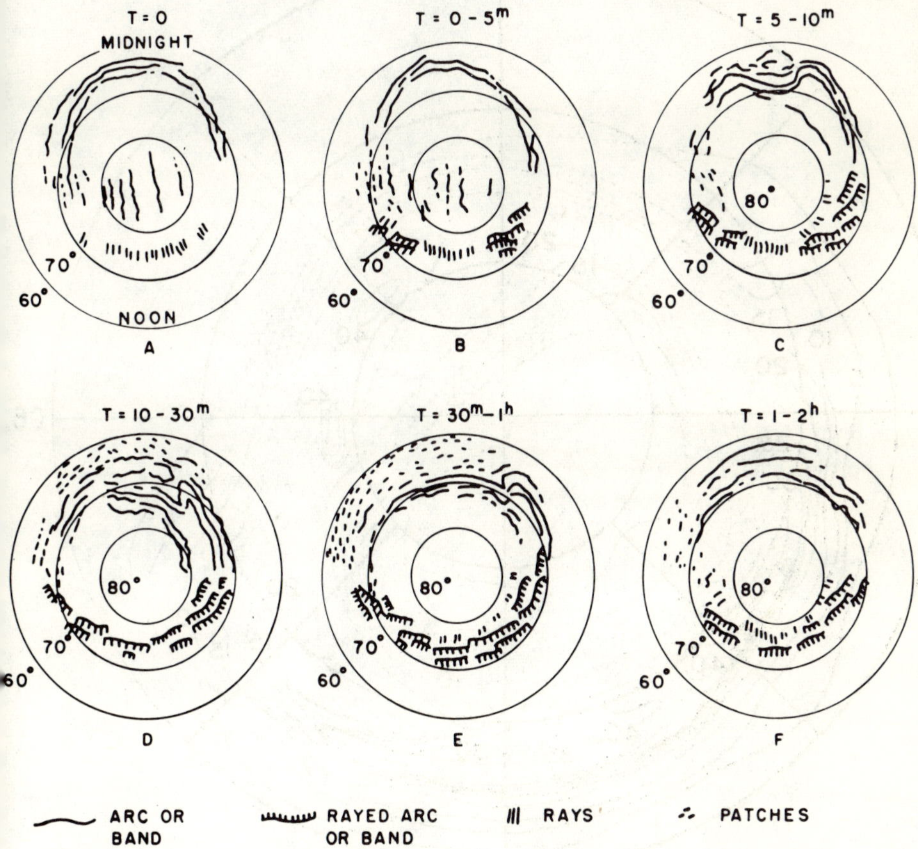

Figure 2. A schematic diagram of the discrete visible aurora illustrating the development of a typical auroral substorm from onset time T. The aurora returns to the original quiet condition in time T = 2-3 hours. Coordinates are corrected geomagnetic latitude and time. (Akasofu, 1965) and (Starkov and Feldstein, 1967b).

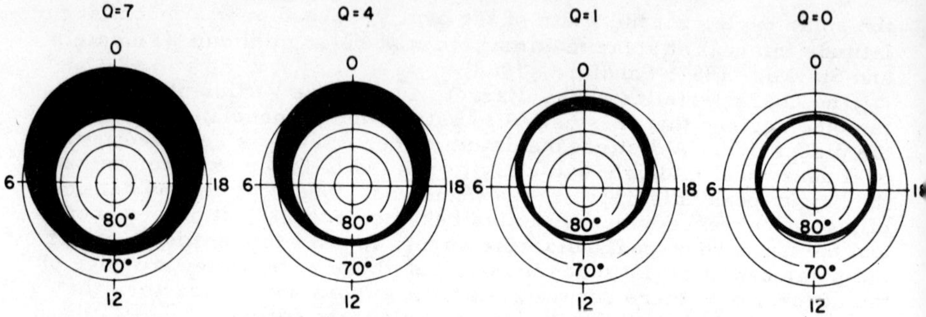

Figure 3. The mean band within which the discrete visible aurora is observed in magnetic calm (Q = 0), moderate (Q = 4), and disturbed (Q = 7) conditions. (Starkov and Feldstein, 1967a).

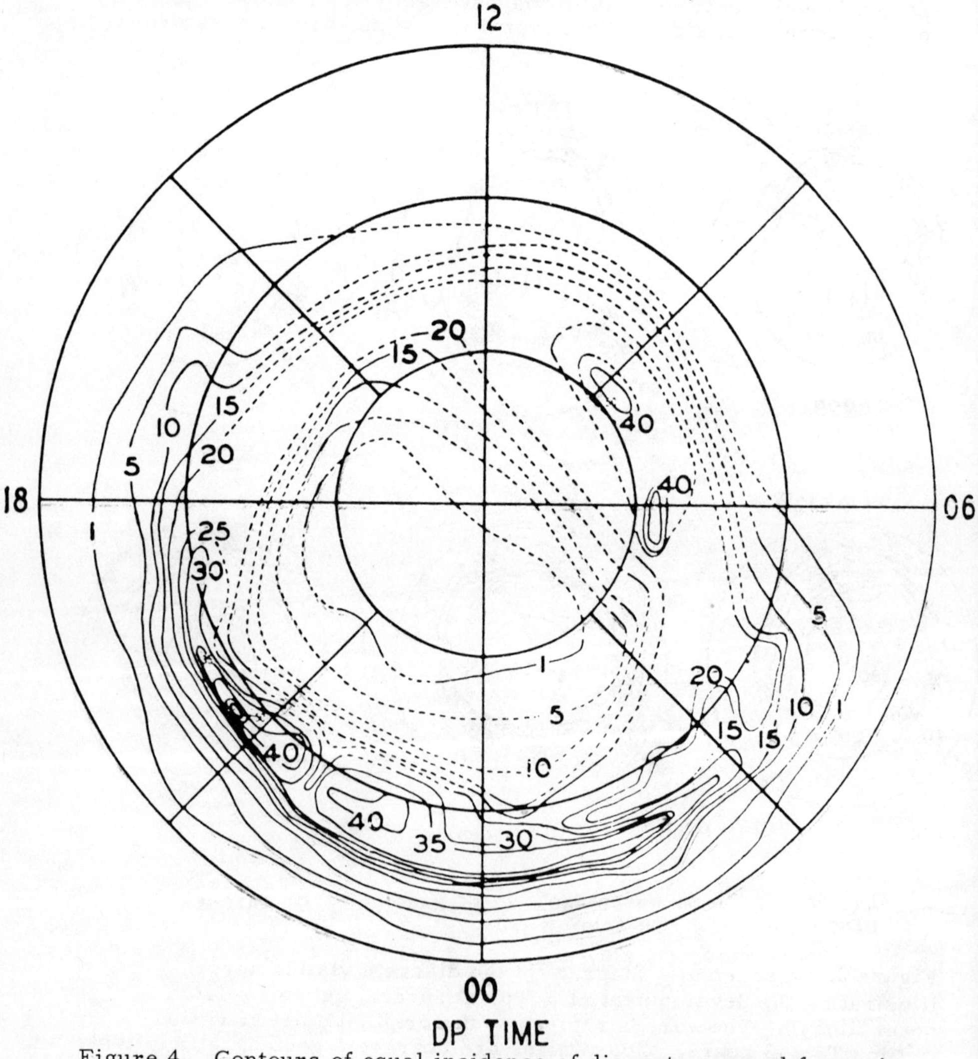

Figure 4. Contours of equal incidence of discrete auroral forms in the IQSY in coordinates of dipole latitude and dipole time from Alaskan stations. (Stringer and Belon, 1967).

develops about every four hours on the average (Davis and Sugiura, 1966) and the discrete aurora then becomes disturbed in the midnight region as illustrated in Figure 2, becoming more active and intense, and developing over a region both poleward (Akasofu, 1966) and equatorward (Pudovkin et al., 1968) of the original quiet arcs. The activity expands until the whole oval from dusk to dawn is filled with active and intense aurora with enhanced activity completely around the auroral oval. A substorm normally lasts about one hour, slowly dying down after it reaches its peak.

The total area covered by an auroral disturbance is greatest when the magnetic activity is greatest. An example of this behavior is presented in Figure 3 from Starkov and Felstein (1967a) which shows the region defined by the mean north and south boundaries of the discrete aurora for various levels of magnetic disturbance.

(ii) Day-peak discrete aurora. Detailed maps of auroral occurrence like that shown in Figure 4 exhibit a distinct maximum of occurrence in the morning around 8 to 10 hours magnetic time and between corrected geomagnetic latitudes 75 - 80° (Lassen, 1967; Stringer and Belon, 1967). Another maximum is sometimes evident at these latitudes shortly after noon, but it has not been established as a permanent feature (Lassen, 1967; 1969).

The day-peak discrete aurora has structure, diurnal variations and a magnetic dependence that differs significantly from the night-peak discrete aurora. Many authors have noted these differences (Lassen, 1967; Stringer and Belon, 1967; Sandford, 1968; Pudovkin et al., 1968) which lead to the concept some time ago of the "inner auroral zone" (Lassen, 1959) and its relationship to the night-peak discrete aurora is not clear. The magnetic field lines from this region map out to the neutral point on the front side of the magnetosphere, leading to the idea that this phenomenon may be associated with the direct entry of solar wind particles into the magnetosphere (Pudovkin et al., 1968; Sandford, 1968).

The typical appearance of the day-peak aurora has been described by Starkov and Feldstein (1967b) and is illustrated in Figure 2. During quiet magnetic conditions faint ray structures and fragmented arcs are observed. Following an auroral substorm in the night-peak aurora, ray structure develops into rayed arcs and there will be some increase in activity if the substorm is sufficiently large. It has been suggested by Sandford (1968) that the day-peak phenomenon has a negative correlation with magnetic disturbance, the aurora disappearing from the morning sky as magnetic activity increases. However, at the same time, with the increasing magnetic disturbance, the night-peak aurora develops completely around the oval, invading the region of maximum occurrence of the day-peak aurora. As yet there is no positive way to distinguish between the two auroral types and thus to test this concept.

Starkov (1968) reports that the height of most of the day-peak aurora is between 145 and 180 km, while Lassen (1969) reports a most frequent occurrence at 120 km, however, both observe a large spread of heights in comparison to night-peak aurora. A close correlation of day-peak aurora and riometer absorption is sometimes observed by Mukherjee (private communication) indicating that quite energetic precipitation sometimes occurs. He also notes evidence of a great variability in the precipitating electron energy spectrum.

Although the frequency of occurrence of the day-peak and the night-peak aurora are comparable, Starkov (1968) has noted that the energy of precipitation of the day-peak aurora is ten times less than that of the night-peak aurora. In both 1959 and 1963 Sandford (1968) observed that the region of peak occurrence of the day-peak aurora was not associated with a peak in intensity of 3914 and 5577 Å emissions from all sources. He concluded that the overall contribution of emissions from discrete aurora in this locality was only a few percent of the total emission which mainly arises from diffuse types of aurora like those discussed later in connection with the important enhancements of the 6300 Å emission.

(iii) Polar cap discrete aurora. At corrected geomagnetic latitudes greater than 80° the occurrence of discrete aurora is relatively low, Figure 4 (Lassen, 1967; Stringer and Belon, 1967; Sandford, 1968). The aurora in this region consists of weak isolated rays and rayed arcs which are aligned, such that the lengths of the arc points toward the sun's azimuth as shown in Figure 5 from Gustafsson (1967). These aurora exhibit a strong negative correlation with magnetic activity (Davis, 1963), particularly on the day-side of the polar cap as shown in Figure 6 from Sandford (1968).

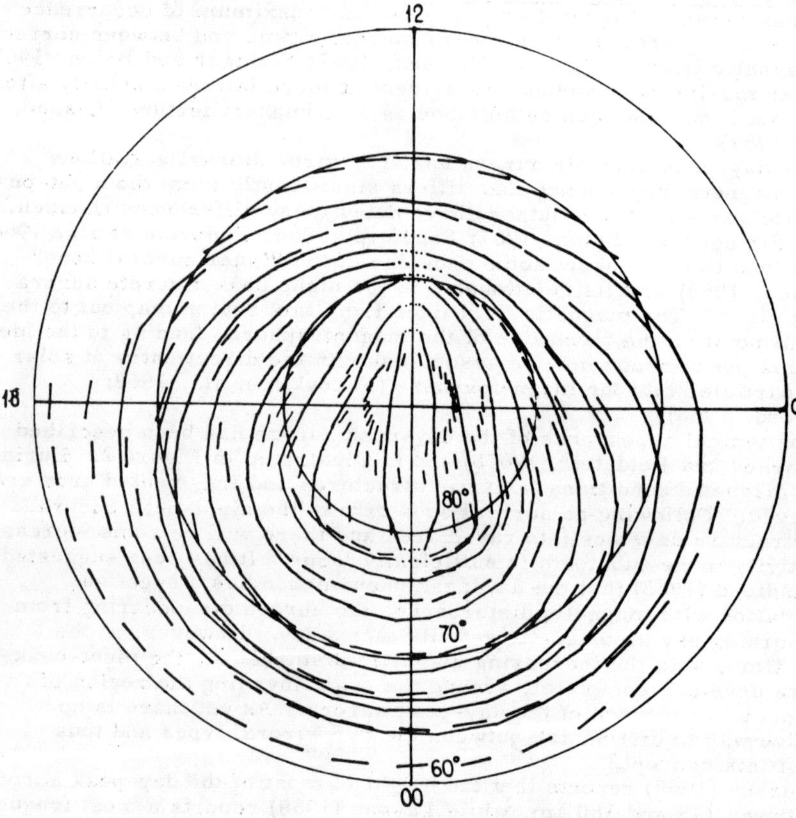

Figure 5. The observed directions in which auroral arcs lie in corrected geomagnetic latitude and time. (Gustafsson, 1967).

The height of the polar cap discrete aurora has not been reported, however, Buchau (private communication) finds considerable variability in the height of the polar cap sporadic Es. Passes by a satellite over the polar cap indicate that this region is often very active with regard to less than 10 keV electron precipitation (Hoffman and Evans, 1968), and Hoffman (1969) reports a great variability in the precipitating electron energies from one event to another. These and other features

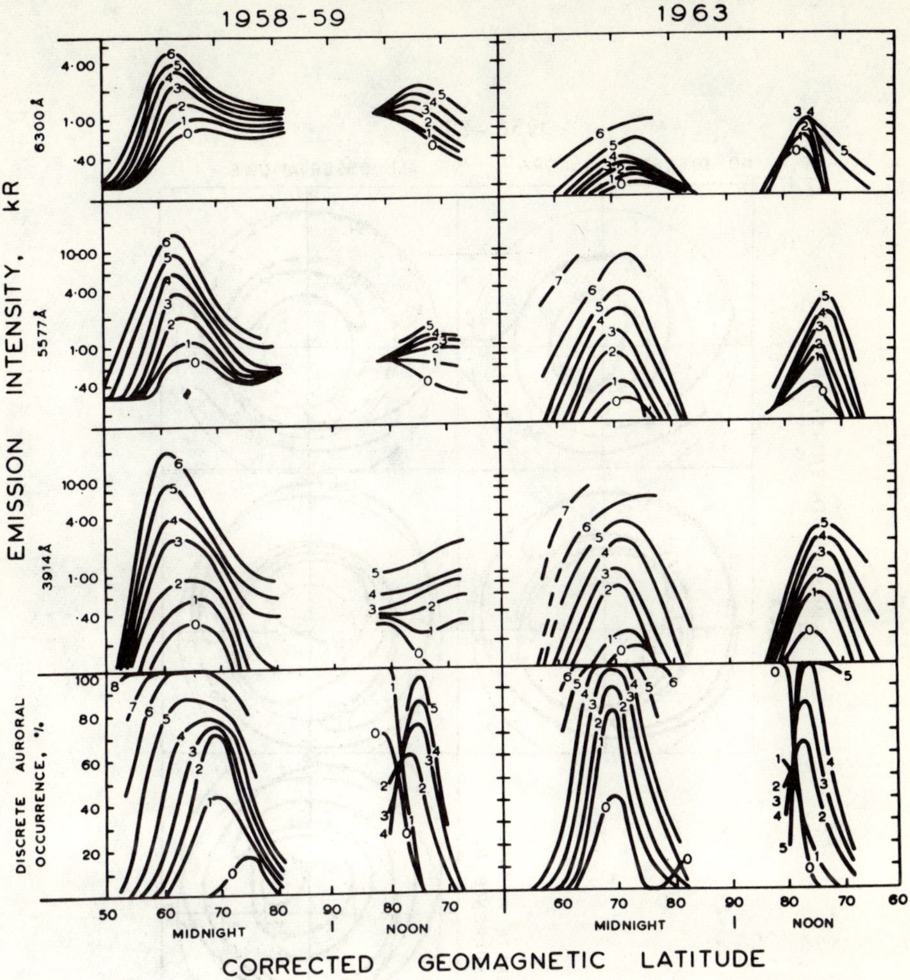

Figure 6. The percentage occurrence of discrete visual aurora and median intensities of 3914, 5577 and 6300 Å emissions as a function of local magnetic K-index and corrected geomagnetic latitude along the noon-midnight corrected geomagnetic time meridian in 1958-59 and 1963. (Sandford, 1968).

discussed by Sandford (1968) indicate that the polar cap discrete aurora is in many ways similar to day-peak discrete aurora.

3.2 Mantle Aurora

In an extensive study of spectrographic data from a number of Antarctic stations in a year of high solar activity Sandford (1964) observed that approximately four times as much auroral emission at 3914 and 5577 Å was generated from a diffuse non-visual glow as was emitted by discrete visible aurora. This emission was generated in a belt which he called the "mantle" aurora, shown in Figures 1 and 7a at approximately constant corrected geomagnetic latitude between

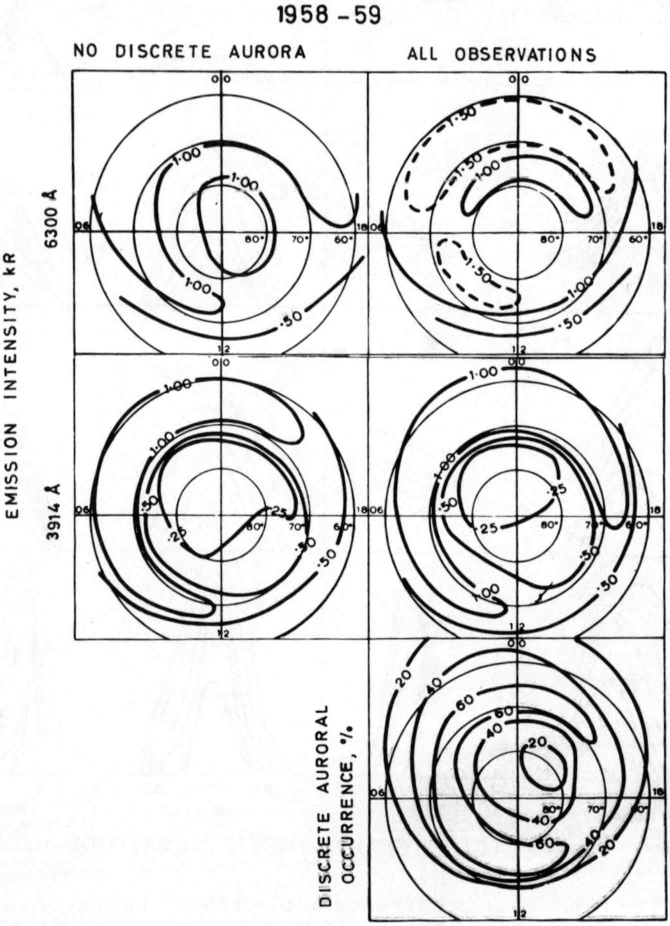

Figure 7a. The diurnal variation of discrete auroral occurrence and intensities of 3914 and 6300 Å emissions in the absence of discrete aurora and for all observations in 1958-59. Contours of percentage occurrence and median intensity in kR in corrected geomagnetic latitude and time. (Sandford, 1968).

60° and 70° with a maximum intensity near midnight and a second more intense maximum in the morning hours. The zone of mantle aurora was subsequently shown by Hartz and Brice (1967) to be a region where energetic electrons of > 20 keV are precipitated.

In a further study from a year of low solar activity Sandford (1968) observed the belt of mantle auroral activity to have shifted about 5° nearer to the pole and to be reduced in intensity by a factor of four as shown in Figure 7b. This reduction made it comparable in intensity

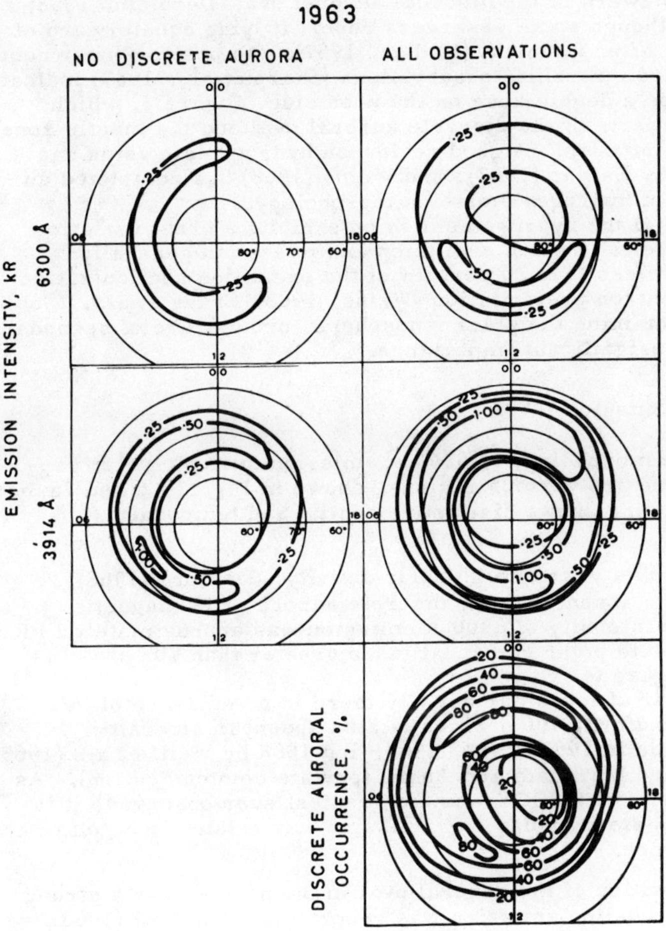

Figure 7b. The diurnal variation of discrete auroral occurrence and intensities of 3914 and 6300 Å emissions in the absence of discrete aurora and for all observations in 1963. Contours of percentage occurrence and median intensity in kR in corrected geomagnetic latitude and time. (Sandford, 1968).

to the discrete visible aurora which exhibited no overall change in intensity over the same period.

The intensity of the mantle auroral emission is closely related to magnetic disturbance, there being a good correlation between the intensity of the morning mantle aurora and the degree of magnetic disturbance in the midnight sector of the auroral oval (Sandford, 1968).

3.3 Proton Excited Emission

Hydrogen emission generated by precipitating protons is observed primarily in an oval zone, shown in Figure 1, which lies equatorward of the discrete auroral oval before midnight. After midnight it probably lies poleward of the discrete auroral oval (Derblom, 1968; Weins, 1968) although some observers report it lying equatorward of the auroral oval after midnight (Eather, 1967). Its position near noon is not well defined. Satellite observations (Sharp et al., 1967) indicate that there may be a double zone on the noon side, Figure 1, which could be counterparts of the discrete auroral oval and the mantle zone of electron precipitation. A good review on hydrogen emission has been presented by Eather (1967), and Weins (1968) has completed an extensive study of hydrogen emission morphology.

The intensity of the hydrogen emission exhibits a great variation with the solar cycle. Proton excitation and ionization generally represents much less than 20 percent of the excitation and ionization generated by electron precipitation (Weins, 1968; Evans et al., 1966). In terms of maintaining the polar ionosphere, protons are of secondary although not of insignificant importance.

3.4 Red-Line Emission

Studies of the morphology of 6300 Å emission over the polar cap indicate that there are various regions, shown in Figures 6 and 7a-b, in which the red-line shows distinctive spatial and temporal characteristics.

(i) During 1959 a year of high solar activity, Sandford (1968) showed that in the absence of any discrete aurora or of magnetic disturbance, the intensity of 6300 Å emission was approximately 1 kR across the complete polar cap at latitudes greater than 70° as illustrated in Figure 6.

In 1963, a year of low solar activity there is no evidence of this enhancement (Sandford, 1968) which is also absent in aircraft observations made in 1963, 1964, 1967, and 1968 by Weill et al. (1965), Eather (1969), and Markham and Anctil (private communication). As the solar maximum of the IGY was the greatest ever observed, it is possible that this steady polar cap emission was related to a solar controlled influence.

(ii) At the position of the auroral oval in the noon sector a strong enhancement of red-line emission was reported by Sandford (1968). This coincides with a region of high F-region electron density (Oguti and Marubashi, 1966) which Maehlum (1968) identifies with a region of "soft" electron precipitation with energies less than 2 keV. Burch (1968) reports that this belt of soft precipitation lies poleward of the auroral oval, however, his results are taken in the summer and are compared with winter auroral data. Maehlum (1968) shows evidence of large seasonal variations in the position of the oval (discussed later) indicating that the auroral oval may lie 5° nearer the pole in summer months which would invalidate Burch's conclusion.

In December 1967 an Air Force Cambridge Research Laboratories aircraft made a series of flights across the day-side auroral oval making ionosonde and optical measurements (Buchau et al., 1969). In December 1968 further flights were flown giving similar, but more comprehensive data (Buchau, Whalen and colleagues, private communication). As the aircraft flew from 70 to 75° latitude towards the auroral oval, Markham and Anctil (private communication) report a broad region of 3914 and 5577 Å emission which might tentatively be identified as the mantle aurora. At a corrected geomagnetic latitude

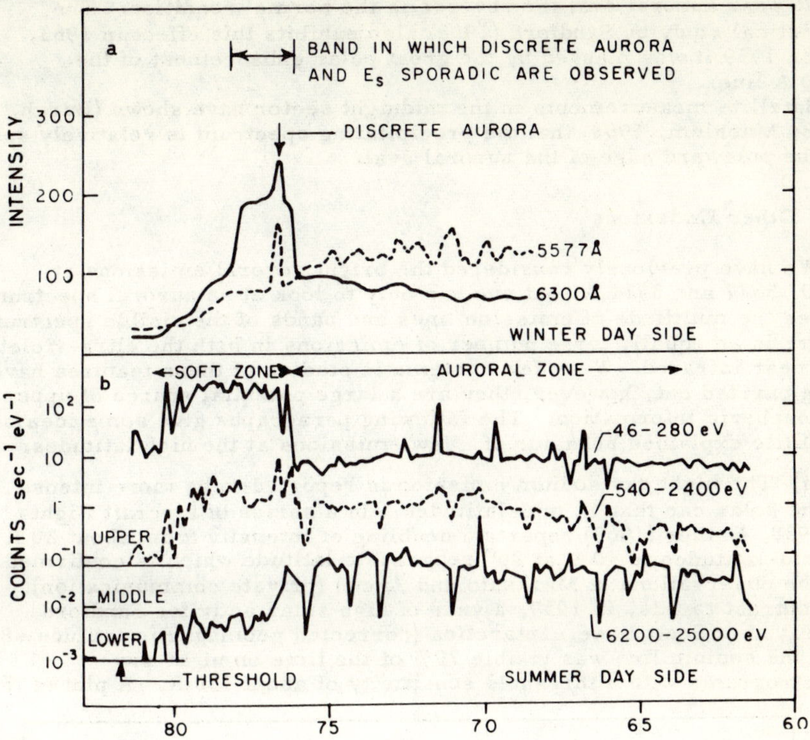

Figure 8. a) A typical sketch of the intensity variation with invariant latitude (similar to corrected geomagnetic latitude) of the 6300 and 5577 Å emissions along the corrected geomagnetic noon meridian in mid-winter showing the band of 6300 Å emission at the position of the auroral oval from Markham and Anctil (unpublished). b) A typical measurement of electron precipitation near the noon corrected geomagnetic meridian in the mid-summer illustrating a well defined band of "soft" electron precipitation in the general region of the day-side auroral oval. (Burch, 1968).

between 75 - 80° a well defined band of 6300 Å emission appeared which may be identified as the day-side auroral oval. Within this band discrete visible aurora and sporadic E at a height of 130 - 140 km was observed. As the aircraft flew beyond the band into the polar cap the optical emissions fell in intensity to extremely low levels. A typical scheme of their results is sketched in Figure 8a.

These results suggest that a relatively narrow band of soft precipitation is occurring within the auroral oval on the day-side and that the soft precipitation has a sharp low and high latitude boundary like that observed by Burch (1968) and reproduced in Figure 8b. Also within the day-side oval there is electron precipitation of a localized nature with a great variability in its electron energy spectrum as discussed in the section of day-peak aurora.

(iii) From dusk until dawn around the poleward edge of the auroral oval Eather (1969), Markham and Anctil (private communication) and Weill et al. (1965) report from airborne studies that 6300 Å emission is frequently more intense than the 5577 Å emission. Within the night-peak auroral oval the reverse is the normal condition. The statistical study by Sandford (1968) also exhibits this effect in 1963, but in 1959 it was masked by the great polar enhancement of the 6300 Å line.

Satellite measurements in the midnight sector have shown (Burch, 1968; Maehlum, 1968) that the precipitating spectrum is relatively soft on the poleward edge of the auroral oval.

3.5 Other Emissions

We have previously considered the bright auroral emissions at 6300, 5577 and 3914 Å, but one has only to look at an auroral spectrum to see the multitude of emission lines and bands of the visible spectrum. There is an equally large number of emissions in both the ultra-violet and near infrared. Very few systematic studies of these features have been carried out, however, they are a large potential source of upper atmospheric information. The following paragraphs give some idea of the little explained behavior of a few emissions at the high latitudes.

(i) The night sky sodium emission is reported to be more intense in the polar cap than at mid-latitudes. In a series of aircraft flights in 1968, Eather (1969) reports a doubling of intensity from about 20 R at mid-latitudes to 40 R at 80° geographic latitude which is confirmed by the observations of Markham and Anctil (private communication). In contrast to this, in 1959, a year of high solar activity, Sandford observed at Scott Base, Antarctica (corrected geomagnetic latitude = 81°) that the sodium line was visible 70% of the time on night-sky spectrograms with a threshold sensitivity of about 150 R. A plot of the

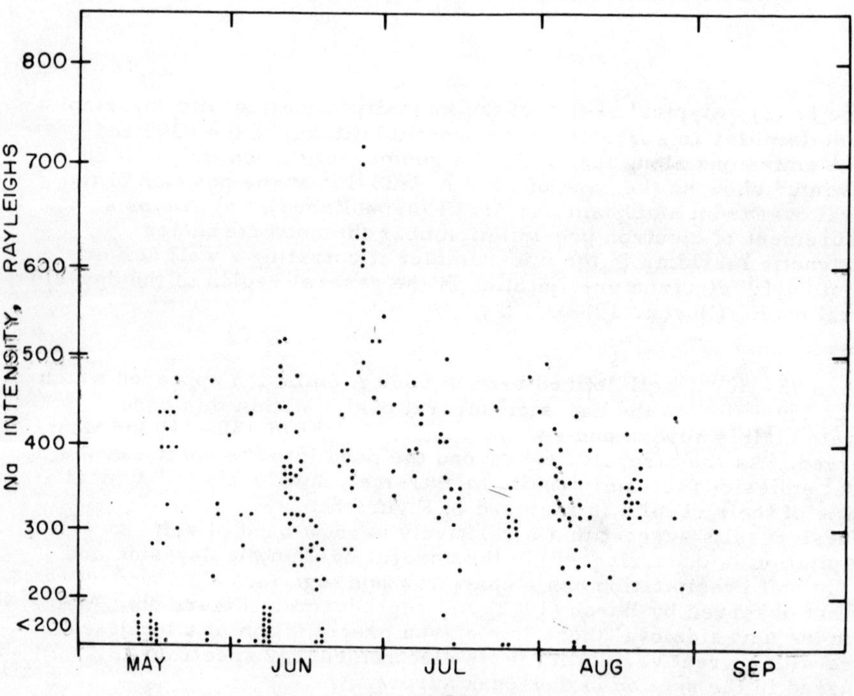

Figure 9. The intensity of polar cap night-sky sodium emission on clear nights in 1959 when the solar depression exceeded 18°, at Scott Base, Antarctica, corrected geomagnetic latitude 81°.

zenith intensities measured on clear nights when the solar depression was greater than 18° (which excludes twilight enhancements) is shown in Figure 9. The highest intensity recorded is about 700 R with a median intensity of 300 R. No correlation with magnetic activity, auroral activity, or time were observed. There is an impression of rapid enhancements followed by a slow decay over a week or so. In 1963 the sodium line was rarely observed on the same spectrograph. This is a further example of an emission exhibiting a great enhancement in the polar cap in the very active year of 1959 followed in 1963 and in the current solar maximum of 1968 (Eather, 1969) by no evidence of such an enhancement.

(ii) An observation of the 1.27 micron infrared atmospheric band of $O_2{}^1\Delta$, by Noxon (private communication) during an aircraft flight across the night time auroral oval produced some interesting results. He observed that the atmospheric bands were enhanced below auroral arcs and that the total energy of emission in these bands alone equalled the total energy, as deduced from 3914 Å measurements, of the precipitating particles producing the aurora. There is adequate energy in the ozone around 85 km to provide an energy source for the infrared emission, but it is not clear how the auroral disturbance could couple into the mesosphere to give rise to this effect. There is evidence that the aurora does have considerable influence on the mesosphere below it, because Fogle and Haurwitz (1966) have observed noctilucent clouds (80 km altitude) to change form or break up about 10 minutes after an auroral form develops above them.

(iii) Satellites have been instrumented to observe the ultra-violet emissions, particularly the Lyman-Birge-Hopfield bands at 1200 - 2200 Å, which are intense in aurora, but cannot be observed from the ground due to atmospheric absorption. On a limited number of passes over the polar cap in November 1965, Johnson et al. (1967) report a UV intensity of at least 300 R across the polar cap for much of the time at wavelengths 1250 - 1800 Å, but the source of this emission was not indicated by them. As UV measurements can be made in daylight the satellite monitoring of UV auroral emissions should prove of great value in polar studies.

(iv) As the airglow emissions at 5577 and 6300 Å are prominent at equatorial and mid-latitudes, they might be expected also to predominate in the polar cap regions at latitudes higher than the auroral oval. From observations of O'Brien and Taylor (1964), Weill (private communication), Sandford (1968), and Eather (1969), it can be said that in the quietest of conditions the average intensities of the 5577 and 6300 Å emissions are comparable to their mid-latitude intensities. The intensities in quiet conditions also exhibit the solar cycle related variations of intensity similar to the mid-latitude variations reported by Hernandez and Silverman (1964). There are some indications (Markham and Anctil, Noxon, private communications) that the polar cap 5577 Å and 1.27 micron emissions may sometimes be noticeably weaker than their mid-latitude intensities for the same night. Further measurements are required to demonstrate the importance of true airglow processes in the polar cap.

4. IONOSPHERIC-OPTICAL RELATIONSHIPS

When energetic electrons are precipitated into the upper atmosphere one of the major energy loss processes is ionizing collisions with the neutral atmosphere. The intensity of 3914 Å emission in a bright stable auroral arc has been used by Walker and

Rees (1968), Stolarski (1968) and Rees et al. (1967) to show that electron densities in excess of 10^6 ions cm^{-3} can be generated over the height range 100 - 250 km. These results are confirmed experimentally by many observers such as Omholt (1960) and Ulwick (1967).

The contribution of this localized auroral ionization to the overall ionization of the polar ionosphere cannot be simply determined. By separating the measurements made in the absence of discrete aurora from those made in the presence of discrete aurora, it can be estimated (Sandford, 1968) that in both 1959 and 1963 the median intensity at 3914 Å emitted by night-peak discrete aurora is at least 250 R around the auroral oval. Its height will be assumed to be between 100 - 110 km (Egeland and Omholt, 1967). From tabulations such as those given by Dalgarno et al. (1965) it can be shown that 250 R at 3914 Å corresponds to precipitating electron flux of the order of 0.3 erg cm^{-2} sec^{-1}, an ion production rate of 10^3 cm^{-3} sec^{-1} and an electron density in excess of 10^4 cm^{-3}. The electron density can only be considered as order of magnitude for a variety of reasons; it is derived from a time average of intense, localized and rapidly varying discrete aurora; winds may play an important role in spreading the ions; the effective recombination coefficient is not precisely known (cf. Omholt, 1960). These results do suggest that a significant amount of E-region ionization can be maintained in the auroral oval by discrete aurora.

The long term average intensity of diffuse emissions is much greater than that of the day-peak discrete aurora which is thus unlikely to contribute a large part in the overall maintenance of the polar ionosphere. An individual display of day-peak aurora does generate localized ionization which is observed as sporadic Es (Buchau et al., 1969). The polar cap discrete aurora is generally faint and less frequent in occurrence than the day-peak aurora and, therefore, may not play a big role in producing polar ionization.

The diffuse auroral emissions generally cover large areas of the sky and are therefore of great importance in the maintenance of the ionosphere. The 3914 Å emission in the mantle aurora in 1959 had an average energetic electron flux of about 1 erg cm^{-2} sec^{-1}, four times greater than the energy of night-peak discrete aurora. The ionization, particularly in the morning hours is mostly generated below 90 km, it being a significant source of ionization density. This ionization is produced at a height which will give high absorption as evidenced by the riometer measurements in the region of mantle aurora (Hartz and Brice, 1967). The much lower intensity of mantle aurora since 1959 means that it is of less importance than discrete aurora as a source of overall ionization, but it will still give rise to important D-region ionization and consequent radio frequency absorption.

Another important diffuse emission is that at 6300 Å from the forbidden transition of atomic oxygen. This line is a good indicator of F-region activity because this transition is very strongly quenched by atmospheric collisions below 150 km (Hunten and McElroy, 1966). The 6300 Å emission is one of the brightest auroral emissions in the visible spectrum above 150 km (Stolarski, 1968), the red-line being generated by four significant processes; direct excitation by secondary electrons; thermal electron excitation which is very efficient when electron temperatures exceed 3200 °K; dissociative recombination of O_2^+ ions and cascading from the ^1S level (green-line) of atomic oxygen. The relative importance of these processes as a function of altitude is shown by Rees et al. (1967) and Stolarski (1968). It is difficult to experimentally determine the relative importance of these processes at any particular time, but with the aid of ionosonde, rocket and satellite data this problem is being solved.

The 6300 Å emission over the polar cap was of the order of 1 kR in 1959 and it fell to only tens of rayleighs by 1963. Maehlum (1969) has noted that during the IGY the F-region electron density was high across the polar cap while ground and satellite studies since 1962 indicate a big depletion of F-region ionization across the polar cap, called the "polar cavity", which suggests a parallel between 6300 Å emission and the F-region ionization. In the winter of 1962-63, Thomas and Andrews (1968) showed that of one in three passes by Alouette I, the F-region ion density was enhanced across the polar cap and from the same satellite Sato and Colin (1969) note that the polar enhancements occur whenever the planetary K-index exceeds 3. They suggest electron precipitation of the order of 1 keV would explain these enhancements. Further evidence of a dynamic excitation process such as particle precipitation is given by Gordon (1968) who reports that the 6300 Å emission showed no correlation between samples taken 6 minutes apart. These results suggest that low energy electron precipitation across the polar cap cannot be ruled out as a source of the red-line and of the polar cap F-region.

Because there had been a lack of evidence of any persistent energetic particle precipitation Cole (1966a, b) put forward the concept that this emission was generated by thermal electrons heated by conduction from a hot tail region of the magnetosphere. He assumed that the source of this heat originated from the interaction between the solar wind and the magnetopause. If a thermal electron temperature of 3200 °K and 4000 °K could be maintained above 200 km about 0.1 kR and 10 kR respectively of 6300 Å emission would be generated (Taylor, 1967). As the thermal excitation at 6300 Å is negligible for electron temperatures below 3200 °K the absence of red emission since the IGY may only indicate that electron temperatures of the polar cap F-region have not since been consistently maintained above this temperature.

The region of 6300 Å enhancement near noon, as discussed earlier, coincides with a region of enhanced F-region electron density and with a band of soft particle precipitation. This emission and the F-region electron density very likely arises from the energetic electron precipitation.

The enhanced region of 6300 Å emission around the poleward edge of auroral oval coincides with a region of soft particle precipitation with an energy influx near midnight in the range of 0.1 - 0.5 ergs cm^{-2} sec^{-1} (Eather, 1969). From the results of Dalgarno et al. (1965), Rees et al. (1967), and Stolarski (1968) it seems that this flux could maintain an F-region electron density in the region of 10^4 - 10^5 cm^{-3}. An F-region enhancement in this general region is reported by Thomas and Andrews (1968) and it suggests that "soft" precipitation may occur completely around the oval connecting with the noon enhancement region discussed above.

Weill et al. (1965) reported a series of arc-like structures in 6300 Å emission in the same region which may be likened to the series of peaks and troughs of F-region ionization reported by Muldrew (1965) from the Alouette top-side sounder.

In summary it can be said that there is considerable evidence of particle precipitation in the high latitude regions which may be sufficient to maintain the polar ionosphere. More correlated studies between optical measurements, ionosondes and satellite records of particle precipitation are needed to indicate the precise relationship.

5. U. T. VARIATIONS AND MAGNETOSPHERIC RELATIONSHIPS

One phenomenon that has been shown to be present in the F-region electron density and in the "soft" electron precipitation is a Universal Time variation. This is believed to arise from the fact that the earth's magnetic axis is inclined to the geographic axis, giving rise to diurnal variations in the magnetospheric configuration (Figure 10). Observations made at different magnetic longitudes at the same local time will show the boundaries of the magnetically related phenomena to occur at different magnetic latitudes as the magnetic field distortion changes through the day. As the inclination of the earth's axis to the ecliptic changes over the year the magnetospheric configuration will also exhibit a seasonal variation.

Ionospheric phenomena, particle precipitation, magnetic activity and auroral absorption have been reported by Maehlum (1968; 1969), Fukushima and Hirasawa (1964) and Zhulina (1968) to exhibit large diurnal and seasonal changes at high latitudes, particularly on the day-side, which can be attributed to the changes in magnetospheric geometry as the inclination of the magnetic axis to the solar wind varies. There have been almost no reports from purely optical measurements of these variations, but few people have looked for them. In their determination of the auroral oval position Feldstein and Starkov (1967) report no evidence of such variations. The "soft" particle precipitation does seem to be closely related to day-peak aurora and 6300 Å emission so that a U. T. variation and seasonal latitude shift of these emissions are to be expected and should be looked for.

The theories that explain the behavior of geophysical phenomena at high latitudes are still in an evolutionary stage and one is referred to reviews such as Piddington (1968) and Shabansky (1968). Further relevant articles are referred to in my paper (Sandford, 1968) which briefly considers magnetospheric-auroral relationships. It is now recognized that the origin of much of the polar cap phenomena are directly related to magnetospheric behavior and I shall briefly indicate the relationship between the magnetosphere as illustrated in Figure 10 and the optical phenomena at high latitudes (Figure 1).

The mantle auroral phenomena is thought to originate from an initial magnetospheric substorm which develops in the midnight sector of the cusp region, B_1, of Figure 10 (Rothwell and Wallington, 1968). This substorm produces energetic particles that drift around the earth in semi-stable orbits at the outer boundary of the trapped radiation belts A, and precipitate to produce mantle aurora. The night-peak discrete aurora is believed to originate from instabilities producing local acceleration in the region B_1 which occur as a result of the magnetospheric substorm. The instabilities in region B_1 spread around the magnetosphere into the day-side region B_2 giving rise to the auroral oval. The day-peak discrete aurora and the noon region 6300 Å enhancement may possibly arise from solar wind interaction with the magnetosphere at the neutral points D_1 (summer) and D_2 (winter). The very different angle of attack between the solar wind and the magnetosphere over the summer and winter poles is clearly shown in Figure 10. The diurnal variation of the angle of attack due to the inclination between the magnetic and spin axes gives rise to the U. T. variations. The polar cap enhancement of 6300 Å in the IGY period and the polar cap region of low activity inside the auroral oval (polar cavity) in the years following the IGY, map out to region E which is the quietest part of the magnetosphere. Interactions at the magnetopause F and between the relatively active plasma sheet C and region E may lead to heating of the plasma in region E. Conduction of this energy into the polar cavity is the process suggested by Cole (1966a) to explain the IGY 6300 Å polar cap enhancement. The region

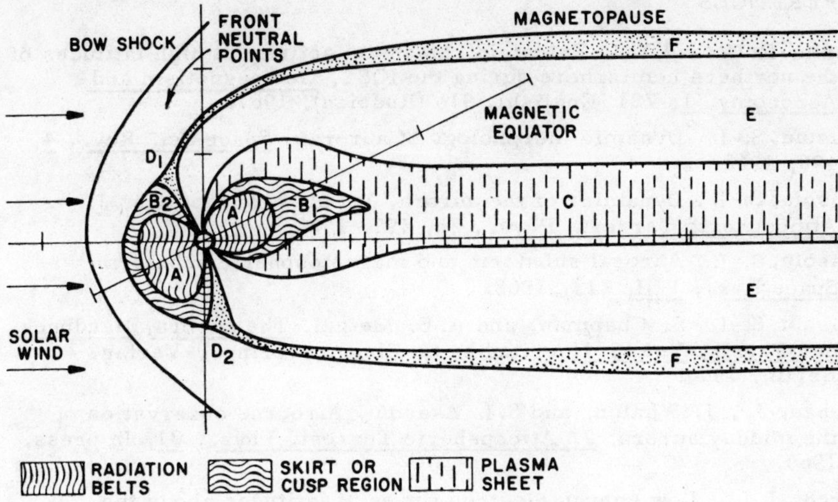

Figure 10. A schematic diagram of the magnetosphere illustrating the distortions caused by the inclination of the earth's magnetic axis to the solar wind.

of "soft" precipitation along the poleward edge of the auroral oval maps out into and may be influenced by the activity on the boundary region F and the boundary between C and E. It is also possible that in the very active IGY period such a disturbed behavior may have filled region E and produced the 6300 Å polar cap enhancement by soft particle precipitation rather than by thermal conduction as suggested by Cole.

There is clearly a close association between the ionospheric activity and the optical emissions at high latitudes. The very great solar maximum of the IGY was associated with great enhancements of both optical and ionospheric phenomena. A great reduction of mantle aurora and polar cap enhancements was evident by the time of the solar minimum of 1964. Although the discrete auroral phenomenon was no less energetic in the IGY than in the IQSY it was confined into a much narrower belt of activity during sunspot minimum in contrast to solar maximum. Evidence from that time until the present rather weak solar maximum shows no return to the great enhancements of the IGY period. The ionosphere seems to exhibit a parallel behavior in the past twelve years.

6. CONCLUSION

The observation of optical phenomena is an extremely useful adjunct to radio frequency observations of the polar ionosphere. It is clear that our understanding of the optical excitation processes in the atmosphere are not well understood in many areas, but it does show a close parallel in overall morphology to ionospheric phenomena.

REFERENCES

Afonia, R. G. and Y. I. Feldstein, Magnetic activity at high latitudes of the northern hemisphere during the IQSY, Geomagnetism and Aeronomy, 1, 784 (English), 916 (Russian), 1967.

Akasofu, S.-I., Dynamic morphology of auroras, Space Sci. Rev., 4, 498, 1965.

Akasofu, S.-I., Dynamics of the aurora, V Poleward motions, J. Atmospheric Terrest. Phys., 28, 497, 1966.

Akasofu, S.-I., Auroral substorm and magnetospheric substorm, Space Res., VIII, 213, 1968.

Akasofu, S.-I., S. Chapman, and A. B. Meinel, The aurora, Handbuck der Physik, Vol 49/1, edited by S. Flugge, Springer-Verlag, Berlin, 1966.

Buchau, J., J. Whalen, and S. I. Akasofu, Airborne observation of the midday aurora, J. Atmopsheric Terrest. Phys., 31, In press, 1969.

Burch, J. L., Low energy electron fluxes at latitudes above the auroral zone, J. Geophys. Res., 73, 3585, 1968.

Chamberlain, J. W., Physics of the Aurora and Airglow, Academic Press, New York, 1961.

Cole, K. D., Theory of some quiet magnetospheric phenomena related to the geomagnetic tail, Nature, 211, 1385, 1966a.

Cole, K. D., Magnetic storms and associated phenomena, Space Sci. Rev., 5, 699, 1966b.

Dalgarno, A., I. D. Latimer and J. W. McConkey, Corpuscular bombardment and N_2^+ radiation, Planetary Space Sci., 13, 1008, 1965.

Davis, T. N., Negative correlation between polar cap visual aurora and magnetic activity, J. Geophys. Res., 68, 4447, 1963.

Davis, T. N., and M. Sugiura, Auroral electrojet activity index and its universal time variations, J. Geophys. Res., 71, 785, 1966.

Derblom, H., On the location of hydrogen emission with respect to auroral forms, Ann. Geophys., 24, 163, 1968.

Eather, R. H., Auroral proton precipitation and hydrogen emissions, Rev. Geophys., 5, 207, 1967.

Eather, R. H., Latitudinal distribution of auroral and airglow emissions: the soft auroral zone, J. Geophys. Res., 74, 153, 1969.

Egeland, A. and A. Omholt, A statistical study of Carl Stormers height measurements of aurora borealis, Aurora and Airglow, edited by B. McCormac, Reinhold, New York, p. 143, 1967.

Evans, J. E., E. G. Joki, R. G. Johnson, and R. D. Sharp, Austral and boreal zone precipitation patterns for low energy protons, Space Res., VI, 773, 1966.

Feldstein, Y. I., Peculiarities in the auroral distribution and magnetic disturbance distribution in high latitudes caused by the asymmetrical form of the magnetosphere, Planetary Space Sci., 14, 121, 1966.

Feldstein, Y. I. and G. V. Starkov, Dynamics of auroral belt and polar geomagnetic disturbances, Planetary Space Sci., 15, 209, 1967.

Fogle, B. and B. Haurwitz, Noctilucent clouds, Space Sci. Rev., 6, 279, 1966.

Fukushima, N., Seasonal change of auroral zone in corrected geomagnetic latitudes, Rep. Ionos. Space Res. Japan, 19, 367, 1965.

Fukushima, N. and T. Hirasawa, Frequent occurrence of local geomagnetic disturbance in high latitudes, J. Geomagn. Geoelect. Japan, 15, 161, 1964.

Gordon, M. A., Night sky photometry of OI, 5577 and 6300 Å inside the southern auroral zone, Planetary Space Sci., 16, 297, 1968.

Gustafsson, G., On the orientation of auroral arcs, Planetary Space Sci., 15, 277, 1967.

Hakura, Y., Tables and maps of geomagnetic coordinates corrected by the higher order spherical harmonic terms, Rep. Ionos. Space Res. Japan, 19, 121, 1965.

Hartz, T. R. and N. M. Brice, The general pattern of auroral particle precipitation, Planetary Space Sci., 15, 301, 1967.

Hernandez, G. J. and S. M. Silverman, A re-examination of Lord Rayleigh's data on the airglow 5577 Å OI emission, Planetary Space Sci., 12, 97, 1964.

Hoffman, R. A., OGO-4 satellite measurements of low energy - high latitude electron precipitation, in The Polar Ionosphere and Magnetospheric processes, edited by G. Skovli, Gordon and Breach, New York, 1969.

Hoffman, R. A. and D. S. Evans, Field aligned electron bursts at high latitudes observed by OGO 4, J. Geophys. Res., 73, 6201, 1968.

Hunten, D. M. and M. B. McElroy, Quenching of metastable states of atomic and molecular oxygen and nitrogen, Rev. Geophys., 4, 303, 1966.

Johnson, R. G., R. E. Meyerott, and J. E. Evans, Coordinated satellite, ground based and aircraft based measurements on auroras, Aurora and Airglow, edited by B. McCormac, Reinhold, New York, p. 169, 1967.

King, G. A. M., Ionospheric observations of auroral activity, Aurora and Airglow, edited by B. McCormac, Reinhold, New York, p. 589, 1967.

Lassen, K., Existence of an inner auroral zone, Nature, 184, 4696, 1959.

Lassen, K., Dirurnal variation of high latitude auroral frequency on magnetically quiet and disturbed days, Nature, 192, 345, 1961.

Lassen, K., Polar cap aurora, Aurora and Airglow, edited by B. McCormac, Reinhold, New York, p. 453, 1967.

Lassen, K., Polar cap emissions, Atmospheric Emissions, edited by B. McCormac, Reinhold, New York, In press, 1969.

Maehlum, B. N., Universal Time control of the low energy electron fluxes in the polar regions, J. Geophys. Res., 73, 3459, 1968.

Maehlum, B. N., On high latitude Universal Time controlled F-layer, J. Atmospheric Terrest. Phys., 31, 531, 1969.

Matsushita, S., Polar daily geomagnetic disturbances, Ann. Geophys., 24, 227, 1968.

McCormac, B. M., Aurora and Airglow, Reinhold, New York, 1967.

Muldrew, D. B., F-layer ionization troughs deduced from Alouette data, J. Geophys. Res., 70, 2635, 1965.

O'Brien, B. J. and H. Taylor, High latitude geophysical studies with satellite Injun 3, IV, Auroras and their excitation, J. Geophys. Res., 69, 45, 1964.

Oguti, T. and K. Marubashi, Enhanced ionization in the ionospheric F_2 region around geomagnetic noon at high latitudes, Rep. Ionos. Space Res. Japan, 20, 96, 1966.

Omholt, A., Ionization by auroral particles, Electromagnetic Wave Propagation, edited by M. Desirant and J. L. Michiels, Academic Press, London, p. 75, 1960.

Piddington, J. H., The magnetospheric radiation belt and tail plasma sheet, Planetary Space Sci., 16, 703, 1968.

Pudovkin, M. I., O. I. Shumilov, and S. A. Zaitzeva, Dynamics of the zone of corpuscular precipitations, Planetary Space Sci., 16, 881, 1968.

Rees, M. H., J. C. G. Walker, and A. Dalgarno, Auroral excitation of the forbidden lines of atomic oxygen, Planetary Space Sci., 15, 1097, 1967.

Rothwell, P. and V. Wallington, The polar substorm and electron islands in the earth's magnetic tail, Planetary Space Sci., 16, 1441, 1968.

Sandford, B. P., Optical studies of particle bombardment in polar cap absorption events, Planetary Space Sci., 10, 195, 1963.

Sandford, B. P., Aurora and airglow intensity variations with time and magnetic activity at southern high latitudes, J. Atmospheric Terrest. Phys., 26, 749, 1964.

Sandford, B. P., Polar-glow aurora, Space Res., VII, 836, 1967.

Sandford, B. P., Variations of auroral emissions with time, magnetic activity and the solar cycle, J. Atmospheric Terrest. Phys., 30, 1921, 1968.

Sato, T. and L. Colin, Morphology of electron concentration enhancement at a height of 1000 km at polar latitudes, J. Geophys. Res., 74, 2193, 1969.

Shabansky, V. P., Magnetospheric processes and related geophysical phenomena, Space Sci. Rev., 8, 366, 1968.

Sharp, R. D., J. E. Evans, and R. G. Johnson, Measurements of particle precipitation at the south pole, Planetary Space Sci., 14, 85, 1966.

Sharp, R. D., R. G. Johnson, M. F. Shea, and G. B. Shook, Satellite measurements of precipitating protons in the auroral zone, J. Geophys. Res., 72, 227, 1967.

Silverman, S. M., F. Ward, and R. Shapiro, The correlation between the 5577 Å night airglow intensity and geomagnetic activity, J. Geophys. Res., 67, 2255, 1962.

Starkov, G. V., Auroral heights in the polar cap, Geomagnetism and Aeronomy, 8, 28 (English), 36 (Russian), 1968.

Starkov, G. V. and Y. I. Feldstein, Change in the boundaries of the oval auroral zone, Geomagnetism and Aeronomy, 7, 48 (English), 62 (Russian), 1967a.

Starkov, G. V. and Y. I. Feldstein, Scheme of an elementary disturbance in auroras on the day side of the earth, Geomagnetism and Aeronomy, 7, 294 (English), 367 (Russian), 1967b.

Stolarski, R. S., Calculation of auroral emission rates and heating effects, Planetary Space Sci., 16, 1265, 1968.

Stringer, W. J. and A. E. Belon, Morphology of the IQSY auroral oval, 1. Interpretation of isoauroral diagrams, J. Geophys. Res., 72, 4415, 1967.

Taylor, J. R., Artificial red aurora II, Effect of cooling on emission, J. Atmospheric Terrest. Phys., 29, 771, 1967.

Thomas, L., Ionospheric implications of aurora and airglow studies, Aurora and Airglow, edited by B. McCormac, Reinhold, New York, p. 93, 1967.

Thomas, J. O. and M. K. Andrews, Transpolar exospheric plasma, 1 Plasmasphere termination, J. Geophys. Res., 73, 7407, 1968.

Ulwick, J. C., Rocket measurements of auroral parameters, Aurora and Airglow, edited by B. McCormac, Reinhold, New York, p. 225, 1967.

Walker, J. C. G. and M. H. Rees, Ionospheric electron densities and temperatures in aurora, Planetary Space Sci., 16, 459, 1968.

Weill, G., Sur une aurore polaire d'un type nouveau, Compte Rendu Acad. Sci., Paris, 254, 3402, 1962.

Weill, G., Une contraction de la zone aurorale antarctique observee au cours de la phase initiale des tempetes geomagnetiques, Compte Rendu Acad. Sci., Paris, 256, 985, 1963.

Weill, G., M. Fafiotte, S. Huille, and J. Delannoy, La structure de la zone auroral arctique en periode de minimum d'activite solaire, Ann. Geophys., 21, 469, 1965.

Weins, R. H., Average time-latitude distribution of hydrogen and OI green-line emission in aurora, Ph. D. Thesis, U. of Saskatchewan, June, 1968.

Zhulina, R. H., Seasonal variations in the position of the auroral absorption zone, Geomagnetism and Aeronomy, 8, 376 (English), 471 (Russian), 1968.

SUBJECT INDEX

REFERENCE INDEX